GEOGRAPHIES OF RURAL CULTURES AND SOCIETIES

Perspectives on Rural Policy and Planning

Series Editors:
Andrew Gilg
University of Exeter, UK
Keith Hoggart
King's College, London, UK
Henry Buller
Cheltenham College of Higher Education, UK
Owen Furuseth
University of North Carolina, USA
Mark Lapping
University of South Maine, USA

Other titles in the series

Geographies of Rural Cultures and Societies

Edited by
LEWIS HOLLOWAY and MOYA KNEAFSEY
Coventry University, UK

Routledge
Taylor & Francis Group
LONDON AND NEW YORK

First published 2004 by Ashgate Publishing

2 Park Square, Milton Park, Abingdon, Oxfordshire OX14 4RN
711 Third Avenue, New York, NY 10017

Routledge is an imprint of the Taylor & Francis Group, an informa business

First issued in paperback 2017

British Library Cataloguing in Publication Data
Geographies of rural cultures and societies. -
    (Perspectives on rural policy and planning)
    1. Rural conditions - Developed countries - Congresses
    2. Rural development - Developed countries - Sociological
    aspects - Congresses 3. Rural geography - Congresses
    4. Agriculture - Social aspects - Developed countries -
    Congresses
    I. Holloway, Lewis II. Kneafsey, Moya
    307.1'412

Library of Congress Cataloging-in-Publication Data
Geographies of rural cultures and societies / edited by Lewis Holloway and Moya Kneafsey.
        p. cm. -- (Perspectives on rural policy and planning)
    Includes bibliographical references and index.
    ISBN 0-7546-3571-6
    1. Sociology, Rural. 2. Rural geography. 3. Great Britain--Rural conditions. I.
Holloway, Lewis. II. Kneafsey, Moya. III. Series.

HT421.G47 2004
307.72--dc22

2004041045

ISBN 13: 978-0-7546-3571-0 (hbk)
ISBN 13: 978-1-138-27510-2 (pbk)

Contents

PART I: THINKING RURALITIES

PART II: RURAL SOCIETIES: INCLUSIONS AND EXCLUSIONS

List of Figures

List of Tables

List of Contributors

Hanne Kirstine Adriansen
Institute for International Studies, Gl. Kongevej 5, 1610 Copenhagen N, Denmark.

Caroline Cozens
Transport and Planning Service, Floor 10, Civic Centre, Royal Parade, Plymouth, PL1 2EW, UK.

Bill Edwards
Institute of Geography and Earth Sciences, University of Wales, Aberystwyth, Llandinam Building, Aberystwyth, SY23 3DB, Wales, UK.

Rob Fish
School of Geography, University of Nottingham, University Park, Nottingham, NG7 2RD, UK.

Graham Gardner
Institute of Geography and Earth Sciences, University of Wales, Aberystwyth, Llandinam Building, Aberystwyth, SY23 3DB, Wales, UK.

Lewis Holloway
Geography Subject Area, Coventry University, Priory Street, Coventry, CV1 5FB, UK.

Annie Hughes
School of Earth Sciences and Geography, Kingston University, Penryhn Road, Kingston upon Thames, Surrey, KT1 1LQ, UK.

Moya Kneafsey
Geography Subject Area, Coventry University, Priory Street, Coventry, CV1 5FB, UK.

Lene Møller Madsen
Danish Centre for Forest, Landscape and Planning, Hørsholm Kongevej 11, DK-2970, Hørsholm, Denmark.

Carol Morris
Centre for Rural Research, University of Exeter, Lafrowda House, St. Germans Road, Exeter, EX4 6TL, UK.

Caitríona Ní Laoire
Irish Centre for Migration Studies, National University of Ireland, Cork, Western Road, Cork, Ireland.

Martin Phillips
Department of Geography, University of Leicester, University Road, Leicester, LE1 7RH, UK.

Susanne Seymour
School of Geography, University of Nottingham, University Park, Nottingham, NG7 2RD, UK.

David Storey
Department of Applied Sciences, Geography and Archaeology, University College Worcester, Henwick Grove, Worcester, WR2 6AJ, UK.

Michael Woods
Institute of Geography and Earth Sciences, University of Wales, Aberystwyth, Llandinam Building, Aberystwyth, SY23 3DB, Wales, UK.

Richard Yarwood
School of Geography and Geology, University of Plymouth, Drake Circus, Plymouth, PL4 8AA, UK.

Acknowledgements

The majority of the chapters in this book are based on papers given at the 'Progress in Rural Geography: Towards 21st Century Geographies of Rurality' conference, held at Coventry University, on the 5th and 6th of July, 2001. The editors would like to thank the Rural Geography Research Group, the Royal Geographical Society (with the Institute of British Geographers), and the Geography Subject Group, Coventry University, for their support for this event. Chapters were subjected to an anonymous refereeing process, and we would like to extend our thanks to all those who gave their time and effort to comment on chapters. Thanks also to Stuart Gill of the Cartography Unit at Coventry University, and to Lucy Holloway, for their assistance with the production of the manuscript. Many thanks to Pam Bertram at Ashgate for providing invaluable technical advice and support.

We are grateful to the Boolavogue Textile Studio, of County Wexford, Ireland, for permission to reproduce the photograph in Figure 10.3.

Chapter 1

Geographies of Rural Cultures and Societies: Introduction

Lewis Holloway and Moya Kneafsey

Introduction

The notion of a 'cultural turn' in rural studies has become a widely accepted part of the story which rural geographers tell concerning the history of their (sub)discipline. This move to engage with matters cultural is usually placed in the 1990s, and is epitomized in edited collections such as *Contested Countryside Cultures* (Cloke and Little, 1997), *Revealing Rural Others* (Milbourne, 1997) and more recently, *Country Visions* (Cloke, 2003), and is also illustrated by the increasing presence of culturally-oriented articles in, for example, *Journal of Rural Studies* and *Sociologia Ruralis*. Taking the idea of a cultural turn as a starting point, this book brings together a collection of chapters with two main aims. First, it examines the position of rural cultural and social geography at the beginning of the twenty-first century, and suggests new agendas in the study of changing cultural and social geographies of rurality in the 'developed' or 'Western' world. In doing so, critical evaluations of theoretical positions and approaches are offered, and suggestions given for the development of new theoretical perspectives. Second, it presents empirical studies which both draw on particular conceptual and methodological positions, and draw attention to some of the key dimensions of cultural and social experience in the contemporary countryside. Here, many chapters focus on the ways in which rural spaces are contested and struggled over, and on how rural social life, far from the idyllic assumptions of many popular understandings of the countryside, is bound up with relations of inequality and exclusion.

The cultural turn within rural studies has commonly been associated with a shift from a concern with the material world towards an interest in including consideration of the immaterial dimensions of social life (for a detailed account of this, see Phillips, 1998a; see also Cloke and Thrift, 1994 and Cloke, 1997). Studies based around the former, materialist, concerns have engaged more or less directly with the material basis of rural areas, either through the empirical mapping of 'rural' entities, land uses, settlements, populations and so on, or through the structuralist examination of production, employment and class relations and rural government exemplified in political economic critical theory. Thus they have, for example, attempted to identify and describe rural spaces on the basis of their association with particular types of activity, people or landscape (epitomized in the idea of an 'Index of Rurality', see Cloke, 1977; Cloke and Edwards, 1986), or have moved towards explanatory modes of work exploring the structural mechanisms associated with the restructuring of rural space and the development of relationships between rural spaces and the capitalist mode of production as a whole (e.g. Marsden, 1988; Marsden et al, 1993; Murdoch and Marsden, 1994) - modes which for some have led to an effacement of the idea of the rural as a meaningful spatial and analytical category (e.g. Hoggart, 1990). The latter, immaterialist, approaches to rural studies have first, refocused attention on the rural, not as a clearly identifiable space, or as a verifiable set of social or economic characteristics, but as a meaningful concept, discursively constructed, understood and related to in different ways by diverse social groups (e.g. Halfacree, 1993; Jones, 1995; Pratt, 1996); and second, drawn attention to the variety of social experiences of the contemporary countryside, through the deployment of the notion of 'otherness' which differentiates the situations of the powerless, the voiceless and the economically, socially and geographically marginalized, from the male, white, heterosexual, middle class standpoint which had been so often assumed in rural studies (see Philo, 1992). From these perspectives, rather than there being a single rural space which people relate to and within which things happen, it has become more common to refer to a multitude of ruralities, as understood and experienced by different groups.

In shifting its emphasis towards the immaterial, contemporary rural studies has engaged in a number of theoretical and methodological changes. First, stimulated by, for example, the contributions of Philo (1992) and Murdoch and Pratt (1993), postmodern conceptualizations of discourse as related to identity, power and knowledge have become increasingly important in turning rural geographers' interest towards the production and consumption of the rural as a meaningful and contested space (e.g. Mordue, 1999). Thus, for instance, representations of rurality in various media have been explored to examine how various imaginings of the countryside have been constituted: here we might mention ideas of the rural as idyllic (e.g. Bunce, 1994), as site for tourism or adventure playground (e.g. Cloke and Perkins, 1998), as cultural memorial to 'traditional' values (e.g. Phillips et al, 2001), or even as horrific (Bell, 1997). Second, and relatedly, there have emerged new bodies of work examining critically the politics of the rural, focusing on power and examples of conflict in and struggle over rural space. For example, Philo's (1992) identification of the existence of rural 'others' has drawn attention to the complex web of power relations in, or affecting, the countryside, and which bolster the privileged position of some, and marginalize or exclude others from full participation in rural politics, economy and society. Here, for example, might be included studies of the rural experiences of people from ethnic minorities (e.g. Kinsman, 1995), women (e.g. Hughes, 1997), children (e.g. Jones, 1997; McCormack, 2002) and the mentally ill (e.g. Philo et al, 2003). Further, continuing heated debates over the future of rural land use in countries like the UK, given supposed crises in the agricultural sector (Drummond et al, 2000) and socio-political wranglings over environmental conservation, access to land for recreation or 'alternative' lifestyles, and the future of controversial practices such as hunting with dogs or the development of new types of land use, illustrate both the symbolic importance of 'traditional' rural activities and the nature of the countryside as an arena for political struggle (e.g. Cresswell, 1996; Halfacree, 1996; Woods, 1997, 2003a, 2003b). Relatedly, the constitution and contestation of rural identities, increasingly seen as fluid, and often structured around particular or contested understandings of rurality and conflicts over rural space, has become an important focus of attention (see, for example, Gray, 1998 and Saugeres, 2000a, 2000b, on farming and masculinity, or Phillips, 1998b on the middle classes). Third, the range and emphasis of the methodologies deployed in rural studies has altered along with these new theoretical and empirical foci. While quantitative methodologies and more conventional modes of survey work are still significant to much research, rural researchers are increasingly utilising intensive, qualitative ways of conducting their research. Without wanting to suggest a straightforward dualism of quantitative and qualitative, or intensive and extensive research methods, many rural researchers have found that the 'depth' afforded by ethnographic, textual and visual research methods has allowed them to analyse more fully the cultural dimensions, and the differentiated social experience, of rurality (see, for example, Hughes et al, 2000).

Yet while the immaterial dimensions of rurality have assumed greater significance in rural studies, it is also possible to point, in addition, to a resurgence of interest in the materialities associated with the countryside. Here, instead of a focus on mapping or describing material entities, or on theorising the production and exchange of material goods, the material substances and things (human and nonhuman; living and non-living) constituting or found in spaces defined as rural have become important in a number of ways, including the following. First, some rural geographers have begun to pay attention to the embodiment of rural subjects, although as Little and Leyshon (2003) suggest, this has as yet received relatively little attention. Studies of rural embodiment recognize the centrality of human corporeality to the formation and contestation of rural identities, to bodily performance and practice, and to the sensual and physical engagement of people with their worlds. Second, and relatedly, attention has increasingly been paid to the bodies of rural animals (for example in their entanglement in industrial food supply systems; see Ufkes, 1998; Stassart and Whatmore, 2003) and to the materiality of non-sentient living entities, such as trees (Cloke and Jones, 2001; Jones and Cloke, 2002). In such cases, nonhuman living things are increasingly being accorded agency in their relationships with human, and other nonhuman, entities (Murdoch, 1997a). Third, rural studies is showing evidence of increased interest in a range of substances associated with the countryside. These include soil (e.g. Ingrams, 2003), 'pollution' (e.g. Seymour et al, 1997) and food (e.g. Fitzsimmons and Goodman, 1998; Goodman, 2001). Substances such as these are studied as part of the complex networks of entities and relationships which are brought into particular configurations (notably the agro-food network) which structure, and are structured by, rural life, and which extend away from specific rural areas into other spaces at a range of scales. This brings us to the final area here, which is to note the increasing role played in rural studies by approaches influenced by actor-network theory. These approaches (see, for example, Murdoch,

1997a, 1997b; Whatmore, 2002) are especially concerned with explanations which emphasize distributed and relational understandings of, for example, power and agency, rather than seeing these as centred in individual human beings or structural agencies such as governmental institutions. Such approaches are also concerned to deconstruct familiar dualisms, such as that of nature and culture, which have tended to structure explanations of rural change. Writing about the rural, from this perspective, focuses on a world of hybrids, accounting simultaneously for the material and the semiotic, the human and the nonhuman, the natural and the cultural, the material and the immaterial, and so on (see, for example, Hinchliffe, 2001, on the UK BSE issue, or MacMillan, 2003, on the regulation of biotechnology).

The range and diversity of contemporary rural studies, which we have begun to illustrate above, draws attention to a flourishing of research in rural geography, characterized by the emergence of challenging theoretical perspectives, a diverse range of subject matter, and more reflexive methodological approaches towards the study of rural change. The chapters which follow, most of which have been developed from papers presented at a conference held at Coventry University in 2001, illustrate some of this range and diversity in presenting individual accounts of theoretical reflection and empirical research. While not prescriptive, the chapters indicate some possible future directions for rural research, in terms of both the conceptual and methodological tools which might be developed, and the scope of the empirical themes which might be explored. The work of producing an edited collection, particularly from conference papers, is often an attempt to forge a semblance of continuity and consistency between an eclectic range of contributions. Yet, in assembling the various chapters in this collection the intention has not been to develop the image of a coherent body of work to represent either a 'cultural turn' or 'rural geography'. In this sense, the motivation for the book is not to suggest that the different contributions are necessarily underpinned by a common agenda, strategy, theoretical framework or methodological approach. Suggesting instead a fragmented rural geography, we want to illustrate, and welcome, a continuing diversity of work focusing on often very different aspects of the social and cultural dimensions of rurality. We see such diversity as a strength of rural social and cultural geography, offering constant opportunities for theoretical and methodological innovation and the exploration of new research avenues. In addition, however, it is also increasingly clear (and hinted at above) that the value of maintaining distinctions between the analytical categories of the social or cultural, and other such fields (e.g. the economic, political or natural), is being brought into question by approaches such as actor-network theory which emphasize the contingency and instability of such categories, and are associated with alternative understandings of power, agency etc. It might thus be that a turning away from ideas of culture, society, economy, politics and nature as analytical tools will increasingly produce other forms of rural study, opening up new perspectives on processes of rural change, experiences of rurality, and embodied and subjective forms of being in 'rural' space.

Part and Chapter Summaries

The book is divided into four parts, each focusing on a key set of issues in relation to the broader themes outlined above. Part 1, Thinking Ruralities, concentrates on emerging ways of conceptualising, telling and researching the rural. The four chapters in this part draw on different theoretical and methodological frameworks which allow us to begin to think about and do rural studies in ways which depart from more conventional approaches. Questions are raised concerning the status and apparent solidity of the category 'rural', about how language and practice is involved in the material and semiotic constitution of rural spaces, and about how rural studies practitioners can access different versions of rurality. In Chapter 2, Martin Phillips raises possibilities for further 'cultural turns' for rural geographers, exploring postmodern experiences of rural space and mobilizing the writings of, in particular, Jean Baudrillard. Drawing on metaphors of obesity and pornography to illustrate the profusion of cultural forms associated with rurality, he suggests that there is always a confusing excess of images and meanings, confounding people's attempts to make sense of the rural. This excess, demonstrated in the proliferation of ways in which the rural has been drawn upon in contemporary culture, has, for many, led to a 'stuttering' sense of uncertainty, hesitancy or scepticism in lay understandings of rurality. The chapter concludes by arguing that more attention needs to be paid in rural studies to processes of subjectification, particularly in relation to the obese and pornographic rurals which are identified. Chapter 3, by Robert Fish, stems from similar concerns over the way in which rural spaces are 'told', becoming parts of narratives which display specific structuring mechanisms in order to develop an argument, but which simultaneously produce, or assume, particular understandings of rurality. This chapter takes a rather different perspective, however, in concentrating on how academic writings in the genre of rural studies have embedded rurality within particular narrative modes. The case is made that the effects of narrative structure on understandings of rurality should be given more consideration, and that alternative narrative (or anti-narrative) forms might be deployed by rural geographers in order to open up the possibilities for exploring the rural in different ways. In Chapter 4, Graham Gardner is specifically concerned with the issue of how academic researchers have implicitly or explicitly conceptualized power in studies of rural societies and communities. A shift is identified from concerns with 'traditional' modes of authority in rural communities, to conceptualizations of rural power relations which draw on wider notions of class differentiation and which also recognize the significance of the modern state in the restructuring of rural spaces and communities. Here, relations between central government and local state institutions are seen as crucial. Yet it is suggested that explicit theorizations of power have been lacking from rural studies, and that future research needs to develop approaches which are able to conceptualize relationships between individuals and the structures of power governing rural areas, and which can understand power as a productive, not simply an oppressive, force. The chapter argues that conceptions of power inspired by Foucauldian thought might offer valuable material for an engagement with those power structures affecting rural spaces. The final chapter, in this section, Chapter

5, by Lene Møller Madsen and Hanne Adriansen, focuses on methodological issues. With illustrative examples from rural research in the very different contexts of Senegal, Denmark and Egypt, the authors first point to the apparent gap between rural studies which tend to focus on first world contexts, and which have experienced a turn towards the cultural, and 'development' research which has focused on third word contexts. They go on to argue for research which can transcend this distinction, showing how strategies which explore material practices and cultural values can shed light on the parallels between very different rurals, while maintaining an essential sense of the differences which clearly exist.

Part 2 of the book, Rural Societies: Inclusions and Exclusions, contains three chapters focusing on the political and social processes influencing experiences of struggle and marginalization in the contemporary countryside. Instead of idyllic representations of the countryside, rural space is shown in these chapters to be contested, stratified and open to the social problems more often associated with urban sites. Rural space itself is shown to be a fundamental part of such struggles and problems, and to be reproduced through them. In Chapter 6, Michael Woods examines the emergence of various forms of rural protest in the UK and France, concentrating on the formation of groups purporting to speak and demonstrate on behalf of rural people and the countryside itself, and on the specific modes of protest engaged in at particular sites. The specific nature of such protests is discussed, and related to particular discourses of rurality, relations of conflict and the formation of social identities. Three case studies illustrate how contemporary rural protests, although unique in their circumstances, nevertheless share a concern with defining rurality and controlling rural space, and are often related to perceptions of a threat to established lifestyles and social structures. In such cases, examination of rural power relations, and the power of the rural to stimulate such protest, warrants serious consideration. Turning to a very different example, Annie Hughes explores the situation of rural lone parents in the UK, in Chapter 7. As an example of the 'neglected other', rural lone parents have remained hidden under dominant representations of rurality which associate country life with the model of the nuclear (two parent, heterosexual) family. Social and geographical constraints on the ability of lone parents to live, work and care for their families in the countryside are highlighted. Rural lone parents hence potentially both lack visibility in mainstream imaginings of rural space, and can find that continuing to live in rural areas is problematic. As illustrated by the case studies which underpin this chapter, they may thus be doubly excluded from the countryside. Chapter 8, the last in this section, presents a third example of the ways in which rural spaces are socially contested. Richard Yarwood and Caroline Cozens discuss the policing of rural Britain, challenging assumptions of the rural as being a crime-free space, and illustrating how competing demands on police time, and the specific nature of rural space, influence crime, the fear of crime, and the policing of crime, in the countryside. Presenting detailed results from research with senior police officers in non-metropolitan areas, the authors point to the ways in which the police define and respond to different sorts of rural space in ways which are related to, and may reproduce, existing rural power relations and social structures. Rural policing was understood as in some ways different to urban policing, yet specific problems with the policing of rural areas were recognized, and the policing of rural protest, civil disobedience and urban visitors to the countryside was seen as an emerging issue.

The three chapters which make up Part 3, Community and Governance, share concerns with ideas of community and the role of the rural community in the definition and management of rural space, thinking about how rural space is governed, and the increasing incidence of 'partnership' approaches to the governance of rural development and environmental protection. Chapter 9, written by Bill Edwards and Michael Woods, considers the notion of community, focusing on the specific context of the UK Labour government's rural agenda and its declared emphasis on the creation of 'vibrant communities' as an essential part of rural development. Concentrating on the relationships that this assumes between the state, local institutions and community members, and drawing on the results of their research into participation in local community governance, the authors present a critical reading of the notions of local and community empowerment and vibrancy that such policies presuppose. This emphasizes that participation in community leadership can be motivated by self (or class) interest, and that structures of community organization can exclude the voices of some, while promoting the demands of others. David Storey, in Chapter 10, looks at the specific case of partnership-led, place-centred, rural development strategies, focusing on tourism and place promotion in Ireland. The emphasis placed on partnership activities by the European Union's LEADER rural development programme is used to explore the ways in which particular sites are marketed as tourist attractions, and thus how rural development initiatives have become implicated in the re-imaging and re-shaping of rural places. In the examples drawn upon in this chapter, an essential component of the imaging of place is the notions of heritage constructed around specific, emplaced historical events and their contemporary significance. Yet, the presentation of heritage is not necessarily straightforward, as first, the construction of a marketable heritage also has implications for how local community members understand their places and histories, and second, tourists do not simply respond naively to what is presented to them as the history of a place. Instead, the more complex responses of these groups warrant further attention. The final chapter in this part, Chapter 11, by Susanne Seymour, returns to the themes of community, participation and partnership, this time in relation to rural environmental protection and sustainability in a global context favouring participatory governance. The chapter focuses on the operation of community, and on making sense of the ways in which certain individuals, representing certain groups and interests, are able to participate in environmental decision making, while others may in practice be or feel excluded from decision making processes. Case study material from the UK, US, Australia and New Zealand is drawn upon in examining the interactions between ideas of community and rurality, and in assessing the implications these have for a range of initiatives which ascribe environmental protection functions to rural communities.

The final part of the book, Part 4, focuses on Cultures of Farming and Food. The three chapters appearing here recognize the continuing social and cultural importance of farming and food production in the countryside of the 'developed' world to both farming and non-farming groups, despite, and sometimes because of, the declining economic significance of agriculture, the reduced proportion of rural populations involved in farming and the ambivalent feelings held by many in relation to farming: farming can be associated with, on the one hand, a bucolic idyll, and on the other, environmental damage and a series of health scares surrounding food. Carol Morris, in Chapter 12, considers the relevance of the notion of the cultural turn for studies of agriculture. Arguing that, heretofore, the cultural dimensions of farming have received little attention from rural geographers, she discusses the small number of studies which have begun to open up some useful ways of bringing cultural perspectives to bear on agricultural research, and suggests ways in which an engagement with the cultural could be increasingly valuable in future studies of agriculture. These include research into the ways agriculture is represented and discursively constructed, the nature-society relations implied in different modes of practising farming, and the importance of the cultural to studies of food, which increasingly extend away from the farm itself into networks of relationships forming agri-food economies. Picking up on this final idea, Lewis Holloway and Moya Kneafsey, in Chapter 13, examine the different ways in which notions and experiences of 'closeness' between food producers, consumers and food itself, are constituted in four very different sets of 'alternative' food relationships. The emergence of 'quality' food networks, the rise of Farmers' Markets in the UK, the engagement of some in small-scale, self-sufficient modes of food production, and the possibilities for entering into different sorts of food production-consumption relationship in 'virtual' space are examined in an attempt to illustrate the underlying desires for connectedness with food associated with various, seemingly diverse production-consumption contexts. The nature of these contexts' (re)production of rural space is also explored, focusing on how ideas of rurality are central to these different ways of growing and consuming food. In the final chapter, Chapter 14, Caitríona Ní Laoire develops a further dimension of a culturally-oriented perspective on agriculture, in examining the production of complex masculine identities amongst young farmers in Ireland. Against a background of rural restructuring which is making economic survival in farming more difficult, she illustrates the continued importance of masculinized farming identities for those young men remaining in agriculture. However, far from such identities remaining static against a backdrop of change, it is shown that the emergence of discourses of competitiveness and efficiency in farming are resulting in different modes of identification, with farmers being identified as economic 'winners' or 'losers', with repercussions for their attitudes towards farming. The chapter concludes by arguing that examinations of gendered identities in farming should be a valuable focus for future research into the cultural geographies of agriculture.

References

Bell, D. (1997), 'Anti-idyll: Rural Horror', in P. Cloke and J. Little (eds), *Contested Countryside Cultures*, Routledge, London, pp. 94-108.

Bunce, M. (1994), *The Countryside Ideal: Anglo-American Images of Landscape*, Routledge, London.

Cloke, P. (1977), 'An Index of Rurality for England and Wales', *Regional Studies*, Vol. 11, pp. 31-46.

Cloke, P. (1997), 'Country Backwater to Virtual Village? Rural Studies and the "Cultural Turn"', *Journal of Rural Studies*, Vol. 13, pp. 367-375.

Cloke, P. (ed) (2003), *Country Visions*, Pearson Education, Harlow.

Cloke, P. and Edwards, G. (1986), 'Rurality in England and Wales 1981: A Replication of the 1971 Index', *Regional Studies*, Vol. 20, pp. 289-306.

Cloke, P. and Jones, O. (2001), 'Dwelling, Place and Landscape: an Orchard in Somerset', *Environment and Planning A*, Vol. 33, pp.649-666.

Cloke, P. and Little, J. (eds) (1997), *Contested Countryside Cultures*, Routledge, London.

Cloke, P. and Perkins, H. (1998), 'Cracking the Canyon with the "Awesome Foursome": Representations of Adventure Tourism in New Zealand', *Environment and Planning D: Society and Space*, Vol. 16, pp. 185-218.

Cloke, P. and Thrift, N. (1994), 'Refiguring the Rural', in P. Cloke, M. Doel, D. Matless, M. Phillips and N. Thrift *Writing the Rural*, Paul Chapman, London, pp. 1-5.

Cresswell, T. (1996) *In Place: Out of Place: Geography, Ideology and Transgression*, University of Minnesota Press: Minneapolis.

Drummond, I., Campbell, H., Lawrence, G. and Symes, D. (2000), 'Contingent or Structural Crisis in British Agriculture', *Sociologia Ruralis*, Vol. 40, pp. 111-127.

Fitzsimmons, M. and Goodman, D. (1998), 'Incorporating Nature: Environmental Narratives and the Reproduction of Food', in B. Braun and N. Castree (eds), *Remaking Reality: Nature at the Millennium*, Routledge, London, pp. 194-220.

Goodman, D. (2001), 'Ontology Matters: the Relational Materiality of Nature and Agro-Food Studies', *Sociologia Ruralis*, Vol. 41, pp. 182-200.

Gray, J. (1998), 'Family Farms in the Scottish Borders: A Practical Definition by Hill Sheep Farmers', *Journal of Rural Studies*, Vol. 14, pp. 341-356.

Halfacree, K. (1993), 'Locality and Social Representation: Space, Discourse and Alternative Definitions of the Rural', *Journal of Rural Studies*, Vol. 9, pp. 23-37.

Halfacree, K. (1996), 'Out of Place in the Country: Travellers and the "Rural Idyll"', *Antipode*, Vol. 28, pp. 42-72.

Hinchliffe, S. (2001), 'Indeterminacy In-decisions - Science, Policy and Politics in the BSE Crisis', *Transactions, Institute of British Geographers*, Vol. 26, pp. 182-204.

Hoggart, K. (1990), 'Let's Do Away With Rural', *Journal of Rural Studies*, Vol. 6, pp. 245-257.

Hughes, A. (1997), 'Rurality and "Cultures of Womanhood": Domestic Identities and Moral Order in Village Life', in P. Cloke and J. Little (eds), *Contested Countryside Cultures*, Routledge, London, pp. 123-137.

Hughes, A., Morris, C. and Seymour, S. (eds)(2000), *Ethnography and Rural Research*, Countryside and Community Press, Cheltenham.

Ingrams, J. (2003), 'Farmer Knowledge of Soil Management and Degradation in England', paper presented to the RGS-IBG International Annual Conference, 3-5 September 2003, London.

Jones, O. (1995), 'Lay Discourses of the Rural: Developments and Implications for Rural Studies', *Journal of Rural Studies*, Vol. 11, pp. 35-49.

Jones, O. (997), 'Little Figures, Big Shadows: Country Childhood Stories', in P. Cloke and J. Little (eds), *Contested Countryside Cultures*, Routledge, London, pp. 158-179.

Jones, O. and Cloke, P. (2002), *Tree Cultures: The Place of Trees and Trees in their Place*, Berg, Oxford.

Kinsman, P. (1995), 'Landscape, Race and National Identity', *Area*, Vol. 27, pp. 300-310.

Little, J. and Leyshon, M. (2003), 'Embodied Rural Geographies: Developing Research Agendas', *Progress in Human Geography*, Vol. 27, pp. 257-272.

MacMillan, T. (2003), 'Tales of Power in Biotechnology Regulation: the EU Ban on BST', *Geoforum*, Vol. 34, pp. 187-201.

Marsden, T. (1988), 'Exploring Political Economy Approaches in Agriculture', *Area*, Vol. 20, pp. 315-322.

Marsden, T., Murdoch, J., Lowe, P., Munton, R. and Flynn, A. (1993), *Constructing the Countryside*, UCL Press, London.

McCormack, J. (2002), 'Children's Understanding of Rurality: Exploring the Inter-relationship Between Experience and Understanding', *Journal of Rural Studies*, Vol. 18, pp. 193-207.

Milbourne, P. (ed) (1997), *Revealing Rural Others: Representation, Power and Identity in the British Countryside*, Pinter, London.

Mordue, T. (1999), 'Heartbeat Country: Conflicitng Values, Coinciding Visions', *Environment and Planning A*, Vol. 31, pp. 629-646.

Murdoch, J. (1997a), 'Inhuman/Nonhuman/Human: Actor Network Theory and the Prospects for a Nondualistic and Symmetrical Perspective on Nature and Society', *Environment and Planning D: Society and Space*, Vol. 15, pp. 731-756.

Murdoch, J. (1997b), 'Towards a Geography of Heterogeneous Associations', *Progress in Human Geography*, Vol. 21, pp. 321-337.

Murdoch, J. and Marsden, T. (1994), *Reconstituting Rurality*, UCL Press, London.

Murdoch, J. and Pratt, A. (1993), 'Rural Studies: Modernism, Postmodernism and the "Post-Rural"', *Journal of Rural Studies*, Vol. 9, pp. 411-427.

Phillips, M. (1998a), 'The Restructuring of Social Imaginations in Rural Geography', *Journal of Rural Studies*, Vol. 14, pp. 121-153.

Phillips, M. (1998b), 'Investigations of the British Rural Middle Classes – Part 2: Fragmentation, Identity, Morality and Contestation', *Journal of Rural Studies*, Vol. 14, pp. 427-443.

Phillips, M., Fish, R. and Agg, J. (2001), 'Putting Together Ruralities: Towards a Symbolic Analysis of Rurality in the British Mass Media', *Journal of Rural Studies*, Vol. 17, pp. 1-27.

Philo, C. (1992), 'Neglected Rural Geographies: A Review', *Journal of Rural Studies*, Vol. 8, pp. 193-207.

Philo, C., Parr, H. and Burns, N. (2003), 'Rural Madness: a Geographical Reading and Critique of the Rural Mental Health Literature', *Journal of Rural Studies*, Vol. 19, pp. 259-281.

Pratt, A. (1996), 'Discourses of Rurality: Loose Talk or Social Struggle?', *Journal of Rural Studies*, Vol. 12, pp. 69-78.

Saugeres, L. (2002a), 'The Cultural Representation of the Farming Landscape: Masculinity, Power and Nature', *Journal of Rural Studies*, Vol. 18, pp. 373-384.

Saugeres, L. (2002b), 'Of Tractors and Men: Masculinity, Technology and Power in a French Farming Community', *Sociologia Ruralis*, Vol. 42, pp. 143-159.

Seymour, S., Lowe, P., Ward, N. and Clark, J. (1997), 'Environmental "Others" and "Elites": Rural Pollution and Changing Power Relations in the Countryside', in P. Milbourne (ed), *Revealing Rural Others: Representation, Power and Identity in the British Countryside*, Pinter, London, pp. 57-74.

Stassart, P. and Whatmore, S. (2003), 'Metabolizing Risk: Food Scares and the Un/Remaking of Belgian Beef', *Environment and Planning A*, Vol. 35, pp. 449-462.

Ufkes, F. (1998), 'Building a Better Pig: Fat Profits in Lean Meat', in J. Wolch and J. Emel (eds), *Animal Geographies: Place, Politics and Identity in the Nature-Culture Borderlands*, Verso, London, pp. 241-255.

Whatmore, S. (2002), *Hybrid Geographies: Natures Cultures Spaces*, Sage, London.
Woods, M. (2003a), 'Deconstructing Rural Protest: the Emergence of a New Social Movement', *Journal of Rural Studies*, Vol. 19, pp. 309-325.
Woods, M. (2003b), 'Conflicting Environmental Visions of the Rural: Windfarm Development in Mid Wales', *Sociologia Ruralis*, Vol. 43, pp. 271-288.

PART I
THINKING RURALITIES

Chapter 2

Obese and Pornographic Ruralities: Further Cultural Twists for Rural Geography?

Martin Phillips

Introduction

As the title of this book highlights, rural geography has undergone something of a cultural turn, or perhaps even turns. Cloke (1997, p. 368), for example, identifies three particular rural 'foci of cultural studies' - landscape, hidden others, and the spatiality of nature - and suggests that these have both 'circumnavigated' the boundaries of a rather 'isolationist' rural geography (see also Phillips, 1998; Pratt, 1996) and connected rural geography into a series of debates of central importance within the contemporary social sciences. He adds that although he feels that 'much of the current popularity, excitement and "fizz" of rural studies draws heavily on these themes', these 'cultural geographies' should not be seen as the 'only significant geographies or sociologies being mapped out in rural areas' (p. 369) and that there are a number of 'unresolved tensions' and 'areas of concern' (p. 372) within these geographies. The cultural turn in rural geography, as in other areas of geography and the social sciences (see Barnes, 1995; Barnett, 1998; Chaney, 1994; Sayer, 1994) has not been uncontested and is arguably far from over, not only because of the presence of a range of unresolved tensions and concerns to which Cloke alludes, but also because of the emergence of new and hitherto largely unexplored subjects. This chapter seeks to illustrate the complex, contested and changeable character of contemporary rural cultural geography by exploring the possibility that one possible further twist for, and indeed perhaps on, rural geographers might be towards examining 'obese' and 'pornographic' ruralities.

To avoid confusion, anxiety or even over-excitement, I should perhaps immediately add that I use the terms obese and pornographic in the metaphorical manner employed by Baudrillard (see particularly Baudrillard (1988a). This is not to say that there is not scope for other, more literal interpretations of these terms: studies by, for example, Matless (1994, 1998), Bell and Valentine (1995), Bell (2000a, 2000b), Little (2002) and Little and Leyshon (2001) have all highlighted how rurality has been sexualized in a whole variety of ways, including some hard-core pornographic representations which configure sex in the countryside around such notions as bestiality and unrestraint. Attention might also be drawn *to* associations between obesity and rurality, in that fattened bodies - both human and non-human - are very much part of the rural scene and economy of the affluent world. Modern agriculture and food processing is focused around producing fat, if no longer fatty, animals and these beasts also figure in many idyllic pastoral images. Substantial bodies, animal and human, also figure strongly in both comic and promotional rural imagery, while skeletal bodies, on the other hand, often connote anti-pastoral imagery: such as in the imagery of famine and the silhouetted limbs of burning cattle during Britain's Foot and Mouth 'crisis'.

This chapter will, however, be using the terms obesity and pornography in the manner adopted by Baudrillard for whom they act as metaphors to characterize important aspects of contemporary society. He argues, for instance, that 'it is not the obesity of a few individuals that is at stake, but that of a whole system ... a whole culture' and that 'how sex appears in pornography' is how things stand 'more generally [within] the project of our culture, whose natural condition is obscenity' (Baudrillard, 1987b, pp. 21-22).

Baudrillard's metaphorical usage of terms such as obesity and pornography are complex: full of ambiguities, inconsistencies and not a little innuendo, a feature which is also to the fore in Doel's (1999) take on Baudrillard but which has come under criticism, notably in relation to feminism (Gallop, 1989; Ross, 1989). Philo (1994, p. 526) similarly remarks that there is 'arguably something objectionable about using metaphors of obesity and cancer as vehicles for making theoretical claims'. Whilst having considerable sympathy with this argument, I feel that there is value in working with these metaphors so long as it is done with sensitivity. In particular I want to use them to reflect on how people might relate to signifiers of rurality, and signpost a variety of potential further 'cultural turns' for rural geographers. In doing so, I will draw upon the ideas, results and experiences of conducting research into two, in many senses quite different, rural research topics, namely rural gentrification and the production, circulation and reception of rural television drama series[1].

Baudrillard, Rurality and Rural Geography

Baudrillard's notions of pornography and obesity, and associated notions such as pornogeography (Doel, 1999), have not really been explored in the context of rural studies, despite Baudrillard's own references to having a rural, peasant background, and suggestions that this has played an important role in the formation of his thought:

> My grandfather stopped working when he died: a peasant. My father has stopped well before his time: civil servant ... I never started work, having soon acquired a marginal, sabbatical situation: university teacher. As for the children, they have not had children. So the sequence continues to the ultimate stage of idleness.
>
> This idleness is rural in essence. It is based on a sense of "natural" merit and balance. You should never do too much ... It is from this I derive a vision of the

world which is both extremist and lazy ... I detest the bustling activity of my fellow citizens, detest initiative, social responsibility, ambition, competition. These are exogenous, urban values, efficient and pretentious. They are industrial qualities, whereas idleness is a natural energy (Baudrillard, 1996a, p. 7).

Baudrillard constructs what might be said to be a static/enclosed view of the rural: it is the place where there is no change or, at most, slow change. So, for example, he writes:

For a primitive or a peasant, imagining that something could exist beyond his native space was impossible; they had never even had a premonition that something other could exist; this horizon was mentally impassable (Baudrillard, 1988a, p. 42).

The stasis and enclosure of Baudrillard's rural is contrasted in his commentaries with the urban, which is seen as the place of the quick and the changing:

The landscape, the immense geographical landscape seems a vast, barren body whose very expanse is unnecessary (even off the highway it is boring to cross), from the moment that all events are concentrated in the cities (Baudrillard, 1988a, p. 19).

Baudrillard's ideas on the rural appear in some sense highly positive – the rural is a place beyond industrial /capitalistic values and practices – and at other times highly negative – the rural is the static and the boring. Both viewpoints can be seen as highly problematic, with voluminous literatures from a whole range of perspectives having critiqued both the notion that the rural is immunized from modernization/capitalism and that the rural is unchanging. Baudrillard's work appears oblivious to such literatures[2], which might go some way to accounting for the relatively scant attention paid to him by rural geographers, although it may have something to do with a persistent, albeit arguably diminishing, 'othering' of continental social theory within rural studies (Cloke et al., 1998; cf. Miller, 1996a; Pahl, 1995).

Having said this, some people have sought to connect Baudrillard's works with rural studies (Cloke, 1992, 1993, 1994; Halfacree 1993; 1997; Lawrence, 1995, 1997, 1998), although several seem to pull back from a full embrace of Baudrillard's arguments. Halfacree (1997), for example, claims that conditions of postmodernity do not overwhelm counterurbanizers, reducing them to the '"stunned silence" ... praised by Baudrillard (1983) as the "silent majority's subversion of imposed cultural meaning"' but rather there is a clear 'lack of "silence" on the part of the counterurbanizer', epitomized by the strident voices of rural pressure groups. He also argues that there is little support for the notion of a postmodern revelry in 'the diversity and ephemerality presented in the postmodern condition, in the manner advocated by Baudrillard', in part because 'the "aesthetic" appeal of rural living is not suited to the "anti-auratic"' character of the postmodern cultural encounter', because rural landscapes are, he claims, 'unsuited to mechanical reproduction; ... [m]ore contemplative than distracted in reception; and ... [have] a popularity rooted in "high art" and Culture rather than the mundane features of everyday life' (Halfacree, 1997, pp. 79-80).

These claims about the construction of the aesthetic appeal of rurality are questionable[3], but here I wish to focus on Halfacree's suggestion that counterurbanizers are far from silent as to the meaning of rurality, not least because it does not chime particularly well with my experiences in the 'fields of rurality'[4]. These suggest that many rural residents, including many who might be described as counterurbanizers, have often been surprisingly 'silent', or at least rather hesitant and not a little troubled, when asked, for instance, about the meaning of rurality and the attractiveness, or not, it holds for them.

Stutterings Over Rurality

Yes it is rural ... It has got no street lights, no pavements, umm, not really, except for post-office it has not got any shops, umm, it has a village hall, and it is very green and everyone has got a garden. *So does that make things rural? I don't think that it does.* Very bad bus service, extremely bad bus service ... I think that the lack of transport makes it rural (Middle-aged, female resident, managerial occupation (OPCS Class II), resident in Berkshire village for over 10 years).

...if I can get in car and I can be in Norwich within about 15 minutes, *it doesn't feel particularly rural.* Now having said that, *of course it's rural*, it surrounded by working farms, big working farms. *But it doesn't feel particularly rural* because everything is so urban (Young female resident, professional occupation (OPCS Class I, resident in Norfolk village for under 5 years).

Berkshire *isn't terribly rural* to be honest, is it? *It is sort of 'con-rural', but I suppose, yes.* Well, it is certainly not urban. It is surrounded by fields and hedges and it is not joined-on to anything that it is a town. So, it can't be urban ... Maybe, maybe I still have this feeling that rural should be farming, should be more farmers, and ...there is none, there are an awful lot of people who don't have anything to do with any rural lifestyle, they just live there, I live here just because it is a nice place to live not because there is anything, not because of their occupation (Semi-retired, male resident, intermediate occupation (OPCS Class II), resident in Berkshire village for 6 years).

Rural? Well what is rural? You can't really explain the meaning of rural ... and I probably [Pauses] ... when you get older you are *a bit dubious about some of the things that are part of the countryside.* I dread to think about this at the moment. They haven't changed it too much ... because when one day this will all come back, but I shan't live long enough to see that, but things are changing (Retired man, manual occupation (OPCS Class V), living in Norfolk village most of life).

Rural. Umm, [laughs slightly and then pauses]. *Not really*, you know. *I mean it is rural* in so far as it is, you know, a smallish village, and it is almost entirely surrounded by farmland, umm, and there is, in physical terms, you know, a significant break between the city and here. So in that sense *you could kid yourself that it's rural*. But on the other hand, it does have, because of its commuterism, if you like, *it does have a slightly suburban feel*. And I suppose if there is one thing that's disappointing me about the village it is a creeping suburbanization. It's not so much that I feel that the city is coming out to us, but I feel that there are, there are a couple of sort of buffer villages between us and the city, which are very suburban, really... And my concern is that we will go the same way...the more development that goes on, the more the balance of the feel of the village will swing more towards, you know, 'modern' buildings and modern life and all of that, and away from anything that might be old and traditional and having links with the past. And that would be a shame for me. But I appreciate that in saying that, you know, I'm, umm, being slightly hypocritical about it all, because I live in a thoroughly modern commuterist sort of life, but I want to come home at the end of the day to something that sort of smacks of old England and classical lines and warm pints (Young female resident, professional occupation (OPCS Class1), living in Norfolk village for under 5 years).

These comments were made within interviews I was conducting as part of a study of rural gentrification. Although there are a number of other interpretations of the concept of gentrification (see Phillips, 1993, 2001, 2002, forthcoming b), a widely held view is that it involves the in-migration of middle class residents who have a strong cultural predilection for rural living (see Little, 1987; Cloke and Thrift, 1990, Cloke et al, 1995, 1998; Smith and Phillips, 2001). As noted in Cloke et al. (1995), questions might be raised about whether this predilection is restricted to the middle classes, or whether these groups simply have greater opportunity than some other groups to express this view through action (see also Hoggart, 1997). The five quotes may be seen to offer some support to this argument in that, whilst occupying a range of different class positions and settlement histories, the respondents all evidenced some attachment to rurality. Rather more readily apparent, however, is the difficulty they had in expressing what rurality meant. Although all managed to volunteer some statement as to what they understood by rurality, these were often retracted, with the interviewees backtracking and at times contradicting themselves, often quite self-consciously. Their accounts of rurality can be seen as stuttering accounts, involving hesitancy and repetition, and in many cases evident signs of unease which teeter on the brink of reducing the person to silence, at least on the subject of what rurality might mean. Drawing on Baudrillard and speaking more metaphorically, Clarke and Doel (1994, p. 510) claim that the 'stutter' marks social movement into a world of 'metastatic, viral and fractal proliferation', a world which, as will be discussed later, might be characterized as obese and pornographic (see also Doel,1999 p. 169).

Hesitancies and silences with regard to rurality have arguably been largely over-looked by rural researchers, particularly following their cultural turn towards 'giving voice' (Philo, 1992) to diverse and often neglected interpretations of rurality. A series of studies influenced by notions of postmodernism, for example, have sought to foreground hitherto neglected representations of rurality (see Cloke and Little, 1997 and Milbourne, 1997; and Phillips, 1998 for an overview), and there has also been increasing recognition of diversity within mainstream conceptions of rurality (see Bunce, 1994; Cloke et al., 1997; Halfacree, 1995; Matless, 1994; Phillips et al., 2001; Short, 1991). One notable exception to this neglect of the hesitant and the silent is Jones' (1995) discussion of 'Lay discourses of the rural' in the village of Priston, Avon. He notes that while a large majority of people interviewed considered this village to be rural, 'a small percentage did not' and 'those who thought it was rural did so for differing reasons, as did those who though it was not' (p. 42). He also notes that Priston's residents generally did not use the term rural until he, the researcher, had used it, although once it had been mentioned people were quite prepared to use the term albeit it being evident that 'rural meant different things to different people', with people 'forming different views of what is rural, which in some cases would barely overlap' (ibid.). Jones concludes from such findings that 'although rural and its synonyms are commonly recognized terms, it is by no means certain that they are commonly UNDERSTOOD in any coherent fashion' (ibid., emphasis in the original).

Following this argument, Jones logically calls for the study of diverse representations of rurality, a research problematic which has, as noted above, come subsequently to form one of the principal 'cultural turns' within rural geography. This problematic is one which, drawing on the work of Baudrillard (1987a, 1987b, 1987c, 1988a), might be characterized as addressing, and indeed arguably contributing to, the 'obesity' of rurality.

Obese Ruralities

Contemporary societies have, Baudrillard argues, become obese in that they are 'bloated' with signifiers such that they start over-lapping and interfering with each other: signs become 'pregnant, symbolically speaking, with all the objects from which they have been unable to separate' (Baudrillard, 1987c, p. 30). Such a description may be seen to chime closely with at least four concerns within contemporary rural geography. First, there is the long running pursuit of a definition of rurality (see Halfacree, 1993, for a critical overview). Whilst for many years the focus was on a search for a single definitive conception of rurality, it is now widely recognized that a multiplicity of alternative definitions can coexist, even though these definitions may well contradict each other and produce quite different cartographies of the rural. Second, the turn towards cultural representations of rurality has served to foreground a series of differing constructions of rurality: to the rural or pastoral idyll, for instance, one can now add the anti-idyll (Bell, 1997), the anti-pastoral (Short, 1991), the non-idyll (Phillips et al., 2001), queer country (Bell and Valentine, 1995), the eroticized rural (Bell, 2000a), mediated rurality (Phillips et al., 2001), and masculinized and feminized countrysides (Agg and Phillips, 1998; Brandth, 1994; 1995; Brandth and Haugen, 2000; Liepins, 2000b; Longhurst and Wilson, 1999), to cite just a few. Third, the turn towards cultural representations of rurality has also served to deconstruct rurality, revealing how notions of rurality draw upon a range of other signifiers, such as nature (rurality as a 'space of nature' (Cloke et al., 1996) or as the agricultural and thereby in effect the 'space of the cultivation of nature' (Short, 1991)), history (rurality as the traditional, the non-modern (Lawrence, 1998; Murdoch and Pratt, 1993; Wright, 1985)), community (Liepins, 2000a, 2000c; Murdoch and Day, 1998; Tönnies, 1963) and urbanity (the rural as simply 'the non urban' (Short, 1991)). Fourth, the notion of the 'post-rural' (Murdoch and Pratt, 1993) points to how cultural constructions of rurality can be found in all manner of sites, often far removed from any spaces which might be conventionally defined as rural. Several studies of urban gentrification have, for instance, identified rurality and rusticity as significant in the reconstruction of inner city localities (e.g. Bell, 1996; Smith N., 1996), while mass media such as television, film and increasingly the Internet can circulate representations of rurality quite literally to the other side of the world (Fish and Phillips, 1999; Phillips, forthcoming c).

These contemporary concerns of rural geographers can be connected to the stuttering accounts of rurality presented earlier. The accounts were produced in response to questions concerning how people understood and applied the term rurality to their present residential locality. In asking such questions I very much had in mind the work of people such as Halfacree (1993; 1995) and Jones (1995), and the notion that there are lay as well as academic 'definitions' of rurality. These quotes also reveal multiplicity in conceptions of rurality: Table 2.1, for instance, shows how these small numbers of quotes reveal most of the senses of rurality identified by Halfacree (1995). The notions of rurality presented also show how notions of rurality flow into a whole series of other concepts including the agrarian, the historical, the natural, the open, the urban and suburban (see also Phillips, 2002; forthcoming b). Finally, while the comments all came from locations which had been classified academically as rural, for several people these places were not completely, and/or authentically, rural.

As well as posing interesting and hitherto rather neglected research questions in their own right, these arguments may be seen to point to a further rather unexplored 'cultural turn' for rural geography. Within many areas of human geography and the social sciences more generally, movements to address the cultural have often involved careful examination of notions of subjectivity and the human subject (see, for example Pile, 1996; Pile and Thrift, 1995). Whilst there have been some important discussions of various 'research subjects' (e.g. Cloke, 1994; Hughes et al., 2000) and 'gendered subjectivities' (e.g. Agg and Phillips, 1998; Brandth and Haugen, 2000; Hughes, 1997; Little and Austin, 1996) within rural studies, there has been little sustained engagement with wider debates over the status of the human subject, although several of the studies listed earlier as drawing on the work of Baudrillard do make some reference to these debates.

Table 2.1 Senses of rurality within the stuttering accounts

Halfacree's (1995) 'descriptions of rurality'	Associated comments in stuttering accounts
Contextual	'surrounded by working farms', 'surrounded by fields and hedges', 'almost entirely surrounded by farmland'
Named settlement type	'village hall', 'smallish village',
Populations size/density	'a smallish village', not 'developed'
Environmental	'very green', 'surrounded by fields and hedges', 'old and traditional and having links with the past'
Occupational	'big working farms', 'rural should be farming'
Locational	'I can get in car and I can be in Norwich within about 15 minutes', 'not joined-on to anything that it is a town', 'significant break between the city and here', 'buffer villages between us and the city'
Social	'something that sort of smacks of old England and classical lines and warm pints'
Functional	'no street lights, no pavements', 'umm, not got any shops', 'very bad bus service'
Animals	-

The work of Halfacree (1997) and Lawrence (1997), for example, can be seen to recognize connections between human subjectivity and what I have called 'obese rurality'. Lawrence, for instance, draws on Baudrillard (1983) to suggest that 'a redoubling of signs of the rural' may have occurred 'as both cause and effect of a growing indeterminacy of rural space', such that these signs may 'no longer make the previously opaque transparently clear' and people 'become an inert mass incapable and unwilling to choose moment to moment under which ... terms of "meaningfulness" ... to live' (Lawrence, 1997, p. 14). Halfacree (1997) also makes reference to the arguments of Baudrillard (1983), although as noted earlier, he is sceptical about whether the 'postmodern condition' of a 'rubble of signifiers' actually 'over-whelms the individual and reduces him or her to a stunned silence' (Halfacree, 1997, p. 78).

Despite their differences, the studies of Lawrence and Halfacree do serve to highlight how symbolic obesity can impact human subjects, a point which has indeed been widely recognized in urban studies particularly through Jameson's (1991, pp. 42-44) claim, itself constructed at least in part through the mediation of Baudrillard's writings, that 'postmodern hyperspace' - a 'filled space' in which 'you yourself are immersed, without any of that distance that formerly enabled perception of perspective or volume' - has exceeded 'the capacities of the individual human body to locate itself, to organize its immediate surroundings perceptually, and cognitively to map its position in a mappable external world'. According to Baudrillard the human subject has become 'schizophrenic', placed in a position of inability to act by 'a state of terror' induced by being 'open to everything' (Baudrillard, 1988a, p. 27). The schizophrenic Baudrillard argues, 'is not, as generally claimed, characterized by his [sic] loss with reality' but is 'over-exposed' to the world, 'by the absolute proximity to and total instantaneousness with things' (p. 27). According to Baudrillard this experience has become generalized such that:

It is no longer a question today of believing or not believing in the images which pass before our eyes. We refract reality and signs indifferently without believing in them. This is not even incredulity: our images simply pass through our brains without landing on the square marked "belief" (Baudrillard, 1994, p. 92).

The stuttering accounts of rurality displayed earlier in this chapter might hence be presented as being a product of a loss of 'cognitive maps' relating to a growing obesity of rural imageries, to an 'excess of rurality' (cf. Baudrillard, 1988b, p. 155). In other words, people are unsure as to what the rural might be because they recognize a range of possible answers, and somewhat contra Halfacree (1995), also recognize some incompatibilities between them. The rural hence does not become a 'fleeting entirety' to be drawn upon in acts of interpretation but more like a swirling vortex of fragmented signs of rurality in the face of which people stumble and stutter, before attempting to mix together a consumable concoction of rurality. Indeed, for some people this may prove impossible and a further consequence of obese rurality is that people may become reluctant to assign rural status to the material and symbolic spaces of their everyday lives because they feel that these spaces do not live up to the full range of meaning associated with rurality. The material and symbolic spaces where these people resided were, for instance, clearly valued but were also frequently described as not 'fully' or 'truly rural', as 'semi-rural', as 'pseudo-rural', even as 'con-rural', notions which do imply some notion of a real, authentic rurality but which locate it as always elsewhere, in some other place or time (Cloke and Phillips, forthcoming; Cloke et al., 1998).

Porno-rurality

For Baudrillard, contemporary cultures are not simply obese, but can also be described revealingly through series of other metaphors, including as already mentioned, the term pornographic. This final section will briefly explore what he means by this term and whether it might be used to interpret some stuttering accounts of rurality.

Baudrillard describes pornography and the obscene as 'the all too visible' (Baudrillard, 1988a, p. 22). He claims that pornographic sexual imagery 'decomposes bodies into their slightest details' (p. 43), zooming in as close as possible to create a series of 'high fidelity images' (Baudrillard, 1990b) in which there is a 'promiscuity of detail' (Baudrillard, 1988a, p. 43) of sexual acts and organs, but in the process undermining the allusions and seductions which are part of the sexual. Similarly, Baudrillard argues that contemporary culture involves a loss of illusion as more and more detailed information is produced and circulated relating to more and more areas of social life such that 'every-thing appears ... transparent, visible, exposed in the raw and inexorable light of information and communication' (Baudrillard, 1988a, p. 22).

Clearly there are points of connection between Baudrillard's notions of the pornographic and the obese, not least in that both revolve around the expansion of signs. There are, however, important differences, differences that in part relate to the relationship of the subject to these signs. The human subject in obesity often appears in Baudrillard's work to be immobilized by signs, simply swallowing everything that comes within reach: to return to Jones' analysis of rurality, agglomerating as many understandings of rurality together as possible. By contrast, the subject in pornography appears within Baudrillard's writings as an active and self-conscious 'voyeur' (Philo, 1994, p. 528), obsessively searching out more detailed and decontextualized signifiers of the desired object and yet at the same time perhaps dimly recognising that these signs represent the disappearance of the desired 'focal point' (Doel, 1994). Baudrillard characterizes pornography as 'the excessiveness of sex and its disqualification' (Baudrillard, 1988a, p. 33), and also as a 'fascination' with this 'lost frame of reference' (Baudrillard, 1987b, p. 15). Applying the term more generally, the pornographic implies an excessiveness of reality (the hyperreal as the all too real, the 'truer than true' (Baudrillard, 1990a, p. 13), 'the all-too visible, the more-visible-than visible' (Baudrillard, 1988a, p. 22)), an undermining of a sense of the real and a fascination with the real. Using such comments one might identity a rural 'pornogeography' (Doel, 1999, p. 75)[5], a 'pornorurality', which involves an excessiveness of rurality, a loss of faith in the rural, and an obsessive fascination with rurality. Indeed Lawrence could be seen to be describing pornographic rurality when characterizes the film *Miles from Home* as a simulation of:

> first of all ... that we had assumed the rural was supposed to be, second of all ... that we now suspect or cynically shrug it off for being, and third of all ... that we obstinately demand that that it remain (Lawrence, 1998, p. 715).

Interestingly given discussion of how Baudrillard's notion of simulation has widely been interpreted as involving signs without a referent (see Halfacree, 1993; Lawrence, 1995), Lawrence claims that:

> Baudrillard understands ... that simulation is never entirely independent of a relation to reality. He describes the "strategy of the real" as a varied range of tactics whereby every effort is made to reassert referentiality in the face of proliferating simulations, to force context and a more specific mean onto simulations which are otherwise marked by their volatility and chameleon significance (Lawrence, 1998, p. 715).

Caution is needed, however, because Lawrence slides quickly between the real and the 'strategy of the real'. If the real is taken to refer to something beyond signs but to which these signs can be made to refer, then for Baudrillard this notion of reality is generally framed as being illusory: there is 'no ... system of reference to tell us what happened to the geography of things' (Baudrillard, 1987a, p. 512) and the notion of reality 'is itself merely a model of simulation' (Baudrillard, 1996b, p. 17). However, Baudrillard at times appears to recognize that there may be reality beyond simulations – '[t]he real ... is the insurmountable limit of theory' (Baudrillard, 1987a, p. 512; see also Butler, 1999; Cubitt, 2001; Perry, 1998) - and more generally suggests that the notion of the real is a necessary illusion in that people find it unbearable to consider an unrepresentable world and develop a variety of strategies to realize the real:

> the world ... is a radical illusion. That at least is one hypothesis. At all events, it is an unbearable one. And to keep it at bay, we have to realize the world, give it force of reality, make it exist and signify at all costs, take from it its secret, arbitrary, accidental character, rid it of appearance and extract its meaning (Baudrillard, 1996b, p. 16).

These three senses of the real - as pure simulation, as the limit of simulation and as a vital illusion to make the world bearable - may be seen to imply quite different human subjects, which might be characterized as a Baudrillardian subject of 'promiscuous voyeurism', a subject of 'creative engagement' and a subject of 'realist resistance'.

The Baudrillardian Subject of Promiscuous Voyeurism

Human subjects might be characterized as 'Baudrillardian' in the sense that they embrace simulation, letting go of the notions of representation and the 'serious' task of creating meaningful signs in favour of a 'playful' (Doel, 1994, p. 126) 'dance' (Philo, 1994, p. 530) or 'motionless voyaging' (Doel, 1999) through the pornography of the real. Whilst such activity may appear very much as an invitation to 'personal indulgence' and the voyeuristic researching of 'quaint topics of pure esoterica' (Peet, 1998, p. 242; see also Phillips, 1994), a series of studies of the mass media and tourism have identified 'popular' Baudrillardian subjects. Feifer (1985), for example, has posited the existence of 'post-tourists' who, as Urry (1990, p. 11) puts it, 'know there is no authentic tourist experience' but gain pleasure from engaging in a multiplicity of tourist experiences which they view as 'merely a series of games or texts to be played'. Urry (1995) identifies a range of other 'postmodern' activities, including many undertaken in the countryside, which involve adoption of simulated identities and a suspension of criteria of authenticity. Whilst authenticity has long been a major ingredient in cultural constructions of rurality – rural spaces being widely represented as places of 'real' communities and of 'natural' relations between people and nature – many rural spaces are being promoted and constructed though the incorporation of 'spectacularized' and 'simulated' identities and meanings (see Cloke, 1992, 1993, 1994). Whilst for many commentators and consumers such spaces are still criticized for their inauthenticity (see Holloway and Hubbard, 2001), their evident popularity may speak to a disengagement with notions of authenticity in favour of other engagements, such as itinerant pursuit of the spectacular and the novel (see Cloke and Perkins, 1998).

The mass media has long figured prominently in Baudrillard's discussions (see Stevenson, 1995), and while its various elements should not be seen as the only source of simulations in contemporary sign-saturated social worlds, as Kellner (1995, p. 5) has observed, they do form 'an immediate and pervasive aspect of contemporary life ... constituted by and constitutive of larger social and political dynamics', and thereby provide 'an excellent optic to illuminate the nature of contemporary society, politics and everyday life'. Perversely, as noted in Phillips (1994), rural geographers have long paid very little attention to the mass media, although recent years have witnessed growing interest in the role that they may play in the construction of symbolic and material ruralities (see Short, 1991; Cloke and Milbourne, 1992; Bunce, 1994; Jones, 1995; Fish and Phillips, 1999; Lawrence, 1997; Phillips et al, 2001).

A number of studies of the mass media can be seen to embrace the Baudrillardian subject, arguing that mass media audiences often adopt a schizophrenic subjectivity, gaining pleasure through fleeting and uncommitted engagements with all manner of simulations which are flashed before them (e.g. Krocker and Cook, 1986; Munson, 1993). Many of these simulations may be seen as pornographic in that they present more and more aspects of social life in ever expanding detail, and the Baudrillardian audience subject might hence be characterized as a 'promiscuous voyeur', gazing on a plethora of simulations of the real at the expense of making direct and selective engagements. One illustration of this possibility is Lawrence's (1995, p. 302) claim that the proliferation of representations of rural homelessness 'has the spectacular effect of actually distancing us from direct engagement with the facts of homelessness'.

The Creatively Engaged Subject

Some studies of the mass media have expressed reservations about the Baudrillardian subject, arguing amongst other things, that it has often been constructed with little actual engagement with mass media audience, save perhaps of the media theorists' reflections of themselves 'as individuals seated in front of the television screen' (Tulloch, 2000, p. 8). As Tulloch (2000, p. 8) notes, many in media studies influenced by the arguments of the likes of Baudrillard have come to reject 'the "metaphysics of presence"' implied in earlier forms of media research whereby the 'lived experience' of audiences (or indeed producers) was seen as a 'source of authentic knowledge' though which to validate, or undermine, the analysis of the media theorist. If the knowledge of the audience is itself merely a simulation, then recovery of this knowledge may be seen to simply imply 'an infinite regress' (p. 9) whereby one simulation is merely supplemented by others of equal (in)validity, in which case why not simply research one's self.

Whilst there is a lot of value in such 'autoethnography' (for a rural illustration see Cloke, 1994), and as Lawrence (1997) has discussed in the context of rural studies, considerable difficulties and dangers associated with engaging with the 'lived experiences' of others, like Tulloch I feel that there is still much to be gained theoretically by such engagements, not least because without them there is a danger of universalizing and essentializing the theorist's own, necessarily positioned and limited, experiences and subject positions (Tulloch, 2000, p. 20; see also Phillips, 2000). Even if all experiences are seen as simulations, this does not mean that everybody experiences the same simulations. As Perry (1998, p. 1) argues, there is no need to presuppose that 'hyperreality is invariant'. Writing from New Zealand, he argues:

> A location at the edge of the (Western) world is especially conducive to these kind of Delphic observations ... For under such conditions hyperrealities seem in no way exotic. For example, the snow scenes on New Zealand Christmas cards signal the arrival of summer - just as the fertility symbols of Easter (eggs and rabbits) signal its departure (Perry, 1998, pp. 1-2).

He adds that in New Zealand another illustration of simulation becoming the mundane is the presence of images of distant places within the country's television broadcasting: 'the nation has a higher proportion of overseas content on its national television than anywhere else in the world' (Perry, 1998, p. 2), and as he has previously observed (Perry, 1994), much of this television content is also rural. British rural dramas have long figured extensively in broadcasting in New Zealand, with drama series such as *Heartbeat* and *Peak Practice* proving highly successful (see Phillips, forthcoming c). Images of the British countryside may hence contribute as much to New Zealanders' sense of the rural as do images of the New Zealand countryside, and indeed conducting research with New Zealanders watching such programmes produced commentaries which suggest that simulated images of British countryside may be being seen as more real than personal contact with the New Zealand countryside:

> they seem more like what you expect in the country ... *Take the High Road* ... we had that about 5 years ago, and that was brilliant, the scenery and the everyday life of people around there, but it also involved people in a stately home ... It was just an ordinary, everyday thing about working people, but brilliant scenery ... I don't think an urban setting has anything going for it, like a country setting, at all ... [T]he scenery mainly, and the people are different. The people in the country are much more tolerant ... and they ... all know their neighbours, they are all friendly ... things are much more laid back, people are co-operative and friendly (Middle aged woman, nurse, resident in New Zealand all her life).

As discussed in a previous work (Phillips et al., 2001), in the past such accounts may well have been dismissed by critical analysts of the rural as indicative of naively idyllic, false or ideological understandings of rurality. One of the most positive aspects of rural geographers' interest in Baudrillard's notion of simulation may be the prohibitions it places on a simple dualism of signifiers into the true and the false, and a resultant dismissal of some people's accounts of the rural. On the other hand, quite similar dualisms may inhabit Baudrillard's own accounts of simulation. Perry (1998, p. 78), for example, identifies a series of dualisms within Baudrillard's accounts of simulation, including 'contrasts between the deep frozen/warm; infantile/affection; atomized/ sociability', but argues that Baudrillard repeatedly privileges one side over the other. Amongst the consequences of this, Perry (1998, p. 78-9) suggests, is that Baudrillard 'suppresses difference' in order to both 'amplify paradox' and to 'ensure the plausibility of his argument'. The differences suppressed are, Perry claims, those that cannot be effaced by simulation, an argument that is also made by Cubitt:

> Baudrillard ... imagined a world in which individual differences have been erased under the sheer homogenising power of the Code ... Commodity culture, the spectacle and simulation, the transparent and the hyperreal, may be triumphant, but they cannot be total ... some reality remains, but it is structured by its exclusion from the spectacularly null totality of commodity culture (Cubitt, 2001, p. 92).

Cubitt and Perry both explore the subversion of 'the code' within Disneyland. Writing about the mass media rather more generally, Nightingale (1994, p. 4) argues for a recognition of 'performed texts' which although connected to the circulated sign, effectively 'outstrips the broadcast text in both significance and vitality'. Rather than focus exclusively on the symbolic construction of the text, there is, she argues, a need to consider 'audience-text' (and indeed audience-medium, audience-industry (Nightingale, 1984)) relations, and to consider them in relation to self-everyday life: 'The relation between self and everyday life is ... the macro-version of audience-text' (Nightingale, 1996, p. 130). Both self-everyday and audience-text constitute performances whose effects are not necessarily those of the simulation: 'the audience-text relationship encompasses the transformation of the heterogeneous, haphazard particularity of the everyday into personal narratives ... transforming particular potential into examples and vice versa' (Nightingale, 1996, p. 150).

At present Nightingale's discussion of such transformations is largely programmatic but her arguments can be seen to resonate with a series of other emergent and cross cutting theoretical projects in contemporary socio-cultural analysts, including re-theorizations of 'audiencing' (e.g. Abercrombie and Longhurst, 1998; Alasuutari, 1999; Ang, 1996; Tulloch, 2000), the spatialities of representation and spaces of representation (Lefebvre, 1991; Phillips, 2002; Soja, 1995, 2000), and non-representational theory and performance (Abercrombie and Longhurst, 1998; Hughes-Freeland, 1998; Thrift, 1996, 1999, 2000a); as well as discussions surrounding Baudrillard and the significance of his notion of simulation to understanding contemporary mass media (Grossberg, 1987; Kellner, 1995; Nelson, 1997; Tulloch, 2000). Elements in all these various streams of writing not only allude to the creation of meaning other than those of the simulated texts, but also to use of these simulations as part of an attempt to make sense of the everyday and the position of the self there. Nelson (1997), for instance, argues that the British rural television drama *Heartbeat* can be interpreted as a postmodern pastiche of simulations for which there are no referents, but then suggests that he cannot go 'the whole way with Baudrillard to deny any reference points in a more grounded semiosis or the reality of lived experience' (p. 87). He claims that

Heartbeat both draws on 'middle distance' conceptions of the world which are widely used by people in their everyday lives, and also 'by offering other potential worlds' intersects 'in other ways with social actuality', being 'taken by some viewers as a critique of the harsher world of historical reality' (p. 123). Likewise, drawing on group discussions with rural residents and viewers of programmes like *Heartbeat*, Phillips et al (2001, p. 14) argue that some people, 'whilst acknowledging some artificiality in the image ... still connect it with their own experience of the real, whilst also projecting their own particular hopes and desires onto the text'. If 'pornography is there ... in order to prove ... by its grotesque realism ... that there is however some real sex somewhere' (Baudrillard, 1987b), then perhaps, for at least some subjects, simulated pornoruralities are being used less voyeuristically and more creatively to point to the possibilities of other unsimulated ruralities?

The Realist Subject

Baudrillard's third sense of the real is as a vital illusion, a strategy of avoiding life in the simulated world by continuing to believe in the possibility of finding and living by reality principles. The very description of this as an illusion recreates dualisms such as fiction/reality and truth/falsity, and thereby acts to effectively distance Baudrillard from this subject position. It appears that while other people might live by this illusion because they cannot bear living in the reality of unreality, Baudrillard is a much stronger subject capable of 'putting the illusion of the world to death' (Baudrillard, 1996b, p. 16).

Such claims invite criticisms of elitism and masculinism, as well as perhaps elements of 'performative contradiction' (Phillips, 1994, p. 112) in that Baudrillard seems able to discern reality whilst arguing for the impossibility of knowing the real (see Best and Kellner, 1991). However, in the present context such concerns will be left to one side in favour of highlighting the significance of Baudrillard's claim that, in a world of proliferating signs, concern for referencing the real also proliferates. Given these claims, it is perhaps unsurprising that in a study of the production and audience readings of four British rural television series, it was apparent that many people were reading these fictional, and for Nelson (1997) thoroughly simulated, programmes in a 'self-extricated fashion' repeatedly 'seeking to separate the real from the imaginary' (Phillips et al., 2001). For many people this was a source of the pleasure they gained from watching these programmes, a claim which appears to resonate with Tulloch's (2000) discussion of 'police drama' viewing in Australia, which suggests that some people use these programmes to expose and probe their own conceptions of the real world. As Nelson comments:

> If realism is illusionistic in more sense than one, its illusions remain potent nevertheless for the human beings who give them credit, and for whom realist narratives are the key means of making sense of their world (Nelson, 1997, p. 110).

For all the emphasis given here to a multiplicity of signs of rurality and the stutterings that may for various reasons accompany them, research in the fields of rurality also leads to encounters with people for whom rurality has very definitive, and in all probability unchangeable, meanings.

Some Points of Conclusion

Stimulated by a number of stuttering accounts of rurality, this chapter has sought to explicate some of the cultural turns that rural geographers have taken in the last few years and signpost a few further turns that I, and perhaps others, might want to pursue in the future. The chapter has focused on the notions of obesity and pornography, drawing on Baudrillard's metaphorical use of these terms although noting that more referential use of these terms might point to some hitherto rather neglected rural geographies. Although there has been widespread examination and exploration of language use within other areas of human geography (e.g. Barnes, 1992; Barnes, 1996a; Barnes and Duncan, 1992; Bondi, 1997; Curry, 1991; Olson, 1980, 1992; Smith J., 1996; Strohmayer and Hannah, 1992), within rural geography there has been very little consideration of such issues (although see Cloke et al., 1994), beyond a concern with how to give voice to others (see Lawrence, 1995, 1997) and make these voices heard in policy context (e.g. Cloke, 1995; Cloke, 1996), together with sporadic complaints about the comprehensibility of theoretical terminology (Gilg, 1985; Miller, 1996a, 1996b; Robinson, 1990). Language, particularly the language used by rural geographers in their talks and writings, is arguably still seen as primarily referential. Whilst new terms and new ways of writing have emerged, they are portrayed largely as better, or perhaps now as simply different, ways to 'capture', 'grasp' or 'portray' reality or ideas about reality. There has as yet been no attempt to document the differing forms of expression used by rural geographers along the lines that Barnes (1992, 1996a, 1996b) has done for economic geography, and this constitutes another cultural turn which rural geographers might potentially take. Rather than attempt such a documentation, I have chosen instead to foreground the metaphorical aspects of language by exploring Baudrillard's notions of obesity and pornography in connection with a number of existing rural studies and my own research in various 'fields of rurality'.

An initial starting point for this exploration was the presence of a series of pauses, retractions and contradictions in accounts as to the meaning of rurality elicited through personal interviews with rural residents. In their various cultural turns exploring and deconstructing multiple definitions and cultural constructions of rurality in rural and non-rural localities, rural geographers have arguably rushed past such hesitancies and silences (although see Jones, 1995). Stopping to consider these 'stutterings' raises a series of questions, including whether the activities of rural geographers might themselves be one of the constituent elements:

Promotion of discourse on the rural which wants to acknowledge the plurality of voices and visions concerned with it in the hopes of empowering different perspectives may inadvertently reproduce Babel instead of broadened communication (Lawrence, 1997, p. 15).

A further dimension of the stuttering accounts is that they may be seen to signpost the neglect within rural geography of a cultural turn towards discussions over the status of the human subject. Although forming a major strand within other areas of human geography it has been largely absent from the rural sub-discipline, although some of the rural geographers who have drawn on Baudrillard's notion of simulation have taken some steps towards the subject. In this chapter, Baudrillard's notions of obesity and pornography are used to take a few further steps down this path. In particular the possibility of an obesity in signifiers of rurality creating a confused, hesitant and in some cases sceptical rural subject are highlighted, while three possible subject positions in pornographic rurality are identified: namely 'promiscuous voyeurism', 'creative engagement' and 'realist resistance'.

In identifying such a range of subject positions I have drawn on the arguments of a range of authors who have both drawn on and expressed some reservations about Baudrillard's notions of simulation. Whilst the notion of rurality as a sign for which there is no original referent, and the associated metaphors of obese and pornographic ruralities, are, for me at least, quite engaging ones, I would concur with Perry's (1998, p. 1) claim that less axiomatic, more contingent, theorizations are required which seek to theorize both the 'plurality of ... determinations and the diversity of ... manifestations'. If rurality has become obese and pornographic for some people, as yet it appears unlikely that it is for everybody, nor that people will live through such ruralities in the same way.

Notes

1. My research on these topics has spanned a number of projects and collaborations but this paper draws particularly upon an ESRC Research Fellowship examining '*The processes of rural gentrification*' (Ref: H53627500695) and an AHRB Research project on '*Globalising British ruralities: the export and reconstruction of socio-spatial identities through British television dramas*' (Ref: AR146000).
2. In this context, it is interesting to note the arguments of Brass (1994) that Baudrillard's 'postmodernism' was prefigured in the work of McLuhan and that McLuhan espoused an 'agrarian populist belief in a mythical/folkloric conception of "primitive"/"natural" (unspoiled) man'.
3. Work by people such as McNaghten and Urry (1998) on 'dwelling', and by Urry (1995b), Harrison (1991) and Cloke (1998) on experiential and emotional aspects of rural recreation, for instance, would all seem to both point to the significance of the mundane as well as question whether the countryside is always approached through quiet contemplation. Similarly, work by people such as Bowler and Everitt (1999), Bunce (1994), Cloke (1992; 1993), Houlton and Short (1995), McNaghten and Urry (1998) and Moir (1964) all point to various mechanical productions of rurality through commodities.

4. Within geography, and also disciplines like anthropology, there has been growing attention paid to the notion of fieldwork (Clifford, 1983, 1997; Clifford and Marcus, 1986; Fox, 1991; Geertz, 1973; Katz, 1992, 1994). Attention has also been drawn, amongst other things, to the way fieldwork has often been used to confer authority to particular knowledge constructs and, somewhat conversely, how research involves working in a variety of 'fields', not all of which are located, to use Murdoch's (1995) words 'out there in the world of others'. A clear example of the former in the context of rural studies is Miller (1996a) who identifies 'a retreat' in rural studies away 'from the awkward realm of "the rural" out there back into the monastic seclusion of reflexivity and derivative theorizing' (p. 17). As I have argued elsewhere (Phillips, 2000), such arguments underplay the significance that theorising and writing play within the practices of so-called field research. They also neglect the extent to which rural researchers are now engaging with a range of new 'fields of research', including published texts, marketing, film and television imagery, and the Internet. As outlined in the text, the present paper draws on experiences and 'texts' from a range of research projects, including transcripted interviews and focus groups with people living in rural areas and with people who 'consume' rural imagery but live in urban areas.

5. Pornorurality might be seen as one form of 'pornogeography' to which a multitude of others could be added. David Clarke and Marcus Doel might be said to have already outlined two pornogeographies in their discussions of 'transpolitical geography' and 'transpolitical urbanism' (Doel and Clarke, 1997 and Clarke and Doel, 1994 respectively). There may well be close parallels with ideas of 'postnationalism' (Anderson, 2000), 'postculture' (Bannet, 1993; Kahn, 1995; Phillips, forthcoming a) and 'post-development' (Crush, 1995; Rahnema and Bawtree, 1997). The notion of the pornographic might also be very applicable to nature.

References

Abercrombie, N. and Longhurst, B. (1998), *Audiences: a Sociological Theory of Performance and Imagination*, Sage, London.

Agg, J. and Phillips, M. (1998), 'Neglected Gender Dimensions of Rural Social Restructuring', in M. Boyle and K. Halfacree (eds), *Migration into Rural Areas: Theories and Issues*, Wiley, London.

Alasuutari, P. (1999), *Rethinking the Media Audience*, Sage, London.

Anderson, K. (2000), 'Thinking "Postnationally": Dialogue Across Aulticultural, Andigenous, and Settler Spaces', *Annals, Association of American Geographers*, Vol. 90, pp. 381-391.

Ang, I. (1996), *Living Room Wars: Rethinking Mass Media Audiences for a Postmodern World*, Routledge, London.

Bannet, E. (1993), *Postcultural Theory: Critical Theory After the Marxist Paradigm*, Macmillan, London.

Barnes, T. (1992), 'Reading the Texts of Theoretical Economic Geography: the Role of Physical and Biological Metaphors', in T. Barnes and J. Duncan (eds), *Writing Worlds: Discourse, Text and Metaphor in the Representation of Landscape*, Routledge, London, pp. 118-35.

Barnes, T. (1995), 'Political Economy 1: the Culture, Stupid', *Progress in Human Geography*, Vol. 19, pp. 423-431.

Barnes, T. (1996a), *Logics of Dislocation : Models, Metaphors, and Meanings of Economic Space*, Guildford Press, London.

Barnes, T., and Duncan, J. (1992), (eds) *Writing Worlds:Discourse, Text and Metaphor in the Representation of Landscape*, Routledge, London.

Barnett, C. (1998), 'Cultural Twists and Turns', *Environment and Planning D: Society and Space*, Vol. 16, pp. 631-634.

Baudrillard, J. (1983) *In the Shadow of the Silent Majorities*, Semiotext(e)), New York.

Baudrillard, J. (1987a), 'Forget Baudrillard: an Interview with Sylvère Lotringer', in J. Baudrillard and S. Lotringer (eds), *Forget Foucault and Forget Baudrillard*, Semiotext(e)), New York, pp. 65-137.

Baudrillard, J. (1987b), 'Forget Foucault', in J. Baudrillard and S. Lotringer (eds), *Forget Foucault and Forget Baudrillard*, Semiotext(e)), New York, pp. 9-64.

Baudrillard, J. (1987c), *Cool Memories*, Galilee, Paris.

Baudrillard, J. (1988a), *The Ecstasy of Communication*, Semiotext(e)), New York.

Baudrillard, J. (1988b), 'On Seduction' in M. Poster (ed), *Jean Baudrillard: Selected Writings*, Polity, Cambridge, pp. 149-165.

Baudrillard, J. (1990a), *Fatal Strategies*, Pluto, London.

Baudrillard, J. (1990b), *Seduction*, Macmillan, London.

Baudrillard, J. (1994), *The Illusion of the End*, Polity, Cambridge.

Baudrillard, J. (1996a), *Cool Memories II*, Polity, Cambridge.

Baudrillard, J. (1996b), *The Perfect Crime*, Verso, London.

Bell, C. (1996), *Inventing New Zealand: Everyday Myths of Pakeha Identity*, Penguin Books (NZ), Auckland.

Bell, D. (1997), 'Anti-idyll: Rural Horror', in P. Cloke and J. Little (eds), *Contested Countryside Cultures: Otherness, Marginalisation and Rurality*, London, Routledge, pp. 94-108.

Bell, D. (2000a), 'Eroticizing the Rural', in R. Phillips, D. Watt and D. Shuttleton (eds), *Decentring Sexualities: Politics and Representations Beyond the Metropolis*, London, Routledge.

Bell, D. (2000b), 'Farm Boys and Wild Men: Rurality, Masculinity, and Homosexuality', *Rural Sociology*, Vol. 65, pp. 547-562.

Bell, D., and Valentine, G. (1995), 'Queer Country: Rural Lesbian and Gay Lives', *Journal of Rural Studies*, Vol. 11, pp. 113-22.

Best, S., and Kellner, D. (1991), *Postmodern Theory: Critical Introductions*, Macmillan, London.

Bondi, L. (1997), 'In Whose Words? On Gender Identities, Knowledge and Writing Practices', *Transactions, Institute of British Geographers*, Vol. 22, pp. 245-258.

Bowler, I. and Everitt, J. (1999), 'Production and Consumption in Rural Service Provision: the Case of the English Village Pub', in N. Walford (ed), *Reshaping the Countryside: Perceptions and Processes of Rural Change*, CAB International, Wallingford, pp. 147-156.

Brandth, B. (1994), 'Changing Femininity: the Social Construction of Women Farmers in Norway', *Sociologia Ruralis*, Vol. 34, pp. 127-149.

Brandth, B. (1995), 'Rural Masculinity in Transition: Gender Images in Tractor Advertisements', *Journal of Rural Studies*, Vol. 11, pp. 123-33.

Brandth, B. and Haugen, M. (2000), 'From Lumberjack to Business Manager: Masculinity in the Norwegian Forestry Press', *Journal of Rural Studies*, Vol. 16, pp. 343-366.

Brass, T. (1994), 'Post-script: Populism, Peasants and Intellectuals, or What's Left of the Future', *Journal of Peasant Studies*, Vol. 21, pp. 246-286.

Bunce, M. (1994), *The Countryside Ideal: Anglo-American Images of Landscape*, Routledge, London.

Butler, R. (1999), *Jean Baudrillard: the Defence of the Real*, Sage, London.

Chaney, D. (ed) (1994), *The Cultural Turn: Scene Setting Essays on Cultural History*, Routledge, London.

Clarke, D. and Doel, M. (1994), 'Transpolitical Geography', *Geoforum*, Vol. 25, pp. 505-524.

Clifford, J. (1983), 'On Ethnographic Authority', *Representations*, Vol. 2, pp. 132-43.

Clifford, J. (1997), *Routes: Travel and Translation in the Late Twentieth Century*, Harvard University Press, Cambridge, Massachusetts.

Clifford, J., and Marcus, G. (eds) (1986), *Writing Culture: the Politics and Poetics of Ethnography*, University of California Press, Berkeley.

Cloke, P. (1992), '"The Countryside": Development, Conservation and an Increasingly Marketable Commodity' in P. Cloke (ed), *Policy and Change in Thatcher's Britain*, Pergamon, Oxford.

Cloke, P. (1993), 'The Countryside as Commodity: New Rural Spaces for Leisure' in S. Glyptis (ed), *Leisure and the Environment*, Belhaven, London, pp. 53-67.

Cloke, P. (1994), '(En)culturing Political Economy: a Life in the Day of a "Rural Geographer"' in P. Cloke, M. Doel, D. Matless, M. Phillips, and N. Thrift *Writing the Rural: Five Cultural Geographies*, Paul Chapman, London, pp. 149-90.

Cloke, P. (1995), 'Research and Rural Planning: from Howard Bracey to Discourse Analysis' in A. D. Cliff, P. R. Gould, A. G. Hoare, and N. J. Thrift (eds), *Diffusing Geography: Essays for Peter Haggett* (Oxford: Blackwell).

Cloke, P. (1996), 'Rural lifestyles: Material Opportunity, Cultural Experience and How Theory Undermined Policy', *Economic Geography*, Vol. 7, pp. 433-449.

Cloke, P. (1997), 'Country Backwater to Virtual Village? Rural Studies and "the Cultural Turn"', *Journal of Rural Studies*, Vol. 13, pp. 367-357.

Cloke, P., Doel, M., Matless, D., Phillips, M. and Thrift, N. (1994), *Writing the Rural: Five Cultural Geographies*, Paul Chapman, London.

Cloke, P. and Goodwin, M. (1992), 'Conceptualizing Countryside Change: From Post-Fordism to Rural Structured Coherence', *Transactions, Institute of British Geographers*, Vol. 17, pp. 321-336.

Cloke, P. and Little, J. (eds) (1997), *Contested Countryside Cultures*, Routledge, London.

Cloke, P. and Milbourne, P. (1992), 'Deprivation and Lifestyles in Rural Wales - II. Rurality and the Cultural Dimension', *Journal of Rural Studies*, Vol. 8, pp. 359-371.

Cloke, P., Milbourne, P. and Thomas, C. (1996), 'The English National Forest: Local Reactions to Plans for Renegotiated Nature-society Relations in the Countryside', *Transactions, Institute of British Geographers*, Vol. 21, pp. 552-571.

Cloke, P., Milbourne, P. and Thomas, C. (1997), 'Living Lives in Different Ways? Deprivation, Marginalization and Changing Lifestyles in Rural England', *Transactions, Institute of British Geographers*, Vol. 22, pp. 210-230.

Cloke, P. and Perkins, H. C. (1998), '"Cracking the Canyon with the Awsome Foursome": Representations of Adventure Tourism in New Zealand', *Environment and Planning D: Society and Space*, Vol. 16, pp. 185-218.

Cloke, P. and Phillips, M. (forthcoming), *The Persistence and Myth of Rural Culture*, Arnold, London.

Cloke, P., Phillips, M. and Thrift, N. (1995), 'The New Middle Classes and the Social Constructs of Rural Living', in T. Butler and M. Savage (eds), *Social Change and the Middle Classes*, UCL Press, London.

Cloke, P., Phillips, M. and Thrift, N. (1998), 'Class, Colonisation and Lifestyle Strategies in Gower', in M. Boyle and K. Halfacree (eds), *Migration to Rural Areas*, Wiley, London.

Cloke, P., and Thrift, N. (1990), 'Class Change and Conflict in Rural Areas', in T. Marsden, P. Lowe and S. Whatmore (eds), *Rural Restructuring*, David Fulton, London.

Crush, J. (ed) (1995), *Power of Development*, Routledge, London.

Cubitt, S. (2001), *Simulation and Social Theory*, Sage, London.

Curry, M. (1991), 'Postmodernism, Language and the Strains of Modernism', *Annals, Association of American Geographers*, Vol. 81, pp. 210-228.

Doel, M. (1994), 'Something Resists: Reading-deconstruction as Ontological Infestation (Departures from the Texts of Jaques Derrida)', in P. Cloke, M. Doel, D. Matless, M. Phillips and N. Thrift, *Writing the Rural: Five Cultural Geographies*, Paul Chapman London, pp. 127-149.

Doel, M. (1999), *Poststructuralist Geographies: the Diabolical Art of Spatial Science*, Edinburgh University Press, Edinburgh.

Doel, M. and Clarke, D. (1997), 'Transpolitical Urbanism: Suburban Anomaly and Ambient Fear', *Space and Culture*, Vol. 1, pp. 13-36.

Feifer, M. (1985), *Going Places*, Macmillan, London.

Fish, R. and Phillips, M. (1999), 'Explorations in Media Country: Political Economic Moments in British Rural Television Drama', in A. Ferjoux (ed), Environment et Nature dans le Campagnes: Agriculture de Qualité et Nouvelles Fonctions. CNRS, Nantes, pp. 113-132.

Fox, R. G. (ed) (1991), *Recapturing Anthropology: Working in the Present*, School of American Research Press, Sante Fe, New Mexico.

Gallop, J. (1989), 'French Theory and the Seduction of Feminism', in A. Jardine and P. Smith (eds), *Men in Feminism*, Routledge, London.

Geertz, C. (1973), 'Thick Description: Towards an Interpretative Theory of Culture', in C. Geertz (ed), *The Interpretation of Cultures*, Basic Books, New York, pp. 3-27.

Giddings, R. (1978), 'A Myth Riding By', *New Society*, Vol. 46, pp. 558-9.

Gilg, A. (1985), *An Introduction to Rural Geography*, Edward Arnold, London.

Grossberg, L. (1987), 'The In-difference of Television', *Screen*, Vol. 28, pp. 28-46.

Halfacree, K. (1993), 'Locality and Social Representation: Space, Discourse and Alternative Definitions of the Rural', *Journal of Rural Studies*, Vol. 9, pp. 1-15.

Halfacree, K. (1995), 'Talking About Rurality: Social Representations of the Rural as Expressed by Residents of Six English Parishes', *Journal of Rural Studies*, Vol. 11, pp. 1-20.

Halfacree, K. (1997), 'Contrasting Roles for the Post-productivist Countryside: a Postmodern Perspective on Counterurbanisation', in P. Cloke and J. Little (eds), *Contested Countryside Cultures: Otherness, Marginalisation and Rurality*, Routledge, London, pp. 109-122.

Harrison, C. (1991), *Countryside Recreation in a Changing Society*, TMS Partnership London.

Harvey, D. (1989), *The condition of postmodernity*, Blackwell, Oxford.

Hoggart, K. (1990), 'Let's do away with rural', *Journal of Rural Studies*, Vol. 6, pp. 245-57.

Hoggart, K. (1997), 'The Middle Classes in Rural England, 1971-1991', *Journal of Rural Studies*, Vol. 13, pp. 253-273.

Holloway, L. and Hubbard, P. (2001), *People and Place: the Extraordinary Geographies of Everyday Life*, Prentice Hall, Harlow.

Houlton, D. and Short, B. (1995), 'Sylvanian families: the production and consumption of a rural community', *Journal of Rural Studies*, Vol. 11, pp. 367-85.

Hughes, A. (1997), 'Rurality and "Cultures of Womanhood"', in P. Cloke and J. Little (eds), *Contested Countryside Cultures: Otherness, Marginalisation and Rurality*, Routledge, London, pp. 123-137.

Hughes, A., Morris, C. and Seymour, S. (eds) (2000), *Ethnography and Rural Research*, The Countryside and Community Press, Cheltenham.

Hughes-Freeland, R. (ed) (1998), *Ritual, Performance, Media*, Routledge, London.

Jameson, F. (1991), *Postmodernism, or the Cultural Logic of Late Capitalism*, Verso, London.

Jones, O. (1995), 'Lay Discourses of the Rural: Development and Implications for Rural Studies', *Journal of Rural Studies*, Vol. 11, pp. 35-49.

Kahn, J. S. (1995), *Culture, Multiculture, Postculture*, Sage, London.

Katz, C. (1992), 'All the World is Staged: Intellectuals and the Projects of Ethnography', *Environment and Planning D: Society and Space*, Vol. 10, pp. 495-510.

Katz, C. (1994), 'Playing the Field: Questions of Fieldwork in Geography', *Professional Geographer*, Vol. 46, pp. 67-72.

Kautsky, M. (1989), *The Agrarian Question*, Zwan, London.

Kellner, D. (1995), *Media Culture: Cultural Studies, Identity and Politics Between the Modern and the Postmodern*, Routledge, London.

Krocker, A. and Cook, D. (1986), *The Postmodern Scene*, Saint Martin's Press, New York.

Lawrence, M. (1995), 'Rural Homelessness: a Geography Without a Geography', *Journal of Rural Studies*, Vol. 11, pp. 297-307.

Lawrence, M. (1997), 'Heartlands or Neglected Geographies? Liminality, Power, and the Hyperreal Rural', *Journal of Rural Studies*, Vol. 13, pp. 1-17.

Lawrence, M. (1998), 'Miles From Home in the Fields of Dreams: Rurality and the Social at the end of History', *Environment and Planning D: Society and Space*, Vol. 16, pp. 705-732.

Lefebvre, H. (1991), *The Production of Space*, Blackwell, Oxford.

Lewis, G. and Maund, D. J. (1976), 'The Urbanisation of the Countryside: a Framework for Analysis', *Geografiska Annaler*, Vol. 58b, pp. 17-27.

Liepins, R. (2000a), 'Exploring Rurality Through "community": Discourses, Practices and Spaces Shaping Australian and New Zealand Rural "communities"', *Journal of Rural Studies*, Vol. 16, pp. 83-99.

Liepins, R. (2000b), 'Making Men: the Construction and Representation of Agriculture-based Masculinities in Australia', *Rural Sociology*, Vol. 65, pp. 605-621.

Liepins, R. (2000c), 'New Energies for an Old Idea: Reworking Approaches to "Community" in Contemporary Rural Studies', *Journal of Rural Studies*, Vol. 16, pp. 23-35.

Little, J. (1987), 'Rural Gentrification and the Influence of Local Level Planning', in P. Cloke (ed), *Rural Planning: Policy Into Action?*, Harper and Row, London, pp. 185-199.

Little, J. (2002), '(Hetero)sexuality and Constructions of the Rural', paper presented at the Association of American Geographers Annual Conference, Los Angeles.

Little, J. and Austin, P. (1996), 'Women and the Rural Idyll', *Journal of Rural Studies*, Vol. 12, pp. 101-11.

Little, J. and Leyshon, M. (2001), 'Embodied Ruralities: The Performance of Gender Identity in the Countryside', paper presented at 'Restless Ruralities, RGS/IBG Rural Geography Research Group Conference', Coventry University.

Longhurst, R. and Wilson, C. (1999), 'Heartland Wainuiomata: Rurality to Suburbs, Black Singlets to Naughty Lingeries', in R. Law, H. Campbell and J. Dolan (eds), *Masculinities in Aotearoa/New Zealand*, Dunmore Press, Palmerston North.

MacNaghten, P. and Urry, J. (1998), *Contested Natures*, Sage, London.

Matless, D. (1994), 'Doing the English Village, 1945-90: an Essay in Imaginative Geography', in P. Cloke, M. Doel, D. Matless, M. Phillips and N. Thrift *Writing the rural: five cultural geographies*, Paul Chapman, London, pp. 7-88.

Matless, D. (1998), *Landscape and Englishness*, Reaktion Books, London.

Milbourne, P. (ed) (1997), *Revealing Rural 'Others': Diverse Voices in the British Countryside*, Pinter Press, London.

Miller, S. (1996a), 'Class, Power and Social Construction: Issues of Theory and Application in Thirty Years of Rural Studies', *Sociologica Ruralis*, Vol. 36, pp. 93-116.

Miller, S. (1996b), 'Theory, Application and Critical Practice', *Sociologia Ruralis*, Vol. 36, pp. 365-370.

Moir, E. (1964), *The Discovery of Britain: English Tourists 1540-1840*, Routledge and Kegan Paul, London.

Munson, W. (1993), *All Talk: the Talk Show in Media Culture*, Temple University Press, Philadelphia.

Murdoch, J. (1995), 'Middle Class Territory? Some Remarks on the Use of Class Analysis in Rural Studies', *Environment and Planning A*, Vol. 27, pp. 1213-1230.

Murdoch, J. and Day, G. (1998), 'Middle Class Mobility, Rural Communities and the Politics of Exclusion', in P. Boyle and K. Halfacree (eds), *Migration into Rural Areas: Theories and Issues*, Wiley, London.

Murdoch, J. and Pratt, A. (1993), 'Rural Studies: Modernism, Postmodernism and the "Post Rural"', *Journal of Rural Studies*, Vol. 9, pp. 411-27.

Nelson, R. (1997), *TV Drama in Transition: Forms, Values and Cultural Change*, Macmillan, London.

Nightingale, V. (1984), 'Media Audiences - Media Products?', *Australian Journal of Cultural Studies*, Vol. 2, pp. 23-35.

Nightingale, V. (1994), 'Improvising Elvis, Marilyn, and Mickey Mouse', *Australian Journal of Communication*, Vol. 21, pp. 1-20.

Nightingale, V. (1996), *Studying Audiences: the Shock of the Real*, Routledge, London.

Olson, G. (1980), *Birds in Egg/Eggs in Birds*, Pion, London.

Olsson, G. (1992), *Lines of Power/Limits of Language*, University of Minnesota Press, Mineapolis.

Pahl, R. (1966), 'The Rural-urban Continuum', *Sociologica Ruralis*, Vol. 6, pp. 299-327.

Pahl, R. (1995), 'Review: *Writing the Rural* and *Family, Economy and Community*', *Rural History*, Vol. 6, pp. 119-22.

Peet, R. (1998), *Modern Geographical Thought*, Blackwell, Oxford.

Perry, N. (1994) *The Dominion of Signs: Television, Advertising and Other New Zealand Fictions*, Auckland University Press, Auckland.

Perry, N. (1998), *Hyperreality and Global Culture*, Routledge, London.

Phillips, M. (1993), 'Gentrification and the Processes of Class Colonisation', *Journal of Rural Studies*, Vol. 9, pp. 123-140.

Phillips, M. (1994), 'Habermas, Rural Studies and Critical Social Theory', in P. Cloke, M. Doel, D. Matless, M. Phillips and N. Thrift, *Writing the Rural: Five Cultural Geographies*, Paul Chapman, London, pp. 89-126.

Phillips, M. (1998), 'The Restructuring of Social Imaginations in Rural Geography', *Journal of Rural Geography*, Vol. 18, pp. 121-153.

Phillips, M. (2000), 'Theories of Positionality and Researching the Rural', in A. Hughes, C. Morris and S. Seymour (eds), *Ethnography and Rural Research*, The Countryside and Community Press, Cheltenham, pp. 28-51.

Phillips, M. (2001), 'Making Space for Rural Gentrification', in Hernando, F. (ed), *Proceedings of 2nd Anglo-Spanish Symposium on Rural Geography*, University of Valladolid, Valladolid, pp. 1-20.

Phillips, M. (2002), 'The Production, Symbolisation and Socialisation of Gentrification: a Case Study of Two Berkshire Villages', *Transactions, Institute of British Geographers*, Vol. 27, pp. 282-308.

Phillips, M. (forthcoming a), *Cultural Geographies*, Sage, London.

Phillips, M. (forthcoming b), 'Other Geographies of Gentrification', *Progress in Human Geography*.

Phillips, M. (forthcoming c), 'Travelling Ruralities? Globalisation, Colonialism and Post-colonialism in the Broadcasting of British Rural Drama Programmes in Aotearoa/New Zealand', submitted to *Environment and Planning A*.

Phillips, M., Fish, R. and Agg, J. (2001), 'Putting Together Ruralities: Towards a Symbolic Analysis of Rurality in the British Mass Media', *Journal of Rural Studies*, Vol. 17, pp. 1-27.

Philo, C. (1992), 'Neglected Rural Geographies: a Review', *Journal of Rural Studies*, Vol. 8, pp. 193-207.

Philo, C. (1994), 'Political Geography and Everything: Invited Notes on "Transpolitical Geography"', *Geoforum*, Vol. 25, pp. 525-532.

Pile, S. (1991), 'Practising Interpretative Geography', *Transactions, Institute of British Geographers*, Vol. 16, pp. 458-69.

Pile, S. (1996), *The Body and the City: Psychoanalysis, Space and Subjectivity*, Routledge, London.

Pile, S. and Thrift, N. (1995), *Mapping the Subject: Geographies of Cultural Transformation*, Routledge, London.

Pratt, A. (1996), 'Deconstructing and Reconstructing Rural Geographies', *Ecumene*, Vol. 3, pp. 345-350.

Probyn, E. (1993), *Sexing the Self: Gendered Positions in Cultural Studies*, Routledge, London.

Rahnema, M. and Bawtree, V. (1997), *The Post-development Reader*, Zed Press, London.

Robinson, G. (1990), *Conflict and Change in the Countryside*, Belhaven, London.

Ross, A. (1989), 'Baudrillard's Bad Attitude', in D. Hunter (eds), *Seduction and Theory*, University of Illinois Press, Urbana, pp. 214-225.

Sayer, A. (1994), 'Cultural Studies and the Economy, Stupid', *Environment and Planning D: Society and Space*, Vol. 12, pp. 635-637.

Short, J. R. (1991), *Imagined Country: Society, Culture and Environment*, Routledge, London.

Smith, D. and Phillips, D. (2001), 'Socio-cultural Representations of Greentrified Pennine Rurality', *Journal of Rural Studies*, Vol. 17, pp. 457-469.

Smith, J. (1996), 'Geographical Rhetoric: Modes and Tropes of Appeal', *Annals, Association of American Geographers*, Vol. 86, pp. 1-20.

Smith, N. (1996), *The New Urban Frontier: Gentrification and the Revanchist City*, Routledge, London.

Soja, E. (1995), *Thirdspace : A Journey Through Los Angeles and Other Real and Imagined Places*, Blackwell, Oxford.

Soja, E. (2000), *Postmetropolis*, Blackwell, Oxford.

Stevenson, N. (1995), *Understanding Media Cultures: Social Theory and Mass Communication*, Sage, London.

Strohmayer, U. and Hannah, M. (1992), 'Domesticating Postmodernism', *Antipode*, Vol. 24, pp. 29-55.

Thrift, N. (1996), *Spatial Formations*, Sage London.

Thrift, N. (1999), 'Steps to an Ecology of Place', in J. Allen, D. Massey, and P. Sarre (eds), *Human Geography Today*, Polity Press, Cambridge.

Thrift, N. (2000a), 'Afterwords', *Environment and Planning D: Society and Space*, Vol. 18, pp. 213-255.

Thrift, N. (2000b), 'Non-representational theory', in R. J. Johnston, D. Gregory, G. Pratt and M. Watts (eds), *The Dictionary of Human Geography*, Blackwell, Oxford:, p. 556.

Tönnies, F. (1963), *Community and Society*, Harper and Row, New York.

Tulloch, J. (2000), *Watching Television Audiences: Cultural Theories and Methods*, Arnold, London.

Urry, J. (1990), *The Tourist Gaze: Leisure and Travel in Contemporary Societies*, Sage, London.

Urry, J. (1995), 'A Middle-class Countryside?', in T. Butler and M. Savage (eds), *Social Change and the Middle Classes*, Routledge, pp. 205-219.

Wright, P. (1985), *On Living in an Old Country*, Verso, London.

Chapter 3

Spatial Stories: Preliminary Notes on the Idea of Narrative Style in Rural Studies

Rob Fish

No Stories, No Rural

> Our society has become a recited society in three senses: it is defined by *stories*...[ ]...by *citations* of stories, and by the interminable *recitation* of stories (de Certeau, 1984, p.186).

> It is in our nature to dramatize (Mamet, 1998, p.3).

It would be fair to say that when the literary theorist Roland Barthes famously claimed that 'narrative is there, like life itself', he probably did not have the work of rural studies in mind. We can't, however, escape his basic premise. Any account of the rural is invariably an exercise in storytelling. Not storytelling in the popular sense of fictitiousness, lies and falsity, but rather as a resource for imaginative thought and writing, and a way of asserting truths about the world. This chapter is an exploration into the way rural studies comes to be imagined through the idea of narrative and a consideration of the problems and dilemmas that are raised in doing so. It is part of a general project that I am developing that seeks to explore the logic and power of 'spatial stories' in making sense of the rural world, a premise that I borrow from the work of de Certeau and his seminal insights into the practice of everyday life (de Certeau, 1984) and one informed by a recent engagement with popular discourses of rurality in television and film (Fish, 2000; Philips et al, 2001; Fish, forthcoming).

We should note from the outset that narrative is a term with meaning more precise than is often given credit. In contemporary academic study the word is typically found in the lexicon of poststructuralism and postmodernism, often as a casual substitute for the word 'discourse', and one suggestive of a loosening up and opening out of meaning. Yet narrative has a very specific definition. In a general sense it refers to sequences of occurrences (or 'event horizons') around which dilemmas and conflicts are explored. It is the process by which the world is cast in an image of tension, discord and disharmony, and depending on the narrative mode employed, one in which conflict may be progressively overcome. My broad purpose in this discussion is to come to terms with this principle as it relates to the work of rural studies. In particular, through an examination of a longstanding mode of narrative that I term 'classical', I wish to argue that contemporary rural studies, while employing certain stable narrative precepts, is also evolving as a highly ambivalent narrative style in which different narrative logics are actively mixed. I begin my analysis by explaining what is at stake in issues of narrative and space and then go on to outline some of the principles governing classical story design. I then use this framework as the basis for inspecting the narrative style of rural studies. In particular I suggest that the narrative stability of rural studies rests on its uniform use of classical 'causality', 'consistency' and 'linearity', constructs of reality that I argue deny room for the more controversial 'anti-narratives' of poststructuralism and postmodernism to emerge. I then tackle the issue of ambivalence in narratives of the rural, suggesting that it actively mixes classical principles of storytelling with what I term a 'mini-narrative' style; one that simultaneously accepts and reverses classical notions of a 'single' and 'stable' main protagonist pursuing 'external forces of antagonism' in a 'minimal' and 'omniscient' reality to a point of 'irreversible' change. I will explain this complex language of narrative as I go along, but my suggestion is that these two narrative logics are spiraled into one another as rural studies seeks to accommodate its evolving theoretical and methodological complexities. By way of conclusion I evaluate what is lost and gained by this emergent style as it relates to the idea of progressive academic discourse.

No Events, No Stories

> [N]arrative structures have the status of spatial syntaxes. By means of a whole panoply of codes, ordered ways of proceeding and constraints, they regulate changes in space (de Certeau, 1984, p.115).

Coming to terms with the idea of narrative means understanding the role of events in the formation of meaning. Events are the building blocks of any narrative style in rural studies. They refer to the process of change (charting directions in the Common Agricultural Policy for instance, or monitoring movements of populations) and are the means by which the rural is re-produced as an object of inquiry. It is therefore quite mistaken to argue that the category of 'rural' only functions in relation to the idea of narrative by dint of putting story events 'in their place' (i.e. as the material site in which narratives unfold). The 'rural' in rural studies is not simply an indifferent/passive context to the process of storytelling. Rather it is an active category shaped through it. For some now working within the discourse of poststructuralism this basic ingredient of narrative - *the event* -is a signature theme that apparently allows notions of space to function in a processive fashion defying neat categorization (Doel, 1999). That is to say, by claiming geography as a series of events, or changes, it is claimed that space is always in a state of opening up, always, to borrow the phraseology of this work, in a state of ' becoming'. As Doel (1999, p.4) puts it, 'one can never know beforehand how an event will play out, spin out and splay out'. Read this way, a category such as 'the rural' apparently works at the limits of meaning because it is always unfolding, always somewhat *on the move*. Interesting though this logic is, it is partly misleading. The event horizons that poststructural geography posit as part of a project of freeing up meanings has to be squared with the rhythms of familiarity and expectation that also govern and limit them. Or to put this another way, events are also scripted and prepared. They have an architecture. And this architecture is a formal set of principles commonly referred to as narrative structure. So from the beginning I wish to make it clear that while the rural is only made possible by creating events, and these events imply a processive, unfolding, notion of space, I do not accept that this simply means undecidibility. Event horizons are designed according to the architectures of narrative, which prescribe in turn the ways in which the rural must be understood and read.

In this section I wish to outline a common model of narrative, which I term classical, as a basis to explore the different narrative logics of rural studies. To do this I briefly highlight two trajectories of modern narratological thought initially guiding my insights. The first is known as 'syntactic' inquiry (Stam et al, 1992) a logic that has its antecedents in early 20th century Russian Formalism - particularly the work of Vladimir Propp (1928) on the Russian folktale – and which has sought in different ways to develop precise grammars of narrative through which the unfolding meanings of texts can be grasped. Syntactic inquiry emphasizes the inevitably temporal patterning of events in the formation of meaning and is often counter-pointed to a semantic approach which, put crudely, downplays the question of sequencing. Semantic inquiry follows the work of Lévi-Strauss (1968) and Greimas (1988) in arguing that configurations of narrative elements, though recounting stories through ostensibly different patterns of sequencing, are thought to be surface expressions of a static or 'achronic' set of antinomies residing in the deeper text. So for instance, in a discussion of how the future of the Common Agricultural Policy might be played out, a syntactic approach would emphasize the logic by which the problem context is expressed (such as the way in which narrative logic constructs its version of heroes and villains), whereas semantic inquiry would emphasize the enduring themes around which the narrative dilemmas are expressed (such as the narrative counterpoints of 'production' versus 'nature conservation').

While I acknowledge that there is more than a tone of high structuralism about this work - a desire to impose a rather mechanistic grid over the idea of narrative - I believe that, taken together, the insights of these two strands of work tell us much about the way in which the rural is brought into being, and in so doing, reveals some of the problems and dilemmas that are raised in narrating the rural in particular ways. I wish to broach this line of reasoning by first outlining, in the barest detail, a set of syntactic principles commonly labeled 'classical' in design; a mode of storytelling whose antecedents lie most notably in the work of Aristotle on poetic form (see Buckley, 1999). The qualities of such a style have been in outlined in relation to a range of cultural texts in which accounts of visual culture, especially filmic texts, have been central (e.g. Bordwell, 1988; Bordwell and Thompson, 1993; Thompson, 1999). I shall elaborate upon these in a moment but broadly put, this structure is associated with a mode of storytelling which, through the creation of an internally causal, coherent, linear, minimal and omniscient reality, sets a single and stable main protagonist in antagonistic relations with an external world, and which clearly resolves its dilemmas over the course of the telling (McKee, 1998). As I will show, classical narrative is certainly not a default narrative standard for rural studies, but its importance cannot be underestimated. In a recent analysis of these principles, for instance, Lowe (2000, p.3) expresses a sentiment shared by many theorists and practitioners of story art, when he asserts that, 'despite fluctuations in fashion ...[classical design]... has remained the resilient narrative paradigm in Western storytelling to this day'. Indeed, where alternative modes of storytelling are considered, the legacy of classical principles is frequently thought to resonate. They are read as a process of reaction and departure from classical form; one in which classical principles have (often implicitly) been mastered and subverted. It is for this reason that one of its key proponents within film, Robert McKee, has chosen to substitute the term 'classical' with 'archplot', as in, 'eminent above others of the same kind' (McKee, 1998, p.45). I now wish to briefly consider the main dimensions of such a logic as it has commonly been formulated in relation to popular cultural texts and then to contextualize these in relation to rural studies.

As *causal* realities, classical narratives function by inextricably tying each story event, by implication or motivation, to other events (Bordwell, 1988). Events are not proffered that do not find their rationale, their explanation, in relation to other occurrences. To borrow the language of practitioners, they refuse to exhibit 'holes' in meaning (Thompson, 1999, p.13). As *coherent* realities, classical narratives establish and subsequently maintain a self-governing logic by which the story world operates. However profound or odd a narrative reality may seem, the notion of a consistent reality concerns the way unfolding events avoid contradicting the sense on which a reality is already being governed. In so far as contradictions appear to occur, they are always explained and thereby accommodated within the logic. In the words of McKee (1998, p.54), classical narrative, however fantastical the creation, creates a reality 'true to itself'. As *linear* realities, classical narratives proceed through the principle of continuous time. While there may actually be some alteration of time in the telling of the narrative (termed in narratology the *sjuzhet or re/cit*), such as a 'flash-back' , the classical model demands that events can be easily re-ordered and assimilated by the reader as a temporally linear order of events (termed *the fabula* or *historie*). As *minimal* realities, classical design conveys the necessary information to be comprehensible, rather than expanding the possibilities of effective meaning. As Lowe (2000, p.62) puts it, classical story events 'minimize redundancy'. There is, of course, a hierarchy at work here in defining necessary information, as Barthes (1977) and Chatman (1969) for instance show in their distinctions between major events (which they describe as *kernels*), and minor ones (which they describe as *catalysts* and *satellites* respectively). But what can be said about the classical formation is its emphasis on the functionality of events to a particular story being told, and by implication, on its forward progression. Finally, as *omniscient* realities, classical narrative grants the reader/viewer the impression that s/he has all-embracing knowledge of story-events at hand, and by dint of erasing (or at least downplaying) the conditions of production which allow this knowledge to be produced, the sense that s/he has unmediated and objective access to the world.

Working alongside these basic principles is a further set of qualities thought to shape the wider arc of the classical narrative in very specific ways. Most notably classical narrative is associated with a mode of storytelling that instigates and clearly resolves dilemmas over the course of its telling. As McKee (1998, p.47) puts it, in classical narrative, 'all questions raised by the story are answered; all emotions evoked are satisfied'. The formal terms on which this arc of story has been articulated develops from Aristotle's seemingly straightforward premise that dramatic forms should have a beginning, middle and end (see Buckley, 1992). Within the field of modern narratology, insight often proceeds from Todorov's seminal equilibrium/dis-equilibrium/equilibrium model (Todorov, 1977), a narrative logic propelled initially by the transformation of an ordered world into one of disorder, and then, over the course of the narrative, by overcoming forces of disorder to duly restore a new sense of equilibrium. Practitioners of story art commonly describe this in terms of a three-act structure (see Field, 1987 and Mamet, 1998) and have sought to discriminate a series of minor and major dramatic reversals around which its logic unfolds.

Central to these accounts is the idea that any notions of equilibrium and disorder quickly, if not immediately, give way to a major occurrence often referred to as an 'inciting incident'; an incident that contrives a dilemma on the part of its protagonist around which subsequent events will develop. Such an occurrence will typically be divided up into a 'set-up', around which a force of antagonism will be introduced (e.g. the so-called 'smoking gun') and an implication or 'pay-off' of the occurrence (e.g. an investigation of some sort). In classical narrative such forces of antagonism are themselves primarily conceived of as external - dilemma and conflict originating between self and others or wider extra/im-personal threats to the order of things - and are typically thought to initiate conflict for a single and clearly definable main protagonist with a seemingly stable set of character traits (Bordwell, 1988; Thompson, 1999). Indeed, in the classical form, the psychological identity and stability of the single main protagonist is rarely called in to question. It is as predictable as the point of irreversible closure that events are contrived to work towards. What proceeds from the inciting incident, then, is the initiation of a conflict that shapes the overall goals of the main protagonist over the course of the event horizon. While the forces of antagonism give rise to these conflicts - they are motivated towards it - it is worth noting here that conflict refers to the contradictory scenarios that these main protagonists mediate between. For instance, the smoking gun is meaningful insofar as it gives rise to a dilemma such as 'justice prevailing' or 'justice not prevailing', counter-pointed ideas which, over the course of the events, will be ultimately resolved to positive, negative and sometimes ironic effect.

Narrating the Rural

Narrativity insinuates itself into scientific discourse as its general denomination (its title) as one of its parts ('case' studies, 'life stories' or stories of groups etc.) or as its counterpoint (quoted fragments, interviews, 'sayings' etc.). Narrativity haunts such discourse. Shouldn't we recognize its *scientific* legitimacy by assuming that instead of being a remainder that cannot be, or has yet not been, eliminated from discourse, narrativity has a necessary function in it, and that a theory of narration is indissociable from a theory of practices, as its condition as well as its production...[ ]...In this way, the folktale provides scientific discourse with a model, and not merely with textual objects to be dealt with (de Certeau, 1984, p.78).

Classical principles are not a hard and fast set of rules for understanding how academic discourses of the rural are made, but they provide us with a useful point of departure for explaining the modes of storytelling at work and what is at stake in pursuing these. It is to this end that I wish to begin this discussion by attempting some crude pointers towards what unites rural studies in terms of its replication of these principles, and in particular, to highlight a virtually exclusive tendency to employ classical notions of causality, consistency, and linearity. In making this point I wish to suggest that rural studies effectively rules out the possibility of creating what I term 'anti-narratives' of the countryside, a style of storytelling that inverts these three premises and one that is closely akin to the principles of poststructuralism and postmodernism. However, the general argument I go onto make is that the narrative logic of rural studies is more scrambled than it is unified: one that is as much about creating 'multiple, unstable and passive protagonists' as it is about employing classical notions of a 'single, stable and active protagonist'; as much about creating 'expansive' realities as it is about 'minimal' realities; as much a mode of 'restricted' narration as it is of 'omniscient' narration; and as much about 'open' narratives as it is about 'closed' narratives. Such reversals exemplify what I term a 'mini-narrative' style and are mixed with classical principles to accommodate different (and often competing) theoretical and methodological trajectories within rural studies (among other things spatial science, political economy, poststructuralism and ethnography). Indeed, I argue that the narrative implications of being in different systems of thought and practice are never fully overcome, but rather spiraled into each other within the arcs of particular narratives. I caution that the style I am about to outline is crude. It is an initial formulation. But I believe this analysis captures something of the totality, the 'structure of feeling' as Thrift (1994) would have it, defining contemporary rural studies (see Figure 3.1 for a summary).

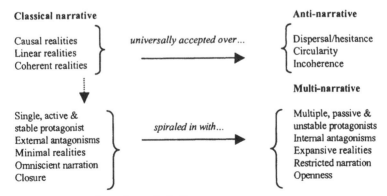

Figure 3.1 Narrative style and rural studies

Let me start, therefore, by saying something of the certainties guiding the narrative style of rural studies, and in particular, to begin by highlighting the enduring role of causality in producing accounts of the rural world. That is to say, one of the typical ways that rural studies holds its realities together is by tying narrative events, directly or by implication, to other story events. So, for instance, a narrative set-up which asserts the importance of agri-environmental schemes to the sustainability of rural areas (story event 1) *then* writes of the importance of understanding land manager attitudes towards conservation (story event 2), *then* conducts a survey of attitudes towards conservation as it is reflected in the uptake of these schemes (story event 3), *then* explains the barriers and constraints to participation (story event 4), and so forth. Expressing this in more abstract terms, occurrences in the arc of a story are always 'motivated' in rural studies. Second, and allied to this, rural studies typically produces internally coherent sets of narrative worlds. As event horizons develop, realities do not contradict the terms on which they have already been constructed. For instance, a story idea boldly claiming that 'rural change is determined by political-economic processes' is not just a statement of intent but a story rule. It governs the terms on which the causal reality is then progressively built. In so far as an author may wish to contravene this logic (e.g. 'we must not dismiss the role of cultural processes') new story rules will be written back into the narrative (e.g. 'rural change is partly determined by political-economic processes') or otherwise existing story rules will be elaborated upon to accommodate the tension (e.g. 'political economic processes do not preclude cultural processes'). In practice, this is one reason why the language of uncertainty pervades rural studies (e.g. 'arguably therefore'; 'potentially we might say'; 'on the basis of this evidence it could be suggested', and so forth). Such caveats are narrative devices allowing consistent realities to be asserted whilst avoiding didactic and polemical views of the world (i.e. that this is how reality *must* be seen to function). Furthermore and finally, rural studies uniformly employs the principle of continuous, linear, time. While accounts of the rural may scramble the unfolding logic of an argument (e.g. the point of arrival in a narrative will frequently precede the empirical basis of that argument) they always allow the reader to re-order and assimilate events in linear fashion (e.g. the point of arrival is shown to be the logical outcome of an empirical inquiry) and will actively guide them towards doing so (e.g. 'on the basis of an extensive survey we will argue that', or ' the survey results therefore show us that').

These components of narrative - causality, coherence and linearity - are not only shared by those propounding modern theoretical discourses of the rural (by which I include political economy, behaviouralism and humanistic thought) but also by those who would perhaps wish to unsettle these seamless qualities of classical narrative (namely postmodernism and poststructuralism). In fact, there is a genuine disjuncture between attitudes towards knowledge that seek to dismantle the verities and correspondences of knowledge and the narrative style by which these directions in rural studies are typically produced and expressed. To all intents and purposes, rural studies engagement with these 'posts' has virtually always been written *through* these classical precepts: a wilful disruption of stable narratives of a world 'out-there' but written through the seamlessly composed narratives of an academic world 'in-here'. To the extent that work swept up by these currents might be thought to betray (or reveal?) their instincts, this is shown at the level of narrative.

So what would rural studies look like if it employed a narrative style that consistently sought to reverse these classical principles? It would surely result in a system of *anti-narrative* where the logic of postmodernism and poststructuralism - in different ways attempts to critique stable and controlling meanings - become not only attitudes to knowledge but also styles by which these critiques narrate the world: 'dispersal' and 'hesitance' disrupting 'causality', 'circularity' overwriting 'linearity', and 'incoherence' unsettling 'coherence'. For instance, it might be said that anti-narrative principles are central to the idea of deconstructive writing, an archetypal strategy of poststructuralism, and one that Doel (1994, p.136) has elegantly described as 'defy[ing] the spirit of gravity, the spirit of reading'. Indeed, Doel's work in this area provides us with useful insight into, and in many respects exemplification of, anti-narrative principles. It is unsurprising, for example, that in attempting to write through the idea of rural change, Doel (1994) book-ends his discussion with the phrase 'find a text to read if you can' (see p.127/p.145), just as it is little wonder that he begins and closes his book on poststructuralist geographies (Doel, 1999) with the phrase 'let us space'(see p.V/p.199). Both are exemplars of the anti-narrative principle of circularity, of arriving at the same point a narrative starts. Rather than create narrative sequences where one discrete story event implies another in the name of narrative progression, deconstructive writing strategically chooses to disperse seemingly discrete events into semiotic infinity. As Doel (1999, p.3) fittingly puts it: 'one will never be finished with the task of doing justice to the event; of reading and re-reading, of thinking and re-rethinking, of repeating and differing'. It is worth noting here that since deconstructive writing downplays this issue of causality and progression, certain problems are raised for the casual claim that poststructuralism is all about the 'speeding up' of meaning. The anti-narrative logic of deconstruction knows nothing of smoothly moving past one narrative event to the next; of creating narrative arcs in which meaning transforms effortlessly across a series of causal *points*. Instead, it embroils itself in the politics of refrain and hesitation; of not being able to proceed past the event. Furthermore, deconstructive writing is less about creating highly coherent realities than about 'writing against itself'. Unlike classical narrative it strategically employs a more circumspect, oblique and duplicitous narrative language to resist stable realities being constructed out of its work. It is less interested in policing realities according to rules and clauses than in playing out its own contradictions and faulty logics.

I shall return to this issue of anti-narrative in the conclusion when I attempt to imagine rural studies in other terms. However, as I have already suggested, my broader purpose is to argue that at the heart of rural studies is a tendency to scramble its narrative lines by mixing classical protocols with what I labeled a *mini-narrative* format; a format where passive and unstable protagonists pursue primarily internal forces of antagonism in expansive realities to largely unresolvable and open effect (see McKee, 1998). This contrary logic will become clear as we go along. In particular I wish to begin this account by briefly and quite simply noting that rural studies slides between a single (i.e. classical) and multi-protagonist (i.e. mini-narrative) format. It is a single protagonist format in the sense that the concerns of rural studies are, according to one process of self-definition, hardwired into reproducing a single meta-concept: 'the rural'. Trajectories of work are recognizable as part of an academic discourse known as 'rural studies' because narrative events are motivated towards journeys of transformation in the experiences or qualities of this general category of concern. However, read this way it is also clear that, for narrative change to be synonymous with the issue of rural change, rural studies must mobilize its events around a much wider set of narrative agents, be they the conflicts of particular social groups (such as 'land managers' or 'children' or the ' service class'), rural sectors (such as 'primary' or 'service' industries in the countryside) or differentiated material spaces (such as 'upland' or 'regional' landscapes). That is to say, these protagonists and their dilemmas effectively take centre stage within particular narrative set-ups (e.g. 'land managers are facing an economic crisis') and are subject to their own particular transformations over the course of story events (e.g. 'land managers are increasingly diversifying their sources of income'), but so too do they allow wider inferences about the state of the single main protagonist – the rural - to be made (e.g. 'rural areas are becoming increasingly diversified').

However, if it is possible to suggest that rural studies mixes a single and multi-protagonist format, the process of narrative transformation it seeks to chart in these protagonist(s) appears far from uniform. By way of the theoretical precepts of postmodernism and poststructuralism it is clear, for instance, that contemporary rural studies increasingly seeks to unbound and breakdown its categories of concern over the course of narrative events, and in so doing, might be said to have gradually veered away from a longstanding predilection with classical narrative's notion of an internally stable main protagonist. Indeed, it might broadly be ventured that rural studies, proceeding from the premises of spatial science and its variants, has exhibited an enduring narrative desire to assert coherence and shape around (otherwise chaotic) categories of concern over the course of story events. Though readily accepting that rural worlds are complex, varied and differentiated (e.g. the frequent claim that 'there are not one but many rurals') the narrative transformations of such work have, in short, been largely motivated towards stabilizing categories (e.g. the creation and analytical use of 'ideal types' of countryside). In theoretical terms, rural studies in this guise has been inclined towards creating 'repetition producing difference' when handling its protagonists: the notion that, out of narrative complexity, variation and so forth, rhythms of constancy and shape start to emerge. In contrast, through the work of postmodernism and poststructuralism, a different image of thought is arguably found, one akin to mini-narrative's principle of instability and exemplified by Doel's (1999, p.18) idea of a 'difference producing repetition'. Whereas the former logic works in the direction of narrative safety, certainty and control, the principle here is reversed. It seeks to show how apparently secure categories of concern can never be adequately held down, but rather affirmed as constantly unstable and shifting configurations of ideas. Protagonists such as 'rural', 'service class', 'farmer', 'child', ' human', 'nature' and so forth therefore become symbolic categories in perpetual motion, reflexively deployed in different ways, at different moments, and for different purposes. Further, according to this narrative style protagonists are both implicated in, and infinitely dispersed through, other categories. The positive terms which allow differences between categories of concern to be expressed and maintained (e.g. human = a, nature = b; rural = x, urban = y) effectively start to collapse and dissolve over the course of events (see Doel, 1999, p.47).

Nonetheless, it is easy to fall into caricatures here. These narrative directions are more a question of tone than discrete and mutually exclusive styles. In practice, rural studies rarely falls into one strict narrative logic or another. Typically one style of work is spiraled into, and then out of, the other. An inclination towards the creation of 'difference producing repetitions', for instance, is not the same thing as adopting wholesale the project of, say, deconstructive writing, which as Doel (1994, p.131) further adds, 'hurls every construction into the air and then perpetually intervenes in an attempt to prevent them from falling back into reconstructions'. More often than not (and I count Doel's 'hollowed out' insight into rural change as a clear exception to this point) narratives of the rural with such an inclination are always typically implicating themselves in, or at the very least cannot escape, the reconstruction of their deconstructions, just as any overt predilection for stable and bounded protagonists in other trajectories of rural studies is more a line of reasoning than a realizable narrative project.

In a similar fashion, rural studies can be said to have an ambivalent approach to the classical principle of an active protagonist, a process that reflects evolving debates over the issue of agency within rural studies. In one sense, it has increasingly affirmed the notion of an active protagonist through its antidotes to the eternal verities of political economy and spatial science where rural protagonists are envisioned as largely passive recipients and exemplars of external logics and forces. To varying extents, and often in competing ways, the contrary notion of dynamic, self-actualizing, human figures set in antagonistic relationships with the world has found expression in the work of behavioural, humanistic and poststructural inquiry, and has evolved alongside their engagements with the practice of ethnography (such as unstructured interviews, participant observation and oral history [see Hughes et al, 2000]). Nonetheless, just as care has to be taken in pigeon-holing work that transforms its protagonists over the course of narrative, so it is true of the simple division I am setting up here. While there is a narrative logic within rural studies that claims, for instance, that its protagonist(s) are not simply open books upon which to unfold extra/im-personal external forces of antagonism, this is not the same thing as overcoming the narrative counterpoints of structure and agency. More often than not the narrative style of 'agency' (i.e. the active protagonist) is always spiraled back into its counter-idea of 'structure' (i.e. the passive protagonist), and vice versa. Therefore, the question is again one of tone than of strict positions. In practice, the narrative style of rural studies is one that evokes an enduring sense of 'mid-way logics' around which these narrative propositions unfold and are debated: note, for instance, behaviouralism's discourse of choice within constraints; poststructuralism's logic of made and re-made worlds; postmodernism and its simultaneous recognition of and resistance to meta-narratives. The narrative style of rural studies is typically to view its narrative protagonists recursively, as simultaneously active and passive, shaped and shaping, cause and outcome.

Following this line of reasoning I furthermore wish to point to a narrative style that moves the nature of conflict as much towards the internal complexities of the narrative agent within mini-narrative as it does towards classical notions of social and extra/impersonal conflict. In particular I wish to point to a current that links the seemingly counter-pointed theories of humanism and poststructuralism within rural studies: that is, a common desire to problematize simple transactions or correspondences between the human agent and the outside world. Whether these logics ultimately reveal a theory of internality (as in humanism), or the social (as in poststructuralism) is not really the issue, for rural studies in this guise has largely been about emphasizing loosely defined complexities and dilemmas of the self, a project that has largely taken shape through ethnographic practice in which the situated voices of different bodies engage their rural worlds in complex ways. To be clear, this bloodline tying seemingly different theoretical projects is not simply a 'retreat within': a narrative style simply counter-pointed to, say, the external forces of 'economy' or ' popular culture' or ' accepted codes of behaviour in rural areas'. Again, they are spiraled into discussions about these structuring agents. External forces of antagonism are the means by which dilemmas are revealed, and conversely, offer important points of departure for understanding our complex (and perhaps more significantly, contradictory) re-engagements and re-makings of the world.

It is arguably through this elevation of active protagonists and internal forces of antagonism within rural studies, and by way of its evolving contract with the practice of ethnography, that narratives increasingly mix classical notions of minimal and functional realities with the more expansive style of mini-narratives. It is, of course, the longstanding promise of social-scientific narratives of the rural to emphasize precision of insight, and the functionality of events. In its various guises the alliance of social science and rural studies has always been motivated towards the issue of minimizing redundancy, whether this is narrating attitudes to the world through the austerity of a structured survey, the precision of numbers or the pristine aesthetic of a figure or chart. Yet for all their grace, these narrative repertoires are now in active negotiation with forces that wish to exceed them: a more embellished style of 'thick' accounts that attempt to do greater justice to each narrative event and turn. The evolving approach within rural studies is increasingly less about classical narrative's emphasis on forward progression than it is about inflecting stories with pauses, departures and all manner of subplots. That said, care must again be taken in creating neat divisions between bodies of work, for what is often found in individual acts of storytelling is a process of social-scientific minimalism working in conjunction with a more expansive ethnographic style. For instance, consider the in-depth semi-structured interview, an endemic feature of this emerging logic. When filtered through the grid of social science a whole panoply of minimizing techniques are often found to be at work: e.g. (1) identify interview 'codes' that can submit and rationalize ethnographic voices to the author's functional meta-narrative; e.g. (2) clip and tuck these voices where appropriate so they don't exceed this meta-narrative; e.g. (3) insert '[...]' where voices become redundant 'noise' in this meta-narrative; and so forth. In short, if the social scientific dream of minimally functional worlds is increasingly being expanded out by ethnographic practices, these practices are themselves being actively contained.

I wish to further highlight a narrative style that shifts between omniscient and restricted modes of narration. To reiterate, narration is the means by which readers are positioned in relation to story events, with omniscient narration the process by which access to story events appears all-embracing and objective, and restricted narration by which it is constructed as limited and partial. While there are many complexions to these ideas, I wish to briefly note the ascendancy of restricted narration as a strategic break on rural studies' general grammar of classical omniscience. Assertions of 'I' within accounts of rural change are a clear expression of this mini-narrative logic for instance, as are the self-reflexive dilemmas expressed around the narrator's ability to speak of/do justice to research subjects. In various ways this logic attempts to mark out the narrator's role in the construction of knowledge (as opposed to using the omnipresent third person, for instance), the inevitable incompleteness of the knowledge constructed (as opposed to, say, claiming 'representativeness' in a study) and allied to this, the impossibility of attaining such completeness (as opposed to the notion of incrementally building towards a 'total' account of a situation). Yet caution must again be taken in using this distinction to draw simple lines around different bodies of work within rural studies. Narratives constructed from the most Olympian of heights will frequently express selectivity and partiality, from explanations over the conceptual framework *chosen* to frame the world – (e.g. 'problem X is approached by drawing upon theory Y') through to the language of uncertainty I mentioned above ('arguably', 'potentially',' perhaps', and so forth). Equally, those who seek to assert the limited transactions taking place between the tripartition of reader, researcher and researched rarely write exclusively in this tone. Typically they strategically emphasize the issue of restricted narration at specific narrative junctures and intervals (such as discussions about positionality and self-reflexivity in the research process) within a broader, more omnipresent, narrative repertoire.

Finally, I wish to highlight a narrative style of rural studies that is as much a process of denying closure as it about resolutions. Unlike classical narrative rural studies, it functions in the format of a continuous series, a structure and mode of narration that, like many other cultural forms, works by developing meaning across temporally extended sequences of output. These outputs - papers, book chapters, presentations and so forth - are effectively 'episodes', 'instalments' or 'bulletins' in a recurring and never ending story. The logic of writing the rural is therefore often as such: (*set-up*) enter into the midst of an already chaotic world (e.g. the rural is culturally contested); (*pay-off*) define an intervention into that world (e.g. in what way is it culturally contested?); (*progressive complication*) explore some of the dilemmas at stake in this intervention (e.g. cultural contestations appear to be defined by class and gender); (*resolution*) create a new sense of equilibrium (e.g. these are important iterations that need further research). In short, the classical notion of closure, of a project fulfilled, never fully comes. While rural studies is often written, implicitly or explicitly, as if it desires major reversals and closure over the course of story events, in practice it is an open narrative mediated through a rather dispersed and segmented form of minor reversals in the order of things, always with the promise of developments: 'avenues of further research' and 'recommendations' that spiral into new debates and trajectories. As a narrative style, rural studies simultaneously closes and opens up.

Rural Studies Otherwise: The Limits of Narrative

In this chapter I have made two key claims regarding the narrative style of rural studies. First, that work is governed by the classical principles of causality, coherence and linearity when constructing its realities; a process that serves to draw a veil over what I termed 'anti-narratives' of the countryside. Second, that the emergent narrative style of rural studies 'spirals in' classical narrative principles with those of a 'mini-narrative' logic. By way of conclusion I wish to speculate on the implications of these two narrative strands of rural studies, and in particular, to suggest that what is at stake here is idea of a progressive academic discourse on rurality.

Let me begin by initially returning to the issue of anti-narrative. According to one line of reasoning, it could be argued that denying room for such a strategy allows rural studies to create for itself something approaching an inter-subjective reality upon which to talk about the world. That is to say, an inclination to create causal, coherent and linear rural realities carries with it the promise of the rural making sense; not 'making sense' in the manner of a final and total capture of the rural, but rather as a basis for shared discourse. Indeed, the argument would go that, for all anti-narrative's potential to scramble and disrupt conventional (classical) constructs of reality, it can never be more than a check on discussions which, in different ways, wish to hold meaning down and proceed as such. With its tendency to endlessly proliferate 'the event', anti-narrative is understood here as the anti-thesis of speech and thereby action, much in the manner of Hall's (1987, p.45) comment that, 'the politics of infinite dispersal is the politics of no action at all'. And yet, we should not deny the possibility of such a style emerging, for if rural studies did, it would simply fall into equally problematical narrative style, that of *anti-conflict*. Anti-conflict is didacticism by another name, most commonly found in propaganda and reflects a desire to preach and silence dissent (Chomsky 2000). It is a narrative style that seeks to assert the singular importance of one particular theory of the world and one particular way of bringing it into being. It is a style that lacks any sort of academic generosity or grace, and through the violence of its words seeks to suffocate the voices of others by claiming the world in its own image (see Miller, 1996, for a perfect exemplar of this). Indeed, if it is possible to discount and dismiss such anti-narrative strategies in the name of their 'anti-progressiveness', or their descent into 'language and game playing', then this arguably comes at the price of another notion of the progressive: an openness to alternative realities, however seemingly absurd, jarring and disconcerting they may be.

Granted all this, rural studies is never as far from the principles of anti-narrative as it may wish to think. Anti-narrative is actually an integral part of the process by which versions of the rural world, in all their causal, coherent and linear glory, finally emerge. Or to put this another way, the whole architecture of producing these rather neat, logical, classical realities is unpinned by completely chaotic, truly monstrous, sets of alternative realities where ideas jar, events don't follow and so forth. Such realities typically reside in the dark underbelly of drafts, re-drafts and re-re-drafts, never seeing the light of day, but nonetheless casting shadows over our work. They exist too in the confidences we share (e.g. 'I can't get this argument *straight*', 'something doesn't *add up*', 'it doesn't *make sense*', 'I'm going around in *circles*', 'it's a complete *mess*') but rarely do they enter into the official discourses of rural studies (i.e. the stories that circulate in formal circuits of communication). Perhaps this is ultimately a good thing, after all, as Hemingway once said, 'the first draft of anything is *shit'*. But I wonder what rural studies would look like if it was simply more self-effacing about how we (and I am surely not alone here) go literally 'round the bend' when attempting to submit the world to this classical grid of expectation, and what is lost in doing so. My guess is that such an approach offers at least another way into a progressive discourse on rurality, something akin to deconstruction's notion of 'writing against writing', of lubricating the threads of our own anti-narratives (for indeed they are there) to see where they lead us.

These issues are played out further when we come to inspect the mixing of classical and mini-narrative formats. There is an argument to make that in constantly ghosting one way of writing the rural (say, a stable and active protagonist operating in minimal realities with closed endings) with another (say, unstable and passive protagonists operating in expansive and realities with open endings), rural studies does avoid the kind of didacticism that I mentioned above. As such it can plausibly entertain the idea of being a progressive discourse because, despite the absence of anti-narratives within official discourse, its storytelling universe is still imbricated with all manner of narrative repertoires with which the rural can be imagined in different and thought provoking ways. Yet, if this is so, it is also clear from the analysis above that rural studies is a rather self-containing narrative discourse, whereby different narrative logics keep each other safely in check. Rather than test the limits of a particular narrative logic (e.g. classical, anti-narrative or minimal), acts of storytelling simply spiral towards points in-between, never quite one thing or the other. This guarded and cautious approach to the practice of storytelling arguably makes rural studies an inherently conservative academic discourse, rarely ambitious enough to the follow the implications of distinctive narrative repertoires and stretch the boundaries of our rural worlds. And while we would have to be careful here of falling into the violence of didacticism I mentioned above, it is in such a practice that the imaginative potential of narrative, largely lost to the work of rural studies, could be found.

References

Barthes, R. (1977), *Image Music Text*, trans. S. Heath, Fontana, London.
Bordwell, D. (1988), *Narration in the Fiction Film*, Routledge, London.

Bordwell, D. and Thompson, K. (1993), *Film Art: An Introduction,* McGraw-Hill, New York.

Buckley, B. (1999), *The Poetics: Aristotle,* Prometheus, Buffalo.

Chatman, S. (1969), 'New Ways of Analyzing Narrative Structure, with an example from Joyce's *Dubliners' Language and Style,* Vol. 2, pp. 3-36.

Cobley, Paul. (2001), *Narrative,* Routledge, London.

Chomsky, N. (2002), *Media Control: The Spectacular Achievements of Propaganda,* Seven Stories Press, London.

de Certeau, M. (1984), *The Practice of Everyday Life,* trans. S. Rendall, University of California Press, Berkeley.

Doel, M. (1994), 'Something Resists: Reading-Deconstruction as Ontological Infestation (Departures from the Texts of Jacques Derrida)', in P. Cloke, M. Doel, D. Matless, M. Phillips and N. Thrift, *Writing the Rural: Five Cultural Geographies,* Paul Chapman, London, pp. 127-48.

Doel, M. (1999), *Poststructuralist Geographies: The Diabolical Art of Spatial Science,* Edinburgh University Press, Edinburgh.

Hughes, A., Morris, C. and Seymour, S. (eds) (2000), *Ethnography and Rural Research,* Countryside and Community Press, Cheltenham.

Fish, R. (2000), *Putting Together Rurality: Media Producers and the Social Construction of the Countryside,* Unpublished PhD thesis, University of Leicester.

Fish, R. (forthcoming), 'Event Horizons: Idyll and Anti-Idyll in the Production of Rural Television Drama', *Environment and Planning D: Society and Space.*

Field, S. (1987), *Screenplay,* Bantam Doubleday, London.

Greimas, A.J. (1988), *Maupassant: The Semiotics of the Text,* trans. Paul Perron, John Benjamins, Philadelphia.

Lévi-Strauss, C. (1968), *Structural Anthropology,* Trans. C. Jacobson and B. Grundfest Schoepf, Allen Lane, London.

Lowe, N. J. (2000), *The Classical Plot and the Invention of Western Narrative,* Cambridge University Press, Cambridge.

Mamet, D. (1998), *Three Uses of the Knife: On the Nature and Purpose of Drama,* Columbia University, New York.

McKee, R. (1998), *Story,* London, Methuen.

Miller, S. (1996) 'Class, Power and Social Construction: Issues of Theory and Application in Thirty Years of Rural Studies', *Sociologia Ruralis,* Vol. 36, pp. 93-116.

Phillips, M., Fish, R. and Agg, J. (2001), 'Putting Together Ruralities: Towards a Symbolic Analysis of Rurality in the British Mass Media', *Journal of Rural Studies,* Vol. 17, pp. 1-27.

Propp, V. (1928), *Morphology of the Folktale,* trans. L. Scott, University of Texas, Austin.

Stam, R. Burgoyne, R. and Flitterman-Lewis S. (1992*), New Vocabularies in Film Semiotics: Structuralism, Post-structuralism, and Beyond,* Routledge: London.

Thompson, K. (1999), *Storytelling in the New Hollywood: Understanding Classical Narrative Technique,* Harvard University Press, London.

Thrift, N. (1994), 'Inhuman Geographies: Landscapes of Speed Light and Power', in P. Cloke, M. Doel, D. Matless, M. Phillips and N. Thrift, *Writing the Rural: Five Cultural Geographies,* Paul Chapman, London, pp. 191-248.

Todorov, T, (1997), *The Poetics of Prose,* trans. R. Howard, Blackwell, Oxford.

Chapter 4

(Re)positioning Power in Rural Studies: From Organic Community to Political Society

Graham Gardner

Introduction

This chapter provides a critical examination of rural studies of power and suggests future agendas regarding the study of power in rural spaces and places. Tracing the emergence of power as an object of interest in UK rural studies, it draws out the relationships between the 'discovery' of power in the countryside, changing theoretical and empirical emphases, and the differing normative attention given to rural society and polity. In particular it pays attention to the (re)positioning of rural 'communities' over the last thirty years, exploring how notions of communities as local social systems governed by 'traditional' modes of authority and status have been replaced by notions of communities structured by wider social divisions and governed by more or less 'hegemonic' class fractions. It also considers recent interest in the configurations and reconfigurations of state power over local social, economic and cultural processes, and in calls for rural researchers to recognize and take account of the socially and discursively constructed nature of the countryside. The chapter concludes by arguing that whilst recent rural studies of power are increasingly critical and reflexive, they have neglected two critical research arenas: the place of individuals in terms of the production, reproduction and contestation of local political spaces, and the notion of power as a constitutive, productive force. These, it suggests, are important foci for future rural research.

The Organic Rural Community – Power as *Authority*

For at least twenty years, between the early 1950s and the early 1970s, the mainstay of research into UK rural society was the 'community study' (Symes, 1981). Such studies were concerned with 'the interrelationships of social institutions in a locality' (Bell and Newby (1971, p.19), seeking either to produce ethnographic accounts of local cultures or (less frequently) to facilitate the construction and elucidation of wider theoretical claims regarding societal structures, processes, relations and mechanisms (Harper, 1989). The majority of studies were also concerned with the impact of technological, social and cultural change on rural localities; particularly the undermining of 'traditional' ways of life as a result of agricultural mechanization, improved rural-urban transport and communication linkages and the in-migration of former urban residents to once relatively closed occupational communities. Rural community studies have declined in popularity since the 1970s, despite being occasionally championed by advocates such as Cohen (1982), although they have by no means entirely disappeared from the academic landscape.

There is little explicit discussion of power in rural community studies. No rural community study, for example, includes interrogation of local political institutions, decision-making or representation to agencies outside a given locality. Nor is attention given to questions of hegemony, contestations of rights to rule, and social and political marginalization. Indeed, the word 'power' rarely appears in such works. Such 'neglect' – at least for one concerned with the issue of power – might be attributed to researchers being unable to see beyond, as Newby (1977, p.12) put it, the 'images of idyllic existence so often conjured up by the words "rural England"'. This can, however be at most only a partial explanation, since the majority of studies were based in Ireland and Wales and often painted rural social life in a far from idyllic light. A more complex but also far more revealing interpretation of an apparent blindness to power is provided by situating studies of the rural community in two wider intellectual landscapes.

(Re)viewing Power in Rural Community Studies (1) – Structural Functionalism

Community was and is a heavily contested concept (Hillery, 1955; Bell and Newby, 1971; Liepins, 1999). The majority of (particularly early) community studies, however, implicitly or explicitly articulated and usually made inseparable three notions of their object of enquiry. First, they perceived communities as geographical areas of residence and social interaction – shared territorial living spaces. Second, they perceived those communities as containing common ways of life – chronically reproduced forms of social interaction. Three, they perceived those communities as reflecting shared states of mind – mutual senses of interdependence, solidarity and morality between individuals[1].

Moreover, communities were envisaged as distinct *local* 'social systems' (Harper, 1989; Lewis, 1986; Stacey, 1969); that is, geographically-bounded social entities, whose features were at least partially causally explainable *only* in terms of their reference to each other through their co-presence (e.g. Frankenburg, 1966, p.162; Williams, 1956, p.174; cf. Parsons, 1960). Research was in most cases informed by either the structural–functionalist perspective dominant in anthropology and sociology at the time (Merton, 1957; Parsons, 1937) or the human ecology perspective then popular in urban community studies (Hawley, 1950; Park, 1916). From such viewpoints, communities are taken to be relatively discrete, self-sufficient, holistic social systems, total structures, or even individual 'organisms'[2]. Description and analysis of them involves, accordingly, ascertaining how the 'parts' of the whole inter-relate and how each part contributes to the functioning of the overall system.

The fates of individuals and social groups are, consequently, only of (theoretical and moral) interest insofar as they illustrate the workings and reproduction of the whole 'community'. Thus communal norms, a common focus of community studies, were of interest not so much in terms of how they advantaged or disadvantaged different individuals or social groups, but in terms of how they maintained systemic integrity. Rees (1996, pp.81-86), for example, in his study of a Welsh village, describes an incident in which a group of adolescent males terrorize a young single female for 'seducing' one of their number and the rest of the village approve their behaviour. Under other circumstances, such 'boisterous behaviour' (Rees's words) would have been frowned on. In this case, however, the action taken by the boys represented the 'moral censure' of the community – regulation of sexual behaviour being seen as critical for maintaining moral cohesion. Similarly, Arensberg (1959, p.68) describes the 'general condemnation' and 'punitive action' against those failing to participate in the reciprocal social arrangements of a farming community in Ireland as the means by which those arrangements were maintained. Community is taken to be *a priori*, and therefore its members are subordinate to it; where their behaviour threatens the reproduction of the whole, it must, if possible, be curbed.

The overall effect of such perspectives, which often drew comparisons between 'communities' and biological entities, was the de-politicization of rural power. By placing the overall 'social system' first in the order of analysis, community studies effectively 'normalized' community structures. 'Internal' social relations, and the relationships between individuals or social groups and the overall community, understood respectively as functional attributes and as fundamentally relationships between a 'whole' and its 'parts', were rendered 'natural'. Alternative interpretations of division and conflict were downplayed, and the legitimacy of dominant distributions of power left unquestioned (cf. Bell and Newby, 1976; Whitehead, 1976).

The structural-functionalist perspective does not, however, preclude discussion of power. On the contrary, power can be seen as an integral and necessary part of the community – a systemic property facilitating collective action and social reproduction (see Parsons, 1986). Indeed, *implicit* in the majority of rural community studies is a *social-integrationist* conception of power, where power is understood as the means of securing the cohesion and reproduction of the community. The absence of any specific reference to power, and any problematization of it, therefore requires further contextualization – that of the conceptualization of the rural as a space of *tradition* rather than *modernity*.

(Re)viewing Power in Rural Community Studies (2) - Tradition and Modernity

Early rural community studies were predominantly studies of the social systems bound up with subsistence or local-market-orientated agriculture, often where the family was the main unit of economic production. Most of the communities studied were situated in agriculturally peripheral regions and had low concentrations of hired labour at the time the research was carried out, the 1930s to early 1960s (Newby, 1977, pp.101-103). 'Capitalist business' farming (Frankenburg, 1966, p.252), involving heavily mechanized production methods and labour relations based primarily on economic contract, such as in much of East Anglia, was not a focus.

This focus was not accidental. Most early community studies, as their authors make clear, are purposeful attempts to identify 'traditional', if not actually 'peasant', social systems (cf. Symes, 1981, pp. 23-24; e.g. Arensburg, 1959; Rees, 1996; Geddes, 1955; Davies and Rees, 1960; Lewis, 1970). They are 'descriptions' of non-capitalist modes of life. In this respect, it is notable that the majority of rural communities studied were located in the Celtic fringe (the West of Ireland and North Wales) or upland England - areas relatively physically isolated from urban centres. 'Rural', in this sense, as a supposed historical-cultural distinction given a geographical location (Weintraub, 1969), was taken to be identical with 'non-modern' (cf. Gusfield, 1975).

The majority of community studies are therefore not just studies of 'communities' in an empirical sense, nor are they simply examples of the use of community study as a methodology. They are, rather, studies of community where 'community' is understood as a *way of life* distinct from other possible modes of existence (cf. Wirth, 1938). As such, they firmly position themselves in regard to a dominant and persistent binary regarding ways of conceptualising social collectives and the inter-subjective relations within them. On one side of this binary is the idea of 'community', referring to social relations based on emotive, affective, non-rational 'bonds' and a 'more or less organized totality' of common sentiments and beliefs (Durkheim, 1964, p.129). On the other side of the binary is 'society', referring to social relations based on impersonal, non-emotive social 'associations' and the individual's occupational position in an internally heterogeneous social body (Weber, 1964b, p.136).

This binary, usually attributed to Tonnies (1957, pp.162-163), is found in almost all West European and North American social theory from the early nineteenth century onwards, as well as in the mainstream sociological tradition and a wider tradition in those cultures. It belongs to, and unifies, a series of linked antitheses running through – and driving – those traditions, distinguishing and setting in opposition the characteristics of 'traditional' or pre-capitalist social relations and 'modern' or capitalist social relations (see Bell and Newby, 1971; Giddens, 1971; Gusfield, 1975; Nisbet, 1970). These antitheses include individualism and alienation versus mechanical solidarity, sacred versus secular frameworks of collective cultural reference, status versus class division and, critically, in terms of the present argument, *authority* versus *power*.

The concept of *power* has specific terms of reference within this set of antitheses. First, it refers to one or a combination of three phenomena: *force* exercised through the military, the legal system or bureaucracy (Nisbet, 1970, pp.107-116); the strength of a *political party* (Weber, 1964a)[3]; and political rule legitimated by a *social contract* (Hobbes, 1962) or some other form of explicit secular rationality (Durkheim, 1972). Second, power is possessed and exercised by a geographical centre: usually the State or an agency of the State (Mann, 1993). These terms of reference are taken to render power as a feature specific to modern capitalist (or at least proto-capitalist) forms of social organization, since pre-capitalist forms of life are taken not to exhibit a unified military (rather than private armies), legal system, bureaucracy and (therefore) State. In other words, within these terms of reference, power is taken to be *absent from pre-capitalist, 'traditional' modes of social organization.*

It is therefore hardly surprising that 'power' was not a focus of early community studies. Power, as a product of modernity, is taken to be alien to community – it does not naturally exist there and it *should not* be there. Accordingly, if in such studies it is acknowledged that power is present, in the form of local political institutions and decision-making, it is not studied, since 'community' (as a way of life) is the subject of enquiry (e.g. Williams, 1963; cf. Bell and Newby, 1976, pp.190-192). Where it *is* discussed, it is in the context of rural communities not having 'enough power' to 'defend' themselves against outside agencies and bureaucracy (e.g. Frankenburg, 1957; cf. Bell and Newby, 1971, pp.218-221). Analytically, power is restricted to formal political decision-making – and rural community studies did no more than mention in passing the existence of local government in the form of parish, district and county councils. Power is thus taken to be exercised only *over* communities or resisted by communities, rather than being a feature of the relations within them.

What *is* taken to 'naturally' exist within – and be an integral part of – community is *authority*. Authority in social theory refers, as does power, to superior ability or strength in a social relation (Sennett, 1993), and, correspondingly, it usually also refers to restraint placed on action. Understood in terms of *society* (modernity), authority is taken to be a form of power (Dahl, 1986), and can be based, like power, on explicit rationality, legal sanction and violence (Weber, 1964a). The authority of *community*, however, is taken to be distinct from power in three key ways. First, it is rooted in *tradition*; it is not exercised or obeyed primarily on the grounds of a rationally-agreed social contract or political rule. Second, it derives from *within* community, rather than being exercised over it from without. Third it derives from the *consent* of those under it; consent based on either conscious or unconscious belief that it exists for the collective good (Nisbet, 1962).

This authority is read, imagined and written in early rural community studies primarily in relation to *status*. Status was the 'worth' of an individual in the eyes of the community. This social position largely determined their *authority* in either the community or in the household – the 'family unit'; whether they should be deferred to, treated as an equal, or could be overruled without conflict. It was largely ascribed – usually primarily on the basis of kin standing (Frankenburg, 1957, p.46; Rees, 1996, p.31) but also on the basis of age (generation), gender, religion, occupation and place in the wider division of labour (Frankenburg, 1957, 1966, p.46; Rees, 1996; Williams, 1963, p.7). Status was also partially interactional, however, particularly in the more 'closed' communities – something achieved, through accomplishments such as craft skill and generosity to others (Arenberg and Kimball, 1961; Emmett, 1964; Rees, 1996, pp.142-146). Correspondingly, it could also be lost, through failure to adhere to communal norms of conduct; in extreme cases this loss amounted to 'social death' – the symbolic or even physical banishment of a community member (Arensberg, 1959, p.69)[4].

Safeguarding 'Authority' in Rural Community Studies

Authority in these terms exists only with the presence of a local social system; it breaks down with the 'intrusion' of wider social relations, since these have not been derived from within. It can therefore be 'found' only where a distinctly local culture can be established. At the same time, if it is 'found', it is in effect evidence of such a system. The significance of this in terms of the dominant social imaginary of much of post-war and rural studies can be seen in the ways in which the spatial and social imaginaries of researchers are mutually informative in many studies of rural communities.

The active purpose of most analysis in rural community studies was the foregrounding of features that nominally allowed a community to be viewed as a local social system with its own distinct institutional frameworks and structures rather than as a mere reflection of the institutions and structures of society at large (Arensberg and Kimball, 1961, pp.309-310; Emmett, 1964, p.ix; Frankenburg, 1957, p.162; Rees, 1996, p.165; Williams, 1956, p.174, 1963, p.xiv and p.208). This *privileging of the local* was, essentially, how community studies discursively created their objects of enquiry. Even where it was accepted that certain aspects of social organization were refractions of the wider social, economic and cultural contexts in which communities were situated (e.g. Williams, 1956; Frankenburg, 1957), and/or that the 'modern' (urban) had in some cases 'intruded' into the 'non-modern' (rural), such features were downplayed in 'explanation'. A clear example is the prioritization of *status* over *class* in explanations of social division within localities.

As community was taken to be the antithesis of society, so status was taken to be the antithesis of class. Status, whether derived from interaction or ascribed, was a measure of individual position and worth according to a relatively undifferentiated 'social economy'. Class, on the other hand, referred to worth in a highly differentiated property, labour and commodity market. As such, it was an intrusion from the outside world - it did not belong to community. Moreover, status is generally highly place-bound, rather than trans-contextual (cf. Plowman et al, 1962), and so emphasising its importance provided further evidence of a system relatively immune to external influences. Issues of class division could not be avoided altogether, particularly in those settlements experiencing high levels of in-migration from former urban dwellers, but class largely remained an empirical *description* of economic stratification (e.g. Williams, 1963, pp.195-199) rather than a theme for further analysis[5]. Even where 'class consciousness' (Williams, 1963) or a 'fundamental division' between 'capital and labour' was identified (Frankenburg, 1957, p.11), its significance was quashed through the privileging of lay-derived definitions of status rather than theoretical categories imposed from 'outside' (Frankenburg, 1966, p.154; Rees, 1996, p.142; Williams, 1956, p.174; cf. Strathern, 1984). In this way, the notion of authority, as opposed to power, was analytically – and, it might be added, ideologically – safeguarded (cf. Bell and Newby, 1976, pp.190-192).

At the same time, many community studies reflected, implicitly or explicitly, a normative agenda that defensively positioned the way of life of rural communities against the way of life of urban society. A consistent theme was the undermining of the morally-structured and 'organic', 'traditional' communities by social, cultural, economic and technological forces that rendered life increasingly sterile, amoral and fragmented. Several authors (e.g. Rees, 1996) went so far as to hold up life as they 'found' it in the communities they studied as the benchmark against which modes of existence and organization elsewhere should be judged. Against such strains of thought, it is not unreasonable to suppose that many authors of such studies considered the structures and relations of authority they observed as inherently legitimate and beneficial, functioning to the benefit of the community and therefore perpetuated – insofar as they *were* being perpetuated in the face of ongoing changes – by virtue of the active consent of a collective communal 'will'. Accordingly, they did not see it as their place to question them.

It is hard to tell precisely where 'description' ends and 'prescription' begins in many rural community studies (Bell and Newby, 1971, 1976) – as it is in much of the social theory those studies consciously or unconsciously reflect (Gusfield, 1975). What is clear, however, is that their lack of engagement with issues of power and the legitimacy of prevailing social and political arrangements, and their apparent approval of structures and relations of 'authority' in 'traditional' rural communities are intimately entangled, as the result of both being a product of the same presuppositions: the idea that local and above all 'rural' social relations could be independent of wider and 'urban' social relations (cf. Lupri, 1967; Pahl, 1965b) and, in the words of Weber (1964a, p.34), a belief in 'the sanctity of the order' as it had 'been handed down from the past.' Correspondingly, the more recent focus on *power* rather than authority in rural studies is the product of, or at least heavily bound up with, challenges to both of these presuppositions – that is, with both ontological and normative critiques of once dominant and largely unquestioned approaches to rural society and space.

Political Society – Power as *Domination*

The first explicit discussion of power in anglophile rural studies came through research by Newby and associates into 'capitalist business farming' in Suffolk and East Anglia (Newby, 1977; Newby et al, 1978; Saunders et al, 1978). This work interpreted rural social and political organization through the framework of Marxist political-economy, focusing on the role of property ownership and the perpetuation of ideology in the construction and reproduction of relations of authority and deference between rural landowners and agricultural workers. Whilst it would be a mistake to think of this work as a direct critique of rural community studies, it did explicitly set out to challenge what Newby (1977) referred to as the idyllic images associated with the idea of 'rural England'.

The exercise of power was identified as critical to the dominance of the agricultural elite, and three inter-related but distinct 'modes of control' (Newby, 1977, p.121) were outlined. First, it was argued that elites perpetuated an 'ideology of property' (Rose et al, 1976) that positioned those who owned land, rather than those who worked on that land, as sources of wealth. This ideology granted employers the 'natural right' to exercise power over their employees in a variety of domains (Newby, 1977, pp.368-381; Newby et al, 1978, p.25). Second, through face-to-face contact with their workers, and the active cultivation of ties of loyalty and deference, employers ensured that conflicts and dissatisfaction remained as far as possible personal – between a *particular* worker and that employer – rather than articulated as a common, class-based grievance (Newby et al, 1978, pp.157-159; and cf. Littlejohn, 1963, pp. 27-36). Associatively, farmers and landowners had influence if not absolute control over the provision of local employment and, often, also housing and welfare, and thus had the power of negative sanction at their disposal should more subtle means of social control fail. Third, village society, even though usually far from the closed community idealized by many commentators on the rural scene, in effect provided – with the encouragement of landowners – a self-contained world for agricultural workers. Socialized within this symbolic, if not physical, 'total institution' (cf. Goffman, 1961), workers were dependent on finding employment within it, and, it was argued, often saw their subordinate position as natural and/or inevitable. If they did question it, they did so only in the terms of the existing 'system', since they lacked the conceptual resources to imagine a social structure organized on radically different principles (Newby, 1977, pp.165-194).

Although this work has been characterized as a sociology of agriculture (Murdoch and Pratt, 1993), it was in fact driven by a more general concern of sociology and political science at the time: the bases of, and the relationships between, class position, class consciousness and class action (Newby, 1977, pp.103-110; cf. Lukes, 1974). Against this context, it explored how the work situation, economic status and class position of farm workers combined with more 'structural' features of rural social organization to reproduce a 'deferential' class consciousness. It was this *class* basis of their analysis which, along with the work of Pahl (1965b) perhaps most of all initially opened up the rural in terms of critical social analyses of power.

Pahl and Newby et al, along with Littlejohn (1963), were keen to stress that key social and political institutions and lines of social stratification in rural areas were local manifestations of, or otherwise allied to, non-local institutional forms and social structure and therefore could not be considered either entirely 'locally-derived' or specifically rural. This focus accompanied a nascent concern within rural studies to come to terms with and re-conceptualize the social and political composition of rural space – and in particular its *re*-composition through out-migration and counterurbanization. As in more recent class analyses and other work on power (see below), the idea of rural communities being closed, organic entities was, if only partially, slowly and in many cases reluctantly, being replaced by the notion of communities being inextricably bound up with the wider worlds in which they were situated and actively constructed, reproduced and contested by their members. Thus, such researchers placed a key emphasis on class analysis, which related to distributions of power within wider society, as opposed to status divisions, which were taken to be restricted to local social contexts, since it enabled relations of power at the local scale to be read in terms of power relations at the national scale (Buttel and Newby, 1980, p.9; Newby, 1980, p.259; Pahl, 1966).

Whilst some argued that this did not mean the death of the study of local institutional arrangements and relations for their own sake (Stacey, 1969), others firmly rejected the idea that the 'community study' as it had traditionally been envisaged was a theoretically informative mode of enquiry. Instead, local milieus were re-envisioned as 'laboratories' (Pahl, 1966), in which the features of national social systems could be identified and explored. This would 'make it impossible to pretend that the interdependence between processes and structures at various geographical levels [i.e. scales] is ... paid lip service to but in practice ignored' (Benvenuti et al, 1975, p.11). Genuinely 'local' features of social organization, if they could be found, were identified as markers of regressive social tendencies – the inability of situated institutions and individuals to 'overcome' spatial constraints – and were taken to be no longer of primary significance.

Power and the Middle-Classes

Early analyses of class in rural areas tended to draw on notions of class that took property and primary labour relations to be the ultimate determinants of social interests, action and power (Newby et al, 1978; Rose et al, 1976). More recent rural class analyses have placed more stress on the relative abilities of different social fractions to consume in the rural commodity market (Cloke, 1990; Marsden et al, 1993; Phillips, 1993) – a Weberian rather than a Marxist notion of class. A distinction can also be drawn between rural class analyses that treat class divisions as a structural given and therefore in themselves explanations of social and political inequalities and those approaches that treat class divisions as reflections of the 'structures and power relations which have *produced* these inequalities' (Murdoch and Marsden, 1994, p.3; original emphasis). Since the early 1980s it has been towards the latter perspective that rural class analysis has increasingly turned.

Through such analyses, power in rural areas (or in parts of rural England, at least) has increasingly been attributed to incoming 'middle class' or 'service class' fractions (cf. Goldthorpe, 1995) challenging and in some cases overcoming the domination of the former agricultural elite. Indeed, rural class analyses since the early 1980s have been pursued almost wholly with regard to the middle class or service class, however defined (Phillips, 1998a, 1998b).

Power has been attributed to the middle class – or rather, certain fractions of the middle class – on three grounds. First, their statistical over-representation in local government and other local statutory bodies, along with their numerical dominance and leadership of amenity groups (Cloke, 1990). Second, their relative success, through collective action, in (re)shaping the physical, social and ideological characteristics of rural space in a manner that matches their economic and aesthetic interests – a process identified as bound up with and helping to further their reproduction (Murdoch and Marsden, 1994). Third, their privileged ability to 'differentiate themselves from other class fractions' (Cloke and Thrift, 1987, p.323) by materially excluding other classes from rural space (Marsden and Murdoch, 1995), symbolically excluding other classes from representations of that space (Thrift, 1989), and constructing symbolically bounded intra-class 'status groups' around shared notions of taste and morality (Phillips, 1998b).

Middle-class power is seen to be based on superior command of and access to various combinations of a number of key resources. These include, most notably, three assets. First, money, signifying the ability to buy into an increasingly economically exclusive rural housing market and, associatively, purchase various forms of 'cultural capital' or 'cultural assets' representing social position (Cloke and Thrift, 1990; Murdoch and Marsden, 1994). Second, skills, qualifications and 'expert knowledges', signifying the ability to mobilize collective action and organize representation at multiple political scales (Day et al, 1989; Urry, 1995). Third, cultural competence, signifying the ability to communicate the possession of 'good taste' to appropriate individuals and social groups and maintain symbolic boundaries against others (Cloke et al, 1995, 1998).

By the late 1990s, the extent of middle class or service class power was – and remains – a significant bone of contention within rural studies. There have been suggestions that the power of certain fractions of the middle class in certain areas is becoming *hegemonic*: as with the elite before them, those fractions are not only politically dominant, but also exercise *leadership* (Cloke and Goodwin, 1992), mobilising and controlling representations of rural space that benefit them whilst remaining fundamentally unchallenged by those they disadvantage (Cloke and Thrift, 1990; Thrift, 1989). Others have challenged this suggestion, arguing that the relationship between a strong middle class presence in the countryside and the translation of 'middle-class interests' into class leadership is far from proven (Hoggart, 1997; Pahl, 1989) and that any middle class power is itself highly contingent on the conjuncture of multiple other forces (Murdoch and Marsden, 1994)[6].

Other strands of thought attempt to further downgrade the importance of any middle class power, by drawing attention to the wider political-economic forces implicated in the making and re-making of rural space. Newby et al (1978), whilst noting the importance of new social fractions in the countryside, saw those fractions as derivative of shifting property relations, and thus placed transformations of property relations as the key causal component of changing power relations (see also Marsden et al, 1993). Rees (1984) and Urry (1984) in turn argued that this perspective marginalized the overall restructuring of the national and international political economy, and the resulting transformations in national spatial divisions of labour (see also Day et al, 1989: 230; Hoggart et al, 1995; Marsden and Flynn, 1993; Marsden et al, 1993)[7]. Drawing on the school of 'regulation theory', Cloke and Goodwin (1992) and Goodwin et al (1995) have sought to contextualize the apparent formation of new power blocs in the countryside against the transformation of the political structures and cultural ensembles associated with securing new modes of capital accumulation. Common to all these perspectives is the situating of rural localities in the context of wider social, political, cultural and technological spaces and forces.

The State, Community Development and Local Governance

The move towards exploring the inter-relations between and interaction of 'local' and 'non-local' social, political and economic processes has generated interest in three other, inter-linked arenas where power is increasingly recognized as critical: *the state*[8], *community development* and *local governance*.

Since the 1970s, rural studies has seen a growing, if limited, interest in the position and activities of the state regarding rural areas. Initially this interest was limited to considering the state as a neutral body arbitrating between different interests in conflicts over land-development and / or as a political actor in its own right (Cloke, 1989), and many commentators continue to work within this framework of interpretation. From the 1980s, however, there has been a rejection of both these notions by many writers, in the light of political-economic analyses of rural restructuring and increasing dissatisfaction with the lack of explanatory leverage offered by empiricist-pluralist conceptions of the state. Instead, attempts have been made to reposition the state as a class actor, chronically favouring dominant social fractions in the planning process and other arenas of interest representation (Cloke and Little, 1990; Little, 2001). Initially, the power of the state as a class actor was viewed in terms of its role as 'midwife', through the planning process, policy and ideology, to farmers and landowners (Newby, 1980; Smith, 1989; Cox and Lowe, 1984; Cox et al, 1986). More recently there have been efforts to explore the representation of 'newer' middle class interests within the state and the complexities of the relationship between the state and those interests (Buchanan, 1982; Cloke and Little, 1990; Murdoch and Marsden, 1994).

Attention has also been called to the relations between central government and local state institutions and agencies (Goodwin, 1998; Woods, 1998). Increasingly the role of the local state in the institution, reproduction, contestation and transformation of rural topographies of power is seen as ambiguous (Goodwin, 1998; Woods, 1998a), in the light of analyses suggesting that any assumption that the 'local state' is essentially a dependent and docile arm of the wider state, simply imposing top-down strategies of rule (cf. Cockburn, 1977) is erroneous (cf. Duncan and Goodwin, 1988). Accordingly, attention has been directed to power relations and contestation within or at the level of the state itself, especially the tensions between the policy objectives of central government and the frequently conflicting aims and objectives of local state actors involved in policy implementation (see especially Cloke, 1987, 1989, pp.186-187; Cloke and Little, 1990; Goodwin, 1998; Woods, 1998a).

Particular attention has also been given to the changing role of the state, and the reconfiguration of patterns and modes of state intervention, in the context of community development and local governance. The blurring of established divisions of responsibility for community regeneration and service delivery, through the encouragement of 'joined up working' between public, private and voluntary agencies at the local scale (Goodwin, 1998; cf. Stoker, 1996), has raised questions over actual and potential reconfigurations of local state power within and over the local rural polity. In the (at least rhetorical[9]) shift from 'government' to 'governance', the formal governmental apparatus of the state becomes just one of many sets of local actors involved in delivering services and identifying 'problems and solutions', whilst former 'state actors' are redefined as agencies in their own right and so cannot be relied on to act as neutral conduits for or extensions of either central or local state power. The local state itself may therefore become increasingly vulnerable to co-option and occupation by other interests (Woods, 1998a; cf. Duncan and Goodwin, 1988), and thus complicate and reconfigure geometries of power at the local scale.

Such concerns have led to the problematization of moves to facilitate 'endogenous' local community development (Ray, 1998, 1999) and community 'self-governance' by encouraging voluntary action, self-organized partnerships for growth and increasing the power of local 'animateurs' (Edwards, 1998; Edwards et al, this volume). Whilst local empowerment is generally taken to be beneficial – a 'good thing', that same 'empowerment' it is argued can also be a cover for the withdrawal of public services, collective welfare provision 'and other support mechanisms' (Murdoch and Abram, 1998, p.42), and thus constitute the state's thinly disguised disavowal of any responsibility for redistributing power within the local polity (Murdoch, 1997). Government enthusiasm for 'local empowerment' has therefore generated questions regarding its contribution to the differential empowerment of different social groups (Burnett, 1998; Murdoch, 1997), the inclusion and exclusion of different interests, the power dependencies and asymmetries between actors and institutions undertaking collective action (Goodwin, 1998), over whether increased power will entail increased responsibility (Edwards et al, 2001b); and over how resultant new modes and distributions of power in the rural might become 'naturalized' and legitimated (Edwards, 1998; Goodwin, 1998).

Attention has also been called to the mechanisms via which state power is exercised 'from a distance' over local populations. Here, power circulating via capillary 'networks' of state, quasi-state and nominally non-state agencies (Murdoch and Abram, 1998) is seen to define and redefine the discursive territories in which actors act (cf. Rose and Miller, 1996). This 'government from a distance', or 'governmentality' has been positioned as one of the most important modes – if not *the* most important mode - of control over social and political life in modern society (Rose, 1999). In rural studies it has been considered primarily in terms of attempts by central government to enforce a 'dominant strategic line' overriding competing definitions of rural problems and their solutions. This has included the enforcement of central rather than local priorities in housing development (Murdoch and Abram, 1998); the 'management' of local populations through the reallocation of powers and responsibilities within local communities (Murdoch, 1997), the reframing of agricultural and environmental policy around technocratic and 'expert' discourses (Murdoch and Ward, 1997), the 'governance' of partnership working (Edwards et al, 2001a) and the reconfiguration of rural development policy (Ward and McNicholas, 1998).

Emerging Agendas for Rural Studies of Power

If there is any common 'thread' uniting the multiple facets of rural studies at the opening of the twenty-first century, it is an emphasis on the ambiguousness and multi-faceted character of the rural and rural change: an 'incredulity towards meta-narratives' (Lyotard, 1984) that attempt to reduce events to any single, underlying structural process (e.g. Cloke, 1997; Cloke and Goodwin, 1993; Murdoch and Pratt, 1997; Philo, 1992). From this standpoint, there have been explicit attempts to contest and transform what some authors identify as unhelpful – or even ethically and politically dangerous – dominant imaginaries of power.

Perhaps most notable in this respect are Murdoch and Pratt (1993, 1994, 1997), who have argued for the adoption of what they term a 'post-modern' conception of power. Here, power is not *a priori* located in and reduced to structure or the state or a pre-defined class. Rather, such entities and agents, and the power they exercise, are taken to be the *effects* of the 'exercise of particular social relations and practices' (Murdoch and Pratt, 1993, p.423). At the same time, a growing number of rural researchers are arguing that power must be seen as an active social and political *construction*, contingent upon the construction, reproduction and transformation of networks of *multiple actors* (Murdoch, 1995; Murdoch and Marsden, 1995; Enticott, 2001). Accordingly, attention is being called to the 'micro-sociologies' through which the alliances and coalitions of actors constituting different networks of power – those actors often situated in different domains of action and operating across multiple scales – are put together and taken apart (Marsden et al, 1993; Murdoch and Pratt, 1994, 1997; Woods, 1998a, 1998b). These theoretical shifts have involved drawing on the 'actor-network theory' of Callon (1986) and Latour (1986) and the notion of 'discourse' developed by Foucault (1980) to draw attention to the active *practices* and *technologies* of power as well as more institutionalized structures and ideologies.

Such an emphasis helps inform a partial 'recasting' of established arenas of power. Class, for example, rather than being taken as a determining, causal power, can be seen as an outcome of multiple, inter-linked social and political processes and their mediation through both local and wider scale discourses and 'cultural textures' (Cloke et al, 1995; Phillips, 1998a). Class thus becomes an effect and marker of power rather than a determinant of it. Similarly, the local state from this perspective does not 'possess' power in its own right, but must achieve it through (re)positioning itself in wider actor-networks (Woods, 1998a, p.25). At the same time, 'the state' itself is not seen as a monolithic entity but a 'chaotic' ensemble of heterogeneous and often competing actors and agencies.

Since the 1990s in particular, rural studies of power have also been characterized by a growing concern with social and political marginalization in rural areas. This has included work on children and youth, women, those on low incomes, the homeless, travellers and homosexuals (Cloke and Little, 1997). Previous work on rural marginalization tended to treat it largely as a function of distance from centres of services, employment and consumption, albeit structured by intervening variables such as age, gender and income, or as a contingency of rural social restructuring (e.g. Moseley, 1979). Increasingly, however, interest in marginalization has centred on the roles of dominant social fractions and ideological and discursive formations in the symbolic and material exclusion of particular groups from rural areas. Particular significance has been attributed to notions of rural space that represent the countryside as somewhere free from social problems, the economic exclusiveness of the rural housing market and the actions of middle-class actors seeking to symbolically and materially exclude those deemed 'out of place' in rural localities. The discursive power at work here, it is further argued, is the product of not only the state, or landowners, or the middle class, but also the very interpretive and conceptual schemas through which rural studies has sought to read, write, imagine and know 'the rural' (Cloke, 1997; Philo, 1992; Murdoch and Pratt, 1993). Most recently, Cloke and his associates (Cloke, 2002; Cloke et al, 2000) have sought to connect questions of power and marginalization to issues of moral responsibility and the politico-ethical imaginaries of rural researchers, calling for an academic engagement that is not only *with* the 'other' but also *for* the 'other'.

Problematizing Power in Rural Studies

Over the past thirty years, imaginaries of power within the field of anglophile rural studies have shifted and multiplied markedly. However, despite these shifts and a growing plurality of viewpoints, two critical aspects of power remain almost wholly neglected in rural studies of power. The remainder of this chapter briefly sketches these out and indicates how they could inform future research agendas.

The Individual and Power

First, little attention has been given to the place of *individual actors* in the production, reproduction and contestation of rural topographies of power. Rural studies of power have been predominantly directed at the *structural* and *institutional* components of power bound up with the ordering and re-ordering of rural society and space. This has included focusing on the broad ideological and discursive (trans)formations dominant in policy arenas, political action and spheres of cultural representation, class (re)composition and the (re)configuration of state power. In such accounts, rural populations tend to be treated primarily as micro-scale expressions, mediators and victims of those structural and institutional forces. Notably, both middle-class in-migrants and the vestiges of the established rural elite, who have been positioned as key agents of power in local and wider rural polities, have largely been considered only as agents subsumed in class action.

As a consequence, there has been a neglect of two key issues regarding the *interface* between individual actors and the wider contexts of which they are an integral part. These are: how, and the extent to which, the political agency of different individuals *informs* the production, reproduction, contestation and transformation of rural topographies of power (and thereby the wider geographies of rural places and spaces); and, how, and the extent to which, the political agency of different individuals is *informed by* the production, reproduction, contestation and transformation of those topographies of power (and thereby those wider geographies).

Foregrounding these two questions is in no way a call to *replace* other, established and emergent approaches to rural power with such a focus. However, it *is* a call to researchers to consider ways of engaging in rural studies of power that take more seriously, and more critically: the *practices* through which such actors may actively *constitute* both their own identities and contexts of action; the *processes* through which the identities of such actors may be actively *constituted by* the practices of others (see below); the *knowledges and understandings* informing those practices; and the *mechanisms and linkages* enabling and constraining the *translation* of interests into action and knowledge into practice.

Such a theoretical move would greatly assist more imaginative and critical thinking in terms of the interconnection between the 'strategic, local and practical concerns' (Clegg, 1989, p.5) of political actors constructing and pursuing particular goals and the *unintended effects* of those 'local and practical concerns'. Here, a greater engagement with some of the suggestions of Foucault (1980, 1981) would be highly worthwhile. Foucault (1981, pp.94-95) suggests that while strategies and practices of power may be explicit and intentional at the level of inter-subjective relations and local collective practice, those strategies and practices tend to become associated and entangled both with one another and with wider-scale and more coherent formations of power – they 'attract and propagate one another'. The resultant 'networks' of power produce outcomes, in multiple, independent domains and scales of experience and practice, that cannot be directly attributed to individual intentional action but at the same time are absolutely dependent on them for their reproduction and proliferation. Thus it becomes possible to draw out how relatively *institutionalized* patterns of advantage and disadvantage, privilege and marginalization, inclusion and exclusion, enablement and constraint and domination and subordination are (re)created and sustained by *individual* ways of thinking and acting that 'can be regarded as reasonable, ethical and even sympathetic' (Cloke, 2002, p.599) towards others.

Through such an imaginary of power it becomes possible to move towards the construction of studies of rural society and space that draw attention to the critical inter-relations between the individual moral agent and the geographies they inhabit, produce, reproduce, and seek to transform.

Power as a Constitutive Force

Second, there has been minimal engagement with the notion of power as *constitutive force*. Since *power* emerged as a key focus for rural researchers, the underlying terms of reference to it, despite engagement with a range of theory and generalization, have remained remarkably consistent and unproblematized: it is taken to be a *negative force*, found where one actor (individual, collective or institutional) exercises power *over* another actor. It is associated with *suppression* and social *conflict* (Newby et al, 1978; Shorthall, 1992); with the political and cultural *dominance* of elites (Cloke, 1990; Woods, 1997); with the *exclusion*, *marginalization* and *silencing* of 'Other' social and cultural groups (Cloke and Little, 1997); with state *control* over the populations in its territory (Murdoch and Abram, 1998; Ward and McNicholas, 1998). Power is seen as *repressive* and *sectional*; power relations are identified as a 'zero-sum game', the power of one actor necessitating the powerlessness of another. This viewpoint is encapsulated by Murdoch and Pratt (1994, p.85), who, despite being amongst the most vocal advocates for the reconstitution of theoretical imaginations in rural studies, maintain that: 'It is the exercise of power that *restricts*...individuals and collectives, and thus *suppresses* or *limits* their potential to find alternative ways of being' (emphases added).

These terms of reference are in keeping with the dominant tradition of thinking about power in twentieth century political science and social theory, where the notion that power is primarily *repressive* and *sectional* has long been a central pre-occupation (cf. Barnes, 1988; Clegg, 1989; Foucault, 1980; Lukes, 1974). However, to conceptualize power only in this manner, as a negative force, is to only partially engage with its nature. A potentially far more radical way of engaging with power is to conceive of it as a *constitutive and productive force*.

To conceive of power as a constitutive and productive force is to reverse the 'repressive hypothesis': 'rather than assuming a generally acknowledged repression...we must begin with...*positive* mechanisms, insofar as they *produce* knowledge, *multiply* discourse...' (Foucault, 1981, p.73; emphases added). This is not to suggest that power is intrinsically productive in a *political* sense. Rather, it is to see it as intrinsically productive in an *ontological* sense. It is to see social and political actors as actively *constructed* and *transformed* through power – through the relations established and reproduced between persons and between persons and institutions, and through the techniques that persons practice on themselves, or induce others to practice on themselves, in order to facilitate new formations of identity and knowledge.

Thinking of power as a constitutive and productive force has the potential to open up a number of critical research agendas in the study of rural power, some of which have already been hinted at by some of the commentators cited above. Two broad concerns can be specified in this respect. First, there could be critical interrogation of current attempts to construct new moral subjectivities and regimes in rural space through the circulation and propagation of political rationalities and the construction and deployment of particular 'technologies of government' (cf. Dean, 1999). One approach here, for example, might be to look at how emergent neo-liberal political regimes favouring the (trans)formation of individuals into 'self-governing' subjects who take responsibility for both their own welfare and that of others within their 'community' (cf. Rose, 1996, 1999) are being translated into policy and practice in the rural arena (see Murdoch, 1997; Edwards et al, 2001b). Second, there is also the potential to consider how, at a more experiential level, the social and political identities and capacities to act of different rural actors may be actively constructed and transformed through their own actions, the actions of others and their entanglement in the wider discursive and material coherences embedded in ways of thinking about and acting towards rural society and space. Principally this would involve exploring the concrete and more or less 'strategic' movements and manoeuvres made by different rural actors in their attempts to gain, reproduce and transform the relations and institutions of rural spaces and places.

Such an imaginary of power would allow rural studies a heightened engagement with the realities and ambiguities of power, in all its forms, as it is practiced, reproduced, experienced and as it impacts upon the multiple actors constituting and reconstituting the rural.

Conclusion

This chapter has sought to critically engage with both historical and present-day concerns with the geographies of power bound up with the constitution and reconstitution of the UK countryside. It has examined and (re)viewed the changing and expanding perspectives on power that rural researchers have been guided by, and indicated how these perspectives are bound up with different ways of conceptualising rural society and space.

A countryside that was once very much seen as apolitical and uncontested is now positioned as an arena of conflict and inequality. Normative visions of power have fundamentally shifted, from perspectives through which traditional structures, relations, discourses and ideologies of power were seen as natural and collectively beneficial, and marginalization, domination and inequality largely hidden from view in the academic landscape, to perspectives through which structures, relations, discourses and ideologies are recognized as contingent political and social constructions and the material realities of poverty and rule are made highly visible and openly critiqued.

At the same time, the theoretical and conceptual tools through which power in rural society and space is interpreted and understood have radically changed and are open to ongoing contestation and revision. Particularly notable here has been the move away from considering rural localities as autonomous spatial containers whose features can be considered in isolation from wider social and political processes. Rural areas are now recognized as integral components of regional, national and international social, political and cultural economies, their past, present and future intimately entangled and mutually informing. Clear also is a desire or at least willingness amongst a growing number of researchers in rural studies to actively engage with and be informed by theoretical literatures and debates that seek to stimulate new ways of conceptualising power and offer the potential to further widen the political, social and ethical imaginations currently articulated within the field of rural studies.

The alternative ways of thinking about power sketched out above are intended as a means of extending, not criticising, this growing concern. The topographies of power evident in the present-day UK countryside, bound up with multiple and vicious political and social inequalities and injustices, require urgent and vigorous interrogation and critique, as do the processes, structures, mechanisms and practices implicated in their ongoing production and reproduction. Such interrogation and critique is only possible with theoretical tools up to the task of de-reifying contemporary political and social arrangements and able to provide the means to think outside the dominant discourses currently governing much policy and practice.

Notes

1. Based on: Arensberg and Kimball (1961); Emmett (1964); Frankenburg (1957); Geddes (1955); Lewis(1970); Rees (1996); Williams (1956, 1963). See also Bell and Newby (1971); Crow and Allen (1994); Frankenburg (1966); Harper (1989); Hillery (1955); Lewis(1986); Stacey (1969).
2. A notion used for example by Rees (1996: 79-80), although his work was not explicitly theoretical.
3. This distinction is slightly blurred, since Weber (1964a, 1964b) and Tonnies (1957) at times talk about social power (status) and economic power (class), but in these cases 'power' is a 'common sense' term rather than a theoretical concept. The key is not the word itself, but the *terms of reference* within which it is theoretically and normatively employed.
4. Communities might also be stratified – or rather, stratify themselves – into more or less distinct *status* groups: several community studies note how residents marked others' social closeness or distance to them by identifying them with particular, normatively categorized 'lifestyles' (e.g. Emmett, 1964; Williams, 1963).
5. The notable exception of the community studies being Littlejohn (1963), who, in contrast to others, was studying 'capitalist' farming.
6. For debate over this, see especially Cloke et al (1995); Hoggart (1997); Murdoch and Marsden (1994, 1995); Pahl (1989).
7. Such contentions are linked to a much wider debate on the significance of localities in terms of social causation (cf. Cox 1993).
8. Both theoretical and empirical work on the (changing) form and role of the state in the rural context has so far been limited (see Cloke and Little, 1990; Goodwin, 1998; Little, 2001).
9. Goodwin (1998) makes the critical distinction between governance as *theory* and governance as *practice*.

References

Arensberg, C. (1959), *The Irish Countryman: An Anthropological Study*, Originally Published in 1937, Peter Smith, Gloucester, Massachusetts.
Arensberg, C. and Kimball, S. (1961), *Family and Community in Ireland*, Originally Published in 1940, Peter Smith, Gloucester, Massachusetts.
Barnes, B. (1988), *The Nature of Power*, Polity Press, Cambridge.
Bell, C. and Newby, H. (1971), *Community Studies: An introduction to the sociology of the local community*, George Allen & Unwin, London.
Bell, C. and Newby, H. (1976), Community, Communion, Class and Community Action: The Social Sources of the New Urban Politics, in D. Herbert and R. Johnston (eds), *Social Areas in Cities: Volume 2. Spatial Perspectives on Problems and Policies*, John Wiley & Sons, London, pp. 189-207.
Benvenuti, B., Galjort, B. and Newby, H. (1975), 'The Current Status of Rural Sociology', *Sociologia Ruralis*, Vol. 15, pp. 3-21.
Buchanan, S. (1982), 'Power and Planning in Rural Areas: Preparation of the Suffolk County Structure Plan', in M. Moseley (ed), *Power, Planning and People in Rural East Anglia*, Centre of East Anglian Studies, University of East Anglia, Norwich, pp. 1-20.
Burnett, K. (1998), 'Local Heroics: Reflections on Incomers and Local Rural Development Discourses in Scotland', *Sociologia Ruralis*, Vol. 38, pp. 204-225.

Buttel, F. and Newby, H. (1980), 'Toward a Critical Rural Sociology', in F. Buttel and H. Newby (eds), *The Rural Sociology of the Advanced Societies: Critical Perspectives*, Croom Helm, London, pp. 1-35.

Callon, M. (1986), 'Some Elements of a Sociology of Translation: Domestication of the Scallops and the Fishermen of St Brieuc Bay', in J. Law (ed), *Power, Action and Belief: A New Sociology of Knowledge?*, Routledge & Kegan Paul, London, pp. 196-233.

Clegg, S. (1989), *Frameworks of Power*, Sage, London.

Cloke, P. (1987), 'Policy and planning in rural areas', in P. Cloke (ed) *Rural Planning: Policy into Action?*, Harper & Row, London, pp. 1-18.

Cloke, P. (1989), 'Rural Geography and Political Economy', in R. Peet and N. Thrift (eds), *New Models In Geography: Volume I*, Unwin Hyam, London, pp. 164-197.

Cloke, P. (1990), 'Community Development and Political Leadership in Rural Britain', *Sociologia Ruralis*, Vol. 30, pp. 305-22.

Cloke, P. (1997), 'Country Backwater to Virtual Village? Rural Studies and "The Cultural Turn"', *Journal of Rural Studies*, Vol. 13, pp. 367-375.

Cloke, P. (2002), 'Deliver us from Evil? Prospects for Living Ethically and Acting Politically in Human Geography', *Progress in Human Geography*, Vol. 26, pp. 587-604.

Cloke, P. and Goodwin, M. (1992), 'Conceptualizing Countryside Change: From Post-Fordism to Rural Structured Coherence', *Transactions, Institute of British Geographers*, Vol. 17, pp.321-336.

Cloke, P. and Little, J. (1990), *The Rural State? Limits to Planning in Rural Society*, Oxford University Press, Oxford.

Cloke, P. and Little, J. (1997), (eds), *Contested Countryside Cultures: Otherness, Marginalisation and Rurality*, Routledge, London.

Cloke, P. and Thrift, N. (1987), 'Intra-class Conflict in Rural Areas', *Journal of Rural Studies*, Vol. 3, pp. 321-333.

Cloke, P. and Thrift, N. (1990), 'Class Change and Conflict in Rural Areas', in T. Marsden, P. Lowe and S. Whatmore (eds), *Rural Restructuring*, David Fulton, London, pp. 165-81.

Cloke, P., Goodwin, M. and Milbourne, P. (1998), 'Inside Looking Out; Outside Looking In: Different Experiences of Cultural Competence in Rural Lifestyles', in P. Boyle. and K. Halfacree (eds), *Migration Into Rural Areas: Theories and Issues*, John Wiley & Sons, London, pp. 134-150.

Cloke, P., Milbourne, P. and Widdowfield, R. (2000), 'Ethics, Reflexivity and Research Encounters With Homeless People', *Ethics, Place and Environment*, Vol. 3, pp. 33-154

Cloke, P., Phillips, M. and Thrift, N. (1995), 'The New Middle Classes and the Social Constructs of Rural Living', in T. Buller, and M. Savage (eds), *Social Change and the Middle Classes*, UCL Press, London, pp. 220-38.

Cockburn, C. (1977), *The Local State: Management of Cities and People*, Pluto Press, London.

Cohen, A. (ed) (1982), *Belonging: Identity and Social Organisation in British Rural Cultures*, Manchester University Press, Manchester.

Cox, G. and Lowe, P. (1984), 'Agricultural Corporatism and Conservation Politics, in T. Bradley and P. Lowe (eds), *Locality and Rurality: Economy and Society in Rural Regions*, Geo Books, Norwich, pp. 147-166.

Cox, G., Lowe, P. and Winter, M. (1986), *Agriculture: People and Policies*, Allen & Unwin, London.

Cox, K. (1993), 'The Local and the Global in the New Urban Politics: A Critical Review', *Environment and Planning D: Society and Space*, Vol. 11, pp. 433-448.

Crow, G. and Allen, G. (1994), *Community Life: An Introduction to Local Social Relations*, Harvester Wheatsheaf, Hemel Hempstead.

Dahl, R. (1986), 'Power as the Control of Behaviour', in S. Lukes (ed) *Power*, Blackwell, Oxford.

Davies, E. and Rees, A. (1960), *Welsh Rural Communities*, Cardiff University Press, Cardiff.

Day, G., Rees, G. and Murdoch, J. (1989), 'Social Change, Rural Localities and the State: the Restructuring of Rural Wales', *Journal of Rural Studies*, Vol. 5, pp. 227-244.

Dean, M. (1999), *Governmentality: Power and Rule in Modern Society*, Sage, London.

Duncan, S. and Goodwin, M. (1988), *The Local State and Uneven Development*, Cambridge: Polity Press.

Durkheim, E. (1964), *The Division of Labour in Society*, Translated by G. Simpson, Collier-Macmillan, London.

Durkheim, E. (1972), 'The Division of Labour and Social Differentiation', in A. Giddens (ed), *Emile Durkheim: Selected Writings*, Cambridge University Press, Cambridge, pp. 141-154.

Edwards, B. (1998), 'Charting the Discourse of Community Action: Perspectives from Practice in Rural Wales', *Journal of Rural Studies*, Vol. 14, pp. 63-77.

Edwards, B., Goodwin, M., Pemberton, S. and Woods, M. (2001a), 'Partnerships, Power and Scale in Rural Governance', *Environment and Planning C: Government and Policy*, Vol. 19, pp. 289 – 310.

Edwards, B., Woods, M., Anderson, J. and Fahmy, E. (2001b), 'Governing Through Communities? Emerging Geographies of Rural Citizenship', Paper delivered to the *IBG/RGS Annual Meeting*, University of Plymouth, January 2001.

Edwards, B., Woods, M., Anderson, J. and Gardner, G. (this volume), 'Mobilising the local: Community, Participation and Governance', in L. Holloway and M. Kneafsey (eds), *Geographies of Rural Cultures and Societies*, Ashgate, Aldershot.

Emmett, I. (1964), *A North Wales Village*, Routledge & Kegan Paul, London.

Enticott, G. (2001), 'Calculating Nature: the Case of Badgers, Bovine Tuberculosis and Cattle', *Journal of Rural Studies*, Vol. 17, pp. 149-164.

Foucault, M. (1980), *Power/Knowledge: Selected Interviews and Other Writings 1972-1977*, edited by C. Gordon, translated by. C. Gordon, L. Marshall, J. Mepham and K. Sopher, The Harvester Press, Brighton, pp. 78-108.

Foucault, M. (1981), *The History of Sexuality: Volume 1: An Introduction*, Penguin, Harmondsworth.

Frankenburg, R. (1957), *Village on the Border: A Social Study of Religion, Politics and Football in a North Wales Community*, Cohen & West, London.

Frankenburg, R. (1966), *Communities in Britain: Social Life in Town and Country*, Penguin, Harmondsworth.

Geddes, A. (1955), *The Isle of Lewis and Harris: A Study in British Community*, Edinburgh University Press, Edinburgh.

Giddens, A. (1971), *Capitalism and Modern Social Theory: An Analysis of the Writings of Marx, Durkheim and Max Weber*, Cambridge University Press, London.

Goffman, E. (1961), *Asylums: Essays on the Social Situation of Mental Patients and Other Inmates*, Anchor Books, Garden City, New York.

Goldthorpe, H. (1995), 'The Service Class Revisited', in T. Butler. and M. Savage (eds), *Social Change and the Middle Classes*, UCL Press, London, pp. 313-329.

Goodwin, M. (1998), 'The Governance of Rural Areas: Some Emerging Research Issues and Agendas', *Journal of Rural Studies*, Vol. 14, pp. 5-12.

Goodwin, M., Cloke, P. and Milbourne, P. (1995), 'Regulation Theory and Rural Research: Theorising Contemporary Rural Change', *Environment and Planning A*, Vol. 27, pp. 1245-1260.

Gusfield, J. (1975), *Community: A Critical Response*, Basil Blackwell, Oxford.

Harper, S. (1989), 'The British Rural Community: an Overview of Perspectives', *Journal of Rural Studies*, Vol. 5, pp. 161-184.

Hawley, A. (1950), *Human Ecology: A Theory of Community Structure*, Ronald Press, New York.

Hillery, J. (1955), 'Definitions of Community: Areas of Agreement', *Rural Sociology*, Vol. 20, pp. 111-123.

Hobbes, T. (1962), *Leviathan, or, the Matter, Forme and Power of a Commonwealth Ecclesiasticall and Civil*, edited by M. Oakeshott, Collier-Macmillan, London.

Hoggart, K. (1997), 'The Middle Classes in Rural England, 1971-1991', *Journal of Rural Studies*, Vol. 13, pp. 253-273.

Hoggart, K., Buller, H. and Black, R. (1995), *Rural Europe: Identity and Change*, Arnold, London.

Latour, B. (1986), 'The Powers of Association', in J. Law (ed), *Power, Action and Belief: A New Sociology of Knowledge*, Routledge & Kegan Paul, London, pp. 264-80.

Lewis, G. (1970), 'A Welsh Rural Community in Transition: A Case Study in Mid-Wales', *Sociologia Ruralis*, Vol. 10, pp. 143-159.

Lewis, G. (1986), 'Welsh Rural Community Studies: Retrospect and Prospect', *Cambria*, Vol. 13, pp. 27-40.

Liepins, R. (1999), 'New Energies for Old: Reworking Approaches to "Community" in Contemporary Rural Studies', *Journal of Rural Studies*, Vol. 16, pp. 23-35.

Little, J. (2001), 'New Rural Governance?', *Progress in Human Geography*, Vol. 25, pp. 97-102.

Littlejohn, J. (1963), *Westrigg: The Sociology of a Cheviot Parish*, Routledge & Kegan Paul, London.

Lukes, S. (1974), *Power: A Radical View*, Macmillan, London.

Lupri, E. (1967), 'The Rural-Urban Variable Reconsidered: The Cross-cultural Perspective', *Sociologia Ruralis*, Vol. 8, pp. 1-7.

Lyotard, J. (1984), *The Postmodern Condition: A Report on Knowledge*, translated by B. Massumi, Manchester University Press, Manchester.

Mann, M. (1993), *The Sources of Social Power: Volume 2: The Rise of Classes and Nation-states 1760-1914*, Cambridge University Press, Cambridge.

Marsden, T. and Flynn, A. (1993), 'Servicing the City: Contested Transitions in the Rural Realm', *Journal of Rural Studies*, Vol. 9, pp. 201-204.

Marsden, T., Murdoch, J., Lowe, P., Munton, R. and Flynn, A. (1993), *Constructing the Countryside*, UCL Press, London.

Merton, R. (1957), *Social Theory and Social Structure*, The Free Press, New York.

Milbourne, P. (1998), 'Local Responses to Central State Restructuring of Social Housing Provision in Rural Areas', *Journal of Rural Studies*, Vol. 14, pp. 167-184.

Moseley, M. (1979), *Accessibility: The Rural Challenge*, Methuen, London.

Murdoch, J (1995), 'The Spatialization of Politics: Local and National Actor Spaces in Environmental Conflict', *Transactions, Institute of British Geographers*, Vol. 20, pp. 368-381.

Murdoch, J. (1997), 'The Shifting Territory of Government: Some Insights from the Rural White Paper', *Area*, Vol. 29, pp. 109-118.

Murdoch, J. and Abram, S. (1998), 'Defining the Limits of Community Governance', *Journal of Rural Studies*, Vol. 14, pp. 41-50.

Murdoch, J. and Marsden, T. (1994), *Reconstituting Rurality: Class, Community and Power in the Development Process*, UCL Press, London.

Murdoch, J. and Marsden, T. (1995), 'Middle Class Territory? Some Remarks on the Use of Class Analysis in Rural Studies', *Environment and Planning A*, Vol. 27, pp. 1213-1230.

Murdoch, J. and Pratt, A. (1993), 'Rural Studies: Modernism, Postmodernism and the "Post-rural"', *Journal of Rural Studies*, Vol. 9, pp. 411-427.

Murdoch, J. and Pratt, A. (1994), 'Rural Studies of Power and the Power of Rural Studies: A Reply to Philo', *Journal of Rural Studies*, Vol. 10, pp. 83-87.

Murdoch, J. and Pratt, A. (1997), 'From the Power of Topography to the Topography of Power: A Discourse on Strange Ruralities', in P. Cloke and J. Little (eds), *Contested Countryside Cultures: Otherness, Marginalization and Rurality*, Routledge, London, pp. 51-69.

Murdoch, J. and Ward, N. (1997), 'Governmentality and territoriality: the statistical manufacture of Britain's "national farm"', *Political Geography*, Vol. 16, pp. 307-324.

Newby, H. (1977), *The Deferential Worker*, Allen Lane, Harmondsworth.

Newby, H. (1980), 'Urbanization and the Rural Class Structure: Reflections on a Case Study', in F. Buttel and H. Newby (eds), *The Rural Sociology of the Advanced Societies: Critical Perspectives*, Croom Helm, London, pp. 255-279.

Newby, H. (1985), *Green and Pleasant Land? Social Change in Rural England*, Third Edition, Wildwood House, London.

Newby, H., Bell, C., Rose, D. and Saunders, P. (1978), *Property, Paternalism and Power: Class and control in rural England*, Hutchinson, London.

Nisbet, R. (1962), *Community and power*, Oxford University Press, New York.

Nisbet, R. (1970), *The Sociological Tradition*, Heinemann, London.

Pahl, R. (1965a), *Urbs in Rure: The Metropolitan Fringe in Hertfordshire*, London School of Economics & Political Science, London.

Pahl, R. (1965b), 'Class and Community in English Commuter Villages', *Sociolgia Ruralis*, Vol. V, pp. 5-23.

Pahl, R. (1966), 'The Rural-urban Continuum', *Sociologia Ruralis*, Vol. 6, pp. 299-327.

Pahl, R. (1989), 'Is the Emperor Naked? Some Questions on the Adequacy of Sociological Theory in Urban and Regional Research', *International Journal of Urban and Regional Research*, Vol. 13, pp. 709-720.

Park, R. (1916), 'The City: Suggestions for the Investigation of Human Behaviour in the Urban Environment', *American Journal of Sociology*, Vol. 20, pp. 577-612.

Parsons, T. (1937), *The Structure of Social Action*, McGraw Hill, New York.

Parsons, T. (1960), *Structure and Process in Modern Societies*, The Free Press, Glencoe, Illinois.

Parsons, T (1986), 'Power and the Social System', in S. Lukes (ed), *Power*, Blackwell, Oxford, pp. 94-143.

Phillips, M. (1993), 'Rural Gentrification and the Processes of Class Colonisation', *Journal of Rural Studies*, Vol. 9, pp. 123-140.

Phillips, M. (1998a), 'Investigations of the British Rural Middle Classes – Part 1: From Legislation to Interpretation', *Journal of Rural Studies*, Vol. 14, pp. 411-25.

Phillips, M. (1998b), 'Investigations of the British Rural Middle Classes – Part 2: Fragmentation, Identity, Morality and Contestation', *Journal of Rural Studies*, Vol.14, pp. 427-43.

Philo, C. (1992), 'Neglected Rural Geographies: a Review', *Journal of Rural Studies*, Vol. 8, pp. 193-207.

Plowman, D., Minchinton, W. and Stacey, M. (1962), 'Local Social Status in England and Wales', *Sociological Review*, Vol. 10, pp.161-202.

Ray, C. (1998), 'Culture, Intellectual Property and Territorial Rural Development', *Sociologia Ruralis*, Vol. 38, pp. 3-20.

Ray, C. (1999), 'Towards a Meta-Framework of Endogenous Development: Repertoires, Paths, Democracy and Rights', *Sociologia Ruralis*, Vol. 39, pp. 521-537.

Rees, A. (1996), *Life in a Welsh Countryside: A social study of Llanfihangel yng Ngwunfa*, Originally Published in 1950, University of Wales Press, Cardiff.

Rees, G. (1984), 'Rural Regions in National and International Economies', in T. Bradley and P. Lowe (eds), *Locality and Rurality: Economy and Society in Rural Regions*, Geo Books, Norwich, pp. 27-44.

Ribchester, C. and Edwards, B. (1998), 'The Centre and the Local: Policy and Practice in Rural Education Provision', *Journal of Rural Studies*, Vol. 15, pp. 49-64.

Rose, N. (1993), 'Government, Authority and Expertise in Advanced Liberalism', *Economy and Society*, Vol. 22, pp. 283-299.

Rose, N. (1996), 'The Death of the Social? Reconfiguring the Territory of Government', *Economy and Society*, Vol. 25, pp. 327-256.

Rose, N. (1999), *Powers of Freedom*, Cambridge University Press, Cambridge.

Rose, N. and Miller, P. (1996), 'Political Power Beyond the State: Problematics of Government', *British Journal of Sociology*, Vol. 43, pp. 173-205.

Rose, R., Saunders, P., Newby, H. and Bell, C. (1976), 'Ideologies Of Property: A Case Study', *Sociological Review*, Vol. 24, pp. 699-730.

Saunders, P., Newby, H., Bell, C. and Rose, D. (1978), 'Rural Community and Rural Community Power', in H. Newby (ed), *International Perspectives in Rural Sociology*, John Wiley & Sons, Chichester, pp. 55-85.

Sennett, R. (1993), *Authority*, Originally Published in 1980, W.W. Norton, London.

Smith, M. (1989), 'Changing Agendas and Policy Communities: Agricultural Issues in the 1930s and the 1980s', *Public Administration*, Vol. 67, pp. 149-165.

Stacey, M. (1969), 'The Myth of Community Studies', *British Journal of Sociology*, Vol. 20, pp. 134-147.

Stoker, G. (1996), 'Public-Private Partnerships and Urban Governance', Mimeo, Department of Government, University of Manchester.

Strathern, M. (1984), 'The Social Meaning of Localism', in T. Bradley and P. Lowe (eds), *Locality and Rurality: Economy and Society in Rural Regions*, Geo Books, Norwich, pp. 181-199.

Symes, D. (1981), 'Rural Community Studies in Great Britain', in J-L. Durand-Drouhn, L-M. Szwengrub and I. Mihailescu (ed.s), *Rural Community Studies In Europe: Trends, Selected Bibliographies, Analyses: Volume 1*, Pergamon Press, Oxford, pp. 17-67.

Thrift, N. (1989), 'Images of Social Change' in C. Hamnett, L. McDowell and P. Sarre (eds), *The Changing Social Structure*, Sage, London, pp. 12-42.

Tonnies, F. (1957), *Community and Association (Gemeinschaft und Gesellschaft)*, Edited and with a Supplement by C. Loomis, Routledge & Kegan Paul, London.

Urry, J. (1984), 'Capitalist Restructuring, Recomposition and the Regions', in T. Bradley and P. Lowe (eds), *Locality and Rurality: Economy and Society in Rural Regions*, Geo Books, Norwich, pp. 45-6.

Urry, J. (1995), 'A Middle-class Countryside?', in T. Butler and M. Savage (eds), *Social Change and the Middle Classes*, UCL Press, London, pp. 205-219.

Ward, N. and McNicholas, K. (1998), 'Reconfiguring Rural Development in the UK: Objective 5b and the New Rural Governance', *Journal of Rural Studies*, Vol. 14, pp. 27-39.

Weber, M. (1964a), *From Max Weber: Essays in Sociology*, Edited by H. Gerth and C. Wright Mills, Routledge & Kegan Paul, London.

Weber, M. (1964b), *The Theory of Social and Economic Organization*, Translated by A. Henderson and Talcott Parsons, Edited and with an Introduction by T. Parsons, The Free Press, Glencoe, Illinois.

Weintraub, K. (1969), 'The Concepts Traditional and Modern in Comparative Social Research – An Empirical Evaluation', *Sociologia Ruralis*, Vol. 9, pp. 23-42.

Whitehead, A. (1976), 'Sexual Antagonism in Herefordshire', in D. Barker and S. Allen (eds), *Dependence and Exploitation in Work and Marriage*, Longman, London, pp. 169-203.

Williams, W. (1956), *Gosforth: The Sociology of an English Village*, Routledge, London.

Williams, W. (1963), *A West Country Village: Ashworthy. Family, Kinship and Land*, Routledge & Kegan Paul, London.

Wirth, L. (1938), 'Urbanism as a Way of Life', *American Journal of Sociology*, Vol. 44, pp. 36-42.

Woods, M. (1997), 'Discourses of power and rurality: local politics in Somerset in the 20th century', *Political Geography*, Vol. 16, pp. 453-478.

Woods, M. (1998a), 'Advocating Rurality? The Repositioning of Rural Local Government', *Journal of Rural Studies*, Vol. 14, pp. 13-26.

Woods, M. (1998b), 'Rethinking Elites: Networks, Space and Local Politics', *Environment and Planning A*, Vol. 30, pp. 2101-2119.

Chapter 5

Constructing Multiple Ruralities: Practices and Values of Rural Dwellers

Hanne Kirstine Adriansen and Lene Møller Madsen

Introduction

While rural geography in the last decade has been enriched by new and challenging theoretical perspectives and methodological approaches inspired by the cultural turn, the empirical focus seems to be limited to certain cultures – it is difficult to find studies concerned with 'Third World' countries. Although our perception of Europe has changed after the fall of the Berlin wall and the division between 'East' and 'West' broken down, there still appears to be a division between what, for want of better terms, can be labelled 'developed' and 'developing' countries[1]. In this chapter, we want to show that this division is an unnecessary boundary precluding us from inspiration and insights gained in 'the other' world. Instead of studying either developed or developing countries, a more fertile approach is to focus on the issues of interest. When the concerns are identities, construction of ruralities, territoriality, and belonging our research questions may well be the same, but the answers are different. We do not advocate comparative studies, as these often reproduce the boundaries; instead we find that the differences in answers can be of inspiration for asking new questions within the respective areas of research.

Based on this premise, we discuss how rural people and their land use can be understood with special emphasis on what can be labelled 'the cultural turn', emphasising identity, discourse, motives, and agency as well as the cultural construction of rurality (e.g. Cloke and Little, 1997; Milbourne, 1997; Philips et al, 2001; Haartsen et al, 2000; van Hoven, 2001). The recent change in rural studies has raised many interesting questions, but we find that in this process, the physical/material aspects of life in rural areas and the use of rural space have been omitted. We want to (re-) turn the attention to these physical aspects, but we want to do it within the light of culturally inspired studies. Hence, the aim is to provide an understanding of rural people and their land use by considering them in a broader context of the social and cultural embeddedness of their actions. In this way, we also hope to contribute to the ongoing debates concerning the use of qualitative methods in human geography (e.g. Winchester, 1999; Johnston, 2000; Crang, 2002).

Consequently, the purpose of this chapter is twofold. At a general level, we want to show that many of the issues raised by the cultural turn transcend the gap between research in so-called developing and developed countries. Most emphasis, however, is given to the second objective, namely, to provide an approach towards the study of rural areas that can combine studies of physical/material and social/cultural issues. In order to do so, we use examples from our respective research from Denmark, Egypt, and Senegal.

The chapter is divided into six sections. First, we provide a short introduction to the case studies. The next section elaborates on the research gap between developed and developing countries and the issue 'sense of belonging' is used to illustrate how concepts can be used for transcending this research gap. Then, attention is drawn to the issue of combining studies of physical/material and social/cultural issues in rural studies; in this case focus is on a combination of the cultural turn with more 'conventional' geographical studies of land use. An analytical approach is developed in the fourth section – as a study of the practices and values of rural dwellers. The applications of the approach developed are discussed in the fifth section, and finally some concluding remarks are given.

Presentation of Case Studies

The discussion in this chapter is the outcome of problems and questions raised during our work with two quite different research projects, one in Denmark and one in Senegal. We found that during this work, we were faced with a number of similar troublesome considerations concerning the integration of different methodological and philosophical approaches and decided to pursue the answers together. This led us to see that many approaches and concepts transcended the boundary between research concerning developed and developing countries, a boundary that is manifest in many research institutions. Since then, we have continued our research in 'different worlds' while maintaining a common interest in the directions of rural geography, of which the last case study from Egypt is an example.

The brief introduction provided here is meant to acquaint the reader with the three studies because these are used as illustrations in the discussion throughout the chapter. The first two concern what could be called 'the use of rural space', while the third from Egypt is more related to discussions of constructions of ruralities.

The study from Denmark concerns the location of private afforestation under EU-regulation 2080/92[2]. The objective is to analyse land use changes resulting from private afforestation and the implementation of the scheme within the Danish planning system. The focus is on the landowners and their role in the afforestation process. On the basis of interviews, map-analysis, aerial photos, administrative data and field registration, it is shown that the location of new woodlands that fulfil the goals of the afforestation programme is more a coincidence than a result of good planning – it is the individual landowner who has the decisive influence on the woodland's contribution to securing the goals of the afforestation programme (Madsen, 2002b). Landowners engaged in afforestation represent a wide range of different practices and values concerning their forest and the complex reasoning behind the location of afforestation areas on their farm cannot be understood through economic reasoning alone (Madsen, 2003). Instead a typology of different landowners is developed, which can be used to explain both the actual location of a new woodland, and why this location is chosen. The study is further described in Madsen (2002a).

The objective of the study from Senegal is to analyse pastoral mobility both in physical terms (length and duration of mobility) and in terms of cultural understanding and importance of the movements and the concept of being mobile. The pastoralists of the northern Senegal are semi-settled, which means they have a rainy-season camp and depending on the dry season they may go on migration with the herds of cattle, sheep, and goats. Contrary to the usual image of nomads as freedom loving people for whom mobility is a necessity, the pastoralists turned out not to be very keen on going on migration. Nevertheless, the livestock are quite mobile due to daily mobility in the vicinity of the rainy-season camp. GPS[3] was used for measuring livestock mobility and mapping the patterns (Adriansen and Nielsen, 2002). This turned out to be a good input for discussion with the pastoralists about their use and perception of mobility. Further information can be found in Adriansen (2002).

The study from Egypt concerns the construction of rural identities in reclaimed desert lands. Land reclamation and the construction of new communities in the desert has been among the Egyptian government's most important policy objectives for decades. While agro-ecological consequences of land reclamation have been widely discussed, little is known about the settlers' ability to create new communities and make a living in these areas. The study shows that the new lands can be seen both as 'spaces of opportunity' and as 'spaces of poverty' and that these issues central to cultural geographers can be used for understanding peoples' perceptions of their life and their constructions of rural livelihoods. Further, it is argued that measures should be taken to ensure that the existing social exclusion is not reproduced and enhanced in the new lands. The study is based on interviews with settlers in the new villages combined with participatory village mapping and ethnographic observations (see Adriansen, 2003).

Transcending the Research Gap Between Developed and Developing Countries

The current division between research in developed and developing countries can be seen as reminiscent of a more political economic approach, according to which the differences between various parts of the world were so great that different lines of reasoning and approaches should be used. Looking back, however, agricultural and to some extent rural geography have been considered subjects that transcend regions. This can be seen in the progress reports on agricultural geography published in the journal *Progress in Human Geography* in the 1980s (Grigg, 1981; 1982; 1983; Bowler, 1984; 1987). With the increasing emphasis on political economy, the focus turned towards developed countries and since then geographers (and other researchers) have tended to 'stick by their region'. Atkins describes the trend this way: 'There has been an unfortunate neglect by agricultural geographers of the rural development process in poor countries' (1988, p.281). A tendency that can be seen in Whatmore's progress reports in *Progress in Human Geography* from 1991 and 1993 (Whatmore, 1991; 1993) and also in Bowler and Ilbery's paper in *Area* from 1987 (Bowler and Ilbery, 1987). We consider this division a futile boundary. The argument is along the lines of thought presented by Crang when he points to 'the need to think of cultures and spaces in ways other than as bounded containers' (Crang, 1998, p.175).

One of the answers to why the cultural turn has not had an impact on agricultural studies in developing countries may be found in Morris (this volume). Here she responds to the criticism that cultural issues are a kind of luxury, out of the question for researchers concerned with salvaging rural life and providing food. We want to reply by asking 'can we salvage rural life without including cultural issues?' There is ample evidence showing that farming has meaning beyond simply providing food. To show sensitivity towards peoples' own perceptions of their life and constructions of livelihoods is one way to avoid providing ethno-centric and mechanistic solutions to rural problems. This is not to say that all areas should be studied in the same way and that context does not matter. On the contrary, context is important for understanding rurality in a given area, but this should not preclude us from establishing communication about concepts, methods, and material used in rural studies world-wide.

The reluctance among (rural) researchers to establish and value communication between researchers of the developed and developing world is not due to mere lack of goodwill or interest. First of all, the gap is constantly reproduced in the sense that we usually attend different conferences, seek different funding, read different journals, and – not least – often have different professional and personal relations, which hamper us from drawing attention to the benefits of research in 'the other world'. Further, opportunities for communications between research methods are neglected due to the fact that many funding opportunities are dependent upon research having direct policy relevance. As Crang notes, the demands tied to funded research often are a 'clear set of predicted outcomes rather than an evolving programme' (Crang, 2002, p.650). This means that opportunities for research relating concepts and methodologies in developed and developing countries are abandoned even before they start.

The issue of 'sense of belonging' can be used to illustrate how concepts can transcend the research gap between developed and developing countries. In textbook material on cultural geography, sense of belonging is often linked with issues of nationality, spatially bound identity, and 'imagined communities' (e.g. Crang, 1998). With regards to more local scale processes, however, the concept is related to ideas of constructing 'place' and 'home' (Morley, 2001) and it can be valuable for understanding processes of land claims, attachment to certain areas, and even farming behaviour. Sense of belonging has a common-sense air to it that we appreciate, because it is part of everyday language. However, it is important to note that sense of belonging is 'inevitably complex, filtered as it is, and frequently reworked, through sedimented layers of memory and lived experience, shaped by the intangible intimacies of history and the pragmatics of the present' (Hammar, 2002, p. 228).

In the study of people resettling in reclaimed desert in Egypt, the concept has been used to understand why people move to the desert and to find factors of importance for relating to the new village communities (Adriansen, 2003). The Egyptian government tries to make 'the greening of the desert' a national project that everybody should feel obliged to participate in. Hence, some see the possibility to be granted desert land as a way to serve their country. Others feel encouraged by the Qur'ân, where they find justification for migration to the desert. Both the national and the religious discourse provide legitimate reasons, though quite different from the accounts found at the individual and family level. For instance, one family came to the new lands because they had nothing in their original village to make them stay. To settle in the new lands was their only way to get to own land.

Many of the inhabitants of the new villages complained that the families had to split up and some move back to the old lands due to the lack of services and infrastructure. This caused a feeling of loneliness and lack of belonging as well as being abandoned by the government which did not provide the services promised. Furthermore, there was an impression that local officials such as doctors and teachers were not interested in living in the new lands, and this led to an image of the new land as a place not worth staying in. On the other hand, they felt that the land originally brought them together and made a community of them.

It takes time to construct a sense of belonging, and time certainly is a key to establishing a feeling of being at home in the desert. This can be seen when interviewing inhabitants of the first new villages. Despite all the difficulties they experienced in the beginning, and for some the feeling of being left to their own devices by a state not caring about its desert inhabitants, most have grown accustomed to living in the margin. Moreover, with the ongoing reclamation of desert land, the majority of the older reclaimed land can no longer be considered frontier. Today they provide the spatial linkage between the old and the new lands.

In the study of afforestation in the Danish countryside, it was shown that the sense of belonging to a given property affected not only the degree of interest in creating woodland, but it also affected the actual planting and the shape of a given woodland (Madsen, 2002a). A critical question for the grant-aided field afforestation programme is the issue of actual planting of woodlands on the fields. The rationale of the implementation of the subsidy scheme is that if a landowner gets a subsidy, they will plant, and if they do not obtain a subsidy, they will not plant (Madsen, 2003). The controlling mechanism associated with the scheme is based on this rationale: the landowner gets the subsidy in portions according to statements of actual planting and maintenance. However, the study shows that this is a simplification of what is actually happening. The relationship between obtaining a grant and actual planting is far from straightforward. In fact a complex relationship between receiving a grant and planting versus not receiving a grant and not planting was found. This complex relationship turned out to be a question of understanding the individual landowner's use of their 'room for manoeuvre', where sense of belonging was one component. As long as the landowner perceives the farm as 'a place of production', the actual planting follows a pattern of economically rational decision-making. If the subsidy is obtained, the field is planted; if the subsidy is not obtained, the field is not planted. On the other hand, when the landowner perceives the farm as 'a place of living', the planting becomes independent of the subsidy: The landowner applies for the subsidy in order to get financial help for the afforestation, but if the subsidy is not obtained, the fields are planted anyway.

Different senses of belonging also led to differences in the shaping of woodlands. While production-oriented landowners often let physical boundaries like field edges define the woodland shape (it was most often square), landowners expressing different types of recreational relationship with their new woodland made an effort to shape their woodland, so for instance, it could be seen from the house or would screen off a trafficked road.

These two examples show how concepts can transcend the research gap between developed and developing countries – how researchers can acquire conceptual inspiration from each other without studying the same subject matter or region.

Combining Different Approaches in Rural Studies

The cultural turn in rural studies has brought much new and interesting material as well as new methods forward and revitalized almost forgotten subject matters within the research field. Within the sub-field of agricultural geography, however, it might be thought that researchers have bypassed the cultural turn (Cloke, 1997; Little, 1999). As Morris (this volume) shows, a number of possible explanations for this can be put forward. Here we want to stress the influence of policy-oriented research. Many of the agricultural studies of the past decades, for instance, have been impact studies of certain agricultural reforms or policies. Policy makers often expect that the methodologies used are standardized and replicable, so results can be applied nationally and allow comparisons over-time. This is valid both for rural researchers involved in studies concerning evaluation of EU agri-environmental schemes and for impact analysis of development aid. The implication for researchers of this 'reality' is often an implicit denial of the ability of qualitative studies to inform policy-processes and thereby a tendency to neglect the benefits of culturally informed studies. This is nicely illustrated by Beedell and Rehman, who point out that 'subjective studies consisting of a small number of farmer interviews are not sufficiently convincing or easily replicable to inform the policy-making reliably' (2000, p.117). We find instead that policy related studies *will* benefit from the perspectives offered by more culturally informed research.

Although the cultural turn may not have had a significant impact on agricultural geography, it is increasingly recognized that studies of farming practices must be widened to include the culture of agriculture. Among others Winter finds that 'future policy reforms should recognize the complex and inter-related factors influencing farmer behaviour' (2000, p.47). Likewise, Morris and Potter (1995) in a review of research on farmers' participation in agri-environmental schemes point to the need for a more behaviourally informed perspective in order to evaluate the effects of the agri-environmental policies on the ground.

It can be argued that the cultural turn within rural geography has caused a detachment of studies of the material/physical world. Not surprisingly, focus on the verbal and materials such as films, texts, or discourse in general has left studies of land use and agricultural practices largely untouched. Within culturally inspired studies 'the rural' is often considered in terms of its social construction (e.g. Halfacree, 1993; Murdoch and Pratt, 1993). However, we find that this does not limit us to focus on the social part solely. Agriculture still takes up a significant proportion of rural land and as noted by Murdoch 'land is a key marker of rurality' (Murdoch, 2000, p.408). The challenge, therefore, lies in using the insights gained through culturally inspired studies to enrich combined studies of both the material/physical and the social/cultural spheres. Hence, the purpose of the rest of this chapter is to show how issues raised by the cultural turn can enrich 'conventional' geographical studies of land use and agriculture.

Some may argue that this is just old wine in new bottles. And yes, among rural researchers, perhaps especially among those concerned with developing countries, combined studies of both cultural and physical phenomena are quite common. But little discussion is found on the methodological and philosophical implications of combining these issues. One exception is Röling (1994; 1997), who suggests a division into human and physical systems, which should then be studied within different frameworks. Physical systems or environments should be studied using an ecosystem approach within a positivist stance. Human systems, on the other hand, should be studied using action research and based on a social constructionist stance. Consequently, Röling believes that these different epistemologies can be integrated. This approach and similar eclectic approaches are used implicitly in many land use studies (e.g. Turner, 1999). However, instead of integrating knowledge created within different epistemologies, we argue for the need to combine methods within the same philosophical framework in order to build knowledge with due respect to the integrity of the differently obtained data. Such an approach will be discussed in the following sections on a practices-values approach and its implications.

Practices and Values – a Combined Approach

As we are interested in 'the use of rural space', which can be seen as a subject matter transcending rural research in both developed and developing countries, we want to be able to combine studies of the material/physical and the social/cultural spheres. Our notion of 'the use of rural space' concerns both the physical land use and the practices and values of individual actors influencing the land use. In order to understand rural people and their land use, they are considered in the broader context of social and cultural embeddedness of their actions. Moreover, the conceptualization of rural space includes more than agricultural activities related to production of food. In many countries, the majority of people living in rural areas and even owning the countryside are not directly involved in agricultural production (see examples in Haartsen et al, 2000). Even in countries where most people living in rural areas are involved in agricultural production, agriculture is not a question of mere production, as recent research on, for instance, rural livelihoods suggests (Ellis, 1998; Bebbington, 1999).

The focus on the embeddedness of peoples' actions does not mean that the limits and possibilities imposed on rural areas by the structures of economy, ecology, politics etc. should not be acknowledged. Instead, we find that these structures are mediated through the complex and often unarticulated processes of decision-making. It is this mediation that should be analysed if the practices leading to a certain use of rural space are to be understood. Further, we emphasize the importance of studying the individual actor and using him/her as the focus for understanding the processes of shared meanings, negotiations, and beliefs that legitimate certain practices and lead to the abandonment of others. Hence, we examine how individuals respond differentially to structure in order to unfold the complex world of the use of rural space. Further, we want to point to the importance of paying attention to the philosophical implications of the methods used and the interplay between methodology and philosophy throughout the research process. Based on these premises the 'practices-values approach' is outlined.

In the analysis of the use of rural space, the rural dweller is used as the point of departure. This means that in order to combine social issues of rurality with physical issues of rural land use, the focus is on the practices and values of individual actors or the group they are member of, e.g. a household. Here, practices should be understood quite literally as actions carried out by individuals, whereas values consist of traditions, preferences, motives, thoughts, and beliefs. We do not perceive the interrelationship between the practices and values of individuals as a linear one-to-one relationship; rather they are interwoven in a complex patchwork. Hence, the relationship between practices and values is not predefined, as this would leave little space for the unexpected. Further, it is stressed that similar practices can be based on different values and similar values may lead to different practices, hence we see the interrelationship as a complex one where one variable cannot be deduced from the other. No claims are made about the novelty of looking at practices and values (see e.g. Kaltoft, 1999; Busck, 2002); here we simply want to suggest an explicit approach towards these issues. The practices-values approach functions as an analytical tool for understanding the interwoven relations between what people do and what they believe in. Further it facilitates an understanding of discrepancies between what people say they are doing and what they actually do. In this way, our approach is different from Kaltoft (1999), who bases her approach towards practices solely on what people say they are doing; this means that it is only a verbalized approach. We find it important to focus not only on a verbalized approach but also on the actual practices that can be observed in a variety of ways. It must be noted that practices and values are not two discrete boxes; they should not be represented as a dualism calling for different research methods e.g. quantitative methods for practices and qualitative for values. Inevitably, it is the researcher who decides what is labelled practices and what is labelled values. Hence, in order to understand the use of rural space, we have to acknowledge the social embeddedness of actions and to gain information on the legitimating process of certain practices.

Actors' practices are context dependent and do not entail a universal rationality. Hence, actions cannot be understood on their own. Instead every individual can be said to be rational within his/her own perception of the world and hence understood in this context (Fay, 1999). Or, as explained by Vayda, 'recognizing that as more is known of their contexts the better are any activities of concern to us understood' (1983, p.272). Consequently, we find that practices and values should be contextualized. This is in contrast to behavioural studies of farmers' goals, values and attitudes conducted in the 1970s and 1980s (e.g. Gasson, 1969; 1988; Ilbery, 1983), which are often considered important for paving the road for today's cultural turn within geography (Philo, 2000; Morris, this volume). We find that the positivist philosophy inherent in some of these behavioural studies creates problems in relation to culturally informed research especially for combining studies of the social/cultural and material/physical spheres. This is an argument that needs some comments.

The discourse leading to behavioural studies emerged in the 1960s and dismissed the notion of the farmer as 'the economic man' (Thom and Woolmington, 1988). Ilbery, who has used this approach in agricultural studies, describes it this way: 'Based on the assumption that there are further sets of influences which affect agricultural decision-making, including farmers' values, aims, motives and attitudes, the behavioural approach recognizes that farmers may not always perceive the environment as it is ... therefore, the objective of the behavioural approach is clear: to reject the notion of the economic man and replace it with a model that is closer to reality' (1986, p. 25-27). By including other factors in the research of farmers' decision-making process, a more comprehensive notion of the farmer as an actor was obtained. Hence, modelling of human behaviour was the response of some behavioural approaches to the shortcomings of the economic man (Walmsley and Lewis, 1984). However, we do not find this satisfactory. One of the problems is that studies were still conducted within a positivistic framework where the behaviour of farmers was often explained using models that linked behaviour to farm and farmer characteristics (see e.g. the work of Ilbery, 1983). As explained by Röling, assumptions used in mathematical models are problematic because 'in making assumptions, the models violate the variability, diversity and the negotiated, contextual, contingent and adaptive nature of human intentionally, and the flux of trade-offs people make among their different goals' (1997, p. 250).

This 'historical' view is important because this approach still influences contemporary research of motives, attitudes, and values (e.g. Potter and Lobley, 1996; Wilson, 1996; Kazenwadel et al, 1998; Beedell and Rehman, 2000). This creates confusion because these issues are often identical with issues central to the cultural turn, but the research process is quite different. The implicit positivist approach means that the obtained data – and especially knowledge about how the data are derived – are often reduced in order to function as inputs in statistical measures. In this process, valuable information about context is usually disregarded and thereby provides no tools for analysing qualitative data and making them a part of the explanation. In contrast to this, the cultural turn has given emphasis to qualitative methods and usually drawn upon post-structuralist frameworks. Here, context is regarded as valuable for knowledge building.

In order to take full advantage of the cultural turn within the field of rural studies, we therefore need to pay attention to the philosophical implications of the practical research, i.e. of the method and theory choices throughout the research process. This is especially an issue of concern when trying to combine 'conventional' geographical studies of land use with culturally informed perspectives, because these studies rely upon a combination of methods and data that can lead to philosophical eclecticism.

When looking at values, we are concerned with understanding and analysing meanings in specific contexts, or as Baxter and Eyles state 'we set out to learn to view the world of individuals or groups as they themselves see it' (1997, p.506). This is important because the practices of people can only be understood in the context of how they construct and perceive their reality. Values do not lead directly to a certain action and are often expressed at a more abstract level than practices. Therefore, practices and values should be understood in context.

The emphasis on context also leads to an investigation of the space where the actions take place whether this space is physical or abstract. Especially interesting are issues of space and place[4] (e.g. Tuan, 1977; Taylor, 1999). Among human geographers, there is an increasing interest in the idea that space is culturally constructed (Unwin, 2000), in the relation between space/place and identity, and in the image of home/place versus outside/space (e.g. Buttimer, 1998; McHugh, 2000). This means that an important part of understanding context is to be aware of both the physical space and the legitimized use of space created through power, negotiations, and values of a given group of individuals. These different types of space delimit, shape, and challenge the individual actors' 'field of legitimate opportunities'; and in that way they are important when studying the use of rural space (see e.g. Liepins, 2000a; 2000b).

Consequently, the practices-values approach outlined here will often lead to a focus on spatial issues. In order to understand the use of rural space, a focus on both the space in which the actions take place and on the spatial distribution of actions will be often be part of the analysis. Hence, not only tangible issues of the use of rural space should be studied, but a spatial approach that is open towards social and cultural phenomena should be used. This has been done by Müller-Mahn, who has studied social change and poverty in rural Egypt. He has used a spatial approach that '... provides a framework within which people organize their lives and within which they act today. In dealing with the link between space and agency, geography presents some analytical tools that can bring about sensitivity to the spatial aspects of social phenomena' (1998, p.274). However, there is a tendency to equate spatial approaches with quantitative methods (e.g. Johnston, 2000). As argued above, we want to apply a broader understanding of spatial approaches that can also include qualitative methods. We find it unhelpful to reduce spatial approaches to quantitative methods and call for suggestions on how methodologically to enrich the more abstract understanding of space that is part of the 'spatial turn' in humanistic and social sciences. Hence, *there is a need for* spatial approaches that do not deny people free will in decision-making and treat them as members of categories rather than individuals (Johnston, 2000, p.132).

When combining social issues of rurality with physical issues of rural land use, a combination of methods and data is an almost inevitable result. Such a combination of quantitative and qualitative data and methods is a challenge for rural researchers. However, few papers deal explicitly with the philosophical and methodological implications of research based on a combination of different data and methods (some exceptions being Yeung, 1997; Pratt, 1995). In the practices-values approach, we advocate studying practices and values using multi-methods in an iterative process. This is what Yeung describes as 'a total method approach in which these different methods are seen as employed as a *coherent whole*' (Yeung, 2000, p.23, emphasis in original). However this has some philosophical implications.

Different ideas about the conduct of research occur, depending on the philosophical point of departure. This means different data collection methods, data handling methods, and evaluation rules, etc. are considered appropriate. The whole idea of being able to choose between methods is based on implicit or explicit assumptions at the philosophical level. There is a risk of eclecticism when working with multi-methods applying a wide range of methods and data types, because data derived from the use of completely different methods cannot just be compared. In order to avoid eclecticism, the integrity of the different methods has to be respected, which points to the importance for connection between philosophy and methodology throughout the research process. However, often it is how the data are used, and not how they are collected, that differs for the various research methods and philosophical stances. Hence, it is how the data are interpreted that matters. The interplay between philosophy and methodology is important, because only philosophical reflections that are used in interplay with methodological considerations can enhance the research process.

Different Ways to Apply the Practices-values Approach

In the following, some reflections concerning the application of the practices-values approach are presented. As the discussion shows, we do not advocate one specific method for collecting empirical material or one way of analysing this – on the contrary. The first section concerns the use of multi-methods for collecting information on both practices and values and to reveal discrepancies between what people do and what they say they are doing. This implies that fieldwork is important. Like Yeung, we find in situ research necessary for establishing context and want to avoid 'armchair theorising' and 'remote sensing empirical studies' (Yeung, 2000). In the next section, we open the array of methods for collecting information on practice and the lived values by including spatial approaches well-known to geographers. Finally, we describe one means to analyse this information and construct abstractions, this is by the use of ideal types.

Multi-methods

The increasing use of qualitative methods in cultural geography has caused the semi-structured interview to gain great importance (Crang, 2002). While this is a powerful tool for collecting information, the emphasis on both practices and values means that semi-structured interviews become too unidirectional. More tools are needed, as we cannot rely on one method for gaining information on the relationship between people's values and what they actually do – their practices. We have to observe what people are doing as well as hear them tell us about their practices and values. Quite often there is a discrepancy between what people say they are doing and what they actually are doing. This is seen both among Danish farmers applying for afforestation grants and Senegalese pastoralists going on migration; therefore these studies have applied multi-methods. Furthermore, when interviews are used, there is a need to analyse not only the interview itself but also the interview location and the power relations between the interviewer and the interviewee. The importance of the interview location for the outcome of the interview has been described by Elwood and Martin (2000) among others. They discuss how researchers can navigate the spatial aspect strategically to situate the interview in its social and cultural context in order to enrich the explanations of the participants. Further, the interview location can be used actively by the researcher to change the power dynamics of the interview itself. A refreshing view on power relations embedded in research based on interviews is given by Bradshaw (2001), who points to research as a fundamentally negotiated process that is translated and transformed by participants and researchers throughout.

The call for multi-methods and observation of people's practice means that our practices-values approach draws upon experiences from ethnographic research emphasising observations. These have been criticized because they are susceptible to bias from the observer's subjective interpretation. But according to Adler and Adler: 'Direct observation ... enhances consistency and validity' (1998, p.90). Observations can be used to establish context for the other data collection methods. As noted by Kvale, 'If the research topic concerns more implicit meanings and tacit understandings, like the taken-for-granted assumptions of a group or culture, then participant observation and field studies of actual behaviour supplemented by informal interviews may give more valid information (1996, p.104). We use the concept 'observations' to cover broad range of methods and not only the ethnographic method, and a number of tools for collecting observations can be suggested e.g. map analysis, physical registration. Hence, we advocate geographical means to collect observations as an addition to ethnographic fieldwork, as will be discussed in the next section.

As mentioned, researchers have to be aware of the risk of eclecticism when working with a wide range of methods and data types, a risk that does not exist when working with only one type of data and method. One way to handle this challenge is through method triangulation, if the different methods are used with due respect to their integrity. Different forms of triangulation, e.g. data-triangulation, investigator-triangulation, method-triangulation exist (see e.g. Denzin, 1970; Mikkelsen, 1995; Roe, 1998). Method-triangulation allows for studying the same phenomenon from different perspectives. As argued by Baxter and Eyles: 'triangulation is one of the most powerful techniques for strengthening credibility of qualitative research and is based on the notion of convergence: when multiple sources provide similar findings their credibility is considerably strengthened' (1997, p.514). However, as this approach is based on the assumption that the different data complement each other in revealing different facets of the social world (Yeung, 1997), the challenge of triangulation lies in ensuring that the data enrich each other and are not simply presented as separate findings.

The emphasis on multi-methods should not be seen as a way to justify qualitative methods by using quantitative methods as a prop – a critique that has been raised by Winchester (1999). Rather, the call for multi-methods is due to our subject matter and the need for ways to uncover the multi-layered patchwork of practices and values. In this way, we disagree with Winchester when she claims that interviews can be used as a stand-alone technique in order to uncover underlying structures and causal mechanisms of social processes. Only by approaching the issue from different angles, the often unarticulated and complex interrelationships between practices and values can be uncovered. We do not believe that one single method can provide the answers. Moreover, it is difficult to know in advance which methods will bring forward new and surprising knowledge. Few would disagree with this, but in practice it is difficult to remain open and follow the unexpected because many studies rely on prescribed – rigorous – methods, for instance prescribed by a funding agency. Nevertheless, surprise can be an important impetus for pushing inquiries as stressed by Vayda (1983).

Spatial Methods

Although ethnographic fieldwork and participatory observation are useful for collecting information on people's practices, there are other means to observe practice and discrepancies between practice and values. As geographers we have learned some of these methods, but according to Johnston (2000), geographers have tended to neglect spatial analysis in recent years. However, as the examples below show, spatial methods can give valuable insight into the lived values and practices of individual actors in rural areas.

In the Danish study of afforestation (Madsen, 2002a), information about landowners' practices were gained by looking at maps where the landowners had indicated the shape and size of the desired woodland. In this way, the spatial method was used in a manner similar to observations. As the practice had not yet occurred, it was impossible to observe the practice 'on the ground', but instead the maps indicated the landowners' intended practice. Besides functioning as observations, the maps provided valuable spatial information. It turned out that the desired shape of the woodland varied among the different 'types' of landowners. For instance, one landowner had a long narrow field located away from the farm that he wanted to afforest. He described the long narrow field as arduous to cultivate due to the distance to the fields and the poor soil quality. Another landowner wanted to use afforestation in order to create shelter on the property. The quality soil of the area was good and there was already a considerable amount of existing woodland on the property. In order to create an attractive property, he himself drew the location and shape of the area intended for afforestation. He wanted as much fringe as possible, so the area had a sinuous shape. Further, he did not want the forest to throw shadows towards the house, so the forest was placed away from the house.

As can be seen in the two examples, the shaping of woodland is related to the values of the landowner. While the former is a production-oriented landowner not interested in the shape of the woodland itself but in its location in relation to the production unit, the latter is a nature-oriented landowner interested in and actively involved in shaping the woodland. The same can be seen for the issue of soil quality: while it is important for the production-oriented landowners, it means nothing for the nature-oriented. Without the spatial information from the maps, the relation between practices and values would not have been revealed.

Recently the availability of GPS has provided a new way to gain spatial information. This was used in the study of pastoral mobility in Senegal (Adriansen and Nielsen, 2002). Here, a number of pastoralists were given a GPS that was used for collecting information on their whereabouts. This could have been done – and has usually been done – by participatory observations in the sense that a researcher follows the pastoralists. Instead of this very extensive fieldwork, the pastoralists did the mapping 'themselves' and this information was combined with interviews and observations both on cultural issues and on mobility practices. Considering the cost-benefits of the different ways of collecting spatial information on pastoral mobility, GPS appeared as a promising solution especially because it was possible to study several pastoral families at the same time. This also gave insights into the power relations and networks among them.

Foremost, the GPS data provided valuable information on the extent and patterns of mobility. From an analytical point of view the GPS data can be used in combination with qualitative information to make method triangulation. The GPS data can be used prior to qualitative interviews to make informed questions about mobility and they can be used after qualitative investigations to illustrate points made or show inconsistencies. In this study, triangulation of methods was used in an iterative process to obtain information about how mobility is practised and why. Moreover, the data was used to expose differences between the local and the researchers' perception and terminology of mobility. This would have been difficult to uncover without the spatial information.

Especially with regard to research in developing countries, various methods of (participatory) mapping have been developed. These are part of the much discussed Rapid Rural Appraisal (RRA) and Participatory Rural Appraisal (PRA) (see e.g. Mikkelsen, 1995). Techniques such as lifeline interviews, ranking, etc. can be useful for rural studies in most countries. However, mapping tools where locals are asked, for instance, to draw the village and fields may seem redundant in countries with extensive national surveying. Nevertheless, these techniques can be used for gaining insights into the layout of villages including social segregation as well as for the local perception of the village space/place. These types of mappings can provide information on lived practices and thereby function as inspirations for further insight into the values legitimating certain spatial practices in rural societies all over the world.

Ideal Types

Finally, we want to stress the use of 'ideal types' as a way of making use of different types of data for understanding the use of rural space through abstractions. Ideal types are 'pure' types and cannot be transferred directly from interviews or questionnaires. Weber (1949) suggested ideal types as a means to understand social action. Burnham, who advocates the use of ideal types for studies of, for instance, pastoralism, explains Weber's idea this way: 'Weber's concept of the ideal type was developed explicitly to escape from ... the notion in the natural sciences of a causality operating outside of history and of actors' subjective consciousnesses' (Burnham, 1987, p.162). Burnham elaborates that the construction of ideal types involves the recognition of how factors operating within pastoral systems are mediated by the actors' subjective understandings of their situations. One of the benefits of ideal types is that they provide 'a datum against which comparisons can be made to advance the appreciation of particular events' (Johnston, 2000, p.366).

The construction of ideal types is a way to organize the analysis of various types of data and they can be used as abstractions allowing analyses and discussions above the concrete level. The following example shows how ideal types can be used to integrate knowledge on both practices and values in order to construct a coherent image of rural dwellers' land use and the underlying reasoning.

In the study of Fulani pastoralists in Senegal, four ideal types of livelihood strategies were developed based on questionnaires and qualitative interviews (Adriansen, 2002). The ideal types were developed and confirmed through iterative abstraction (Yeung, 1997) and method triangulation (Mikkelsen, 1995; Roe, 1998). In practice, the questionnaires were read several times and combined with information from the qualitative interviews. Later, some of these findings were discussed with the pastoralists during a focus group discussion.

The four ideal types are named: 'the agro-pastoralist', 'the Tabaski[5] pastoralist', 'the commercial pastoralist', and 'the non-herding pastoralist'. Each of the types provides a comprehensive idea of the pastoral way of life; an idea that is based on pastoralists' motivations, socio-economic possibilities, biological constraints, etc. and not on functionalist and structuralist perceptions of pastoral societies. It should be noted that the types are described as 'individuals', but this does not mean that they can stand-alone. All pastoralists are members of a household and their actions should be interpreted within this context. However, these ideal types are abstractions that should be used to understand the diversity of strategies employed, and thus the individual, idealized form is used.

Most of the pastoralists live in large families where a combination of strategies is employed and different strategies can co-exist for many years. Moreover, the applicability of the different strategies depends upon the availability of labour, the number of animals, and the preferences of the household. As household and herd sizes can change from year to year, so can the strategy. Although the different strategies can be interpreted as a shift away from pastoralism, the emphasis on both practices and values show that this is not the case. If the various practices revealed in the ideal types are studied without considering the underlying values, motives, and preferences, the society could be seen as very dynamic and identities as constantly changing or reconstructed, and so could the pastoral way of life. For instance, one year a pastoralist can describe his commercial activities as selling Tabaski sheep and the next he spends most of his time in his shop. If, on the other hand, we take issues such as preferences, values, and motives into account, the strategies can be seen as innovative and adaptive to a changing socio-economic environment, but not changing the pastoral way of life *per se*. Although the preferences and motives of the pastoralists vary, cattle represent the 'cultural capital' among the Fulani and this appears not to be changing, even though the ways to acquire this capital have been diversified. Hence, the new activities are not a denial of the 'pastoral way of life'. Instead they can be interpreted as a way to preserve or consolidate the values and identity of the Fulani. The four types are used to conclude that pastoralists in Ferlo have managed to make the most of market opportunities while maintaining their pastoral way of life.

This discussion of methods is by no means a comprehensive discussion of the implications of combining methods and data, but should rather be seen as an illustration of different ways to conduct rural studies that embrace both social/cultural and material/physical aspects and that respect the philosophy-methodology interplay. It should not been seen as an implicit disregard of other methods or that these methods are in themselves new. For instance, it may be noted that our way of establishing understanding and abstractions is similar to grounded theory (Glaser and Strauss, 1967). While we find that grounded theory is a useful methodology, it is only one in a number of possible methodologies, which can be used for understanding the use of rural space. Likewise, our methodology is similar to 'progressive contextualization' (Vayda, 1983). While these methodologies have been of inspiration, we find a need to stress the importance of using multi-methods and emphasize the interplay between methodology and philosophy.

Conclusions

We set out with the purpose of showing that many of the issues raised by the cultural turn transcend the gap between research in so-called developing and developed countries. This was done by illustrating how concepts such as 'sense of belonging' are useful for understanding ruralities in 'different worlds'. The majority of the chapter was devoted to discussing how to combine studies of physical/material and social/cultural issues and to providing an approach towards the study of the use of rural space. We argue that the cultural turn has given rise to much new and interesting research on neglected parts of rural studies as well as drawing attention to new material and knowledge building. Within this new area of research, we see a challenge in combining the cultural turn with its emphasis on identity, discourse, motives, agency, and cultural construction of rurality with more 'conventional' geographical studies of land use.

In order to do so, we have advocated a multi-faceted approach – the 'practices-values' approach – towards the study of rural areas in both developed and developing countries, illustrated by examples from our research in Denmark, and Senegal. In this way, transcending the gap was an underlying issue throughout the chapter. Within the practices-values approach, studies of individual actors are used as the focus for understanding the processes of shared meanings, negotiations, and held beliefs that legitimate certain practices, abandon others, and thereby shape the use of rural space. Practices and values should by no means be considered a dualism; instead they are used as an analytical tool for uncovering and comprehending the complex and often non-verbalized relations between action and lived values. This calls for the use of multi-methods and a (re-)turn to philosophically informed ways of doing research. Finally, attention is drawn towards addressing the 'new and surprising'. In this process, inspiration and the exchange of experience between researchers studying the developed and the developing world will be valuable; a call that may well be difficult to practice within the restricted frames of research funding and request for policy relevance.

Acknowledgements

A special thanks to Professor Keith Hoggart for his encouragement and useful comments in regard to our discussions of the need to transcend the research gap between rural studies of 'developed' and 'developing countries'. Also thanks to Professor Emeritus Sofus Christiansen for sharing his enthusiasm for geography and his never-ending curiosity about rural life.

Notes

1. Both the dichotomies 'developed-developing countries' and 'First-Third World' are embedded in modernist and orientalist discourses encompassing a certain type of (economic) development, ethnocentrism, and 'othering' (see e.g. Said, 1985; Crang, 1998). The 'North-South' demarcation could be used instead. However, within rural studies there has been a number of interesting contributions from the Australian continent and the 'North-South' division therefore appears inappropriate. Nevertheless, for the argument of this paper, we need a term covering (the boundary between) these 'different worlds'. This boundary – as other boundaries – is a social construction, but one that manifests itself in the construction of research departments and journal delimitations, just to mention a few examples. The argument of the paper is that this boundary is problematic and should be deconstructed; the usage of modernist concepts should therefore not be seen as endorsing these discourses.

2. Council Regulation (ECC) no 2080/92 of 30 June 1992 instituting a community aid scheme for forestry measures in agriculture.

3. GPS means Global Positioning System, a device that can provide coordinates for one's location.

4. A discussion of 'space' and 'place' is beyond the scope of this chapter. Suffice it to say that while space is general, place is particular: 'space is everywhere, place is somewhere... place is a space with attitude' (Taylor, 1999, p.10).

5. Tabaski is the Senegalese word for the Muslim feast called *id-al-adha* in Arabic. Tabaski sheep are young male sheep. They are raised for commercial purposes and fed well so they can earn a good price at the right time.

References

Adler, P.A. and Adler, P. (1998), 'Observational Techniques', in N.K. Denzin and Y.S. Lincoln (eds), *Collecting and Interpreting Qualitative Materials*, Sage Publications, Thousand Oaks, pp. 79-109.

Adriansen, H.K. (2002), 'A Fulani Without Cattle is Like a Woman Without Jewellery: A Study of Pastoralists in Ferlo, Senegal', *Geographica Hafniensia, A11*, Institute of Geography, University of Copenhagen.

Adriansen, H.K. (2003), 'Egypt's Desert Lands: New Spaces of Opportunity or New Spaces of Poverty?', Paper submitted to *Development and Change*.

Adriansen, H.K. and Nielsen, T.T. (2002), 'Going Where the Grass is Greener: on the Study of Pastoral Mobility in Ferlo, Senegal', *Human Ecology* Vol. 30, pp.215-226.

Atkins, P.J. (1988), 'Redefining Agricultural Geography as the Geography of Food', *Area* Vol. 20, pp.281-283.

Baxter, J. and Eyles, J. (1997), 'Evaluating Qualitative Research in Social Geography: Establishing 'Rigour' in Interview Analysis', *Transactions, Institute of British Geographers*, Vol. 22, pp. 505-525.

Bebbington, A. (1999), 'Capitals and Capabilities: a Framework for Analyzing Peasant Viability, Rural Livelihoods and Poverty', *World Development*, Vol. 27, pp. 2021-2044.

Beedell, J. and Rehnman, T. (2000), 'Using Social-psychology Models to Understand Farmers' Conservation Behaviour', *Journal of Rural Studies*, Vol. 16, pp. 117-127.

Bradshaw, M. (2001), 'Contracts and Member Checks in Qualitative Research in Human Geography: Reason for Caution?', *Area*, Vol. 33, pp. 202-211.

Bowler, I.R. (1984), 'Agricultural Geography', *Progress in Human Geography*, Vol. 8, pp. 255-262.

Bowler, I.R. (1987), 'Agricultural Geography', *Progress in Human Geography*, Vol. 11, pp. 425-432.

Bowler, I. and Ilbery, B. (1987), 'Redefining Agricultural Geography', *Area*, Vol. 19, pp. 327-332.

Burnham, P. (1987), 'Pastoralism and the Comparative Method', in L. Holy (ed), *Comparative Anthropology*, Basil Blackwell, Padstow, pp. 153-167.

Busck, A.G. (2002), 'Farmers' Landscape Decisions: Relationships Between Farmers' Values and Landscape Practices', *Sociologia Ruralis*, Vol. 42, pp.233-249.

Buttimer, A. (1998), 'Geography's Contested Stories: Changing States-of-the-art', *Tijdschrift voor Economische en Sociale Geografie*, Vol. 89, pp. 90-99.

Cloke, P. (1997), 'Country Backwater to Virtual Village? Rural Studies and 'the CulturalTturn'', *Journal of Rural Studies*, Vol. 13, pp. 367-375.

Cloke, P. and Little, J. (eds) (1997), *Contested Countryside Cultures: Otherness, Marginalisation and Rurality*, Routledge, London.

Crang, M. (2002), 'Qualitative Methods: the New Orthodoxy?', *Progress in Human Geography*, Vol. 26, pp. 647-655.

Crang, M. (1998), *Cultural Geography*, Routledge, London.

Denzin, N.K. (1970), *The Research Act: A Theoretical Introduction to Sociological Methods*, Aldine, Chicago.

Ellis, F. (1998), 'Household Strategies and Rural Livelihood Diversification', *The Journal of Development Studies*, Vol. 325, pp. 1-38.

Elwood, S.A. and Martin, D.G. (2000), 'Placing Interviews: location and Scales of Power in Qualitative Research', *Professional Geographer*, Vol. 52, pp. 649-657.

Fay, B. (1999), *Contemporary Philosophy of Social Science: a Multicultural Approach*, Blackwell, Oxford.

Gasson, R. (1969), 'Occupational Immobility of Small Farmers: a Study of the Reasons Why Small Farmers Do Not Give Up Farming', Occasional Papers No. 13. Farm Economics Branch, Department of Land Economy, Cambridge University, Cambridge.

Gasson, R. (1988), *The Economics of Part-time Farming*, Longman, Hong Kong.

Glaser, B.G. and Strauss, A.L. (1967), *The Discovery of Grounded Theory: Strategies for Qualitative Research*, Aldine, Chicago.

Grigg, D. (1981), 'Agricultural Geography', *Progress in Human Geography*, Vol. 5, pp. 268-276.

Grigg, D. (1982), 'Agricultural Geography', *Progress in Human Geography*, Vol. 6, pp. 242-246.

Grigg, D. (1983), 'Agricultural Geography', *Progress in Human Geography*, Vol. 7, pp. 255-260.

Haartsen, T., Groote, P. and Huigen, P.P.P. (eds) (2000), *Claiming Rural Identities*, Van Gorcum, The Netherlands.

Halfacree, K. (1993), 'Locality and Social Representation: Space, Discourse and Alternative Definitions of the Rural', *Journal of Rural Studies*, Vol. 9, pp. 1-15.

Hammar, A. (2002), 'The Articulation of Modes of Belonging: Competing Land Claims in Zimbabwe's Northwest', in K. Juul and C. Lund (eds), *Negotiating Property in Africa*, Heinemann, USA, pp. 211-246.

Ilbery, B.W. (1983), 'Goals and Values of Hop Farmers', *Transactions, Institute of British Geographers*, Vol. 8, pp. 329-41.

Ilbery, B.W. (1986), 'Theory and Methodology in Agricultural Geography', in M. Pacione (ed), *Progress in Agricultural Geography*, Croom Helm Progress in Geography Series, Croom Helm, Worcester, pp.13-37.

Johnston, R.J. (2000), 'On Disciplinary History and Textbooks: or Where has Spatial Analysis Gone?', *Australian Geographical Studies*, Vol 38, pp. 125-137.

Kaltoft, P. (1999), 'Values About Nature in Organic Farming Practice and Knowledge', *Sociologia Ruralis*, Vol. 39, pp. 39-53.

Kazenwadel, G., van der Ploeg, B., Baudoux, P. and Häring, G. (1998), 'Sociological and Economic Factors Influencing Farmers' Participation in Agri-environmental Schemes', in S. Dabbert, A.Dubgaard, L.Slangen, and M.Whitby (eds), *The Economics of Landscape and Wildlife Conservation*, CAB international, Wallingford, pp. 187-203.

Kvale, S. (1996), *Interviews: an Introduction to Qualitative Research Interviewing*, Sage Publications, USA.

Liepins, R. (2000a), 'New Energies for an Old Idea: Reworking Approaches to 'Community' in Contemporary Rural Studies', *Journal of Rural Studies*, Vol. 16, pp. 23-35.

Liepins, R. (2000b), 'Exploring Rurality Through 'Community': Discourses, Practices and Spaces Shaping Australian and New Zealand Rural 'Communities'', *Journal of Rural Studies*, Vol. 16, pp. 325-341.

Little, J. (1999), 'Otherness, Representation and the Cultural Construction of Rurality', *Progress in Human Geography*, Vol. 23, pp. 437-442.

Madsen, L.M. (2002a), 'Location of Woodlands: the Danish Afforestation for Field Afforestation', *Geographica Hafniensia A10*, Institute of Geography, University of Copenhagen.

Madsen, L.M. (2002b), 'The Danish Afforestation Programme and Spatial Planning: New Challenges', *Landscape and Urban Planning*, Vol. 58, pp. 241-254.

Madsen, L.M. (2003), 'New Woodlands in Denmark: The Role of Private Landowners', *Urban Forestry & Urban Greening*, Vol. 1, pp.185-195.

McHugh, K.E. (2000), 'Inside, Outside, Upside Down, Backward, Forward, Round and Round: a Case for Ethnographic Studies in Migration', *Progress in Human Geography*, Vol. 24, pp. 71-89.

Mikkelsen, B. (1995), *Methods for Development Work and Research, a Guide for Practitioners*, Sage Publications, Delhi.

Milbourne, P. (1997) (ed), 'Revealing rural 'others': diverse voices in the British Countryside', Pinter Press, London.

Morley, D. (2001), 'Belongings: Place, Space and Identity in a Mediated World', *Cultural Studies*, Vol. 4, pp. 425-448.

Morris, C. and Potter, C. (1995), 'Recruiting the New Conservationists: Farmers' Adoption of Agri-environmental schemes in the U.K.', *Journal of Rural Studies*, Vol. 11, pp. 51-63.

Murdoch, J. (2000), 'Networks – a New Paradigm for Rural Development?', *Journal of Rural Studies*, Vol.16, pp. 407-419.

Murdoch, J. and Pratt, A. (1993), 'Rural Studies: Modernism, Postmodernism and the 'Post Rural'', *Journal of Rural Studies*, Vol. 9, pp. 411-427.

Müller-Mahn, D. (1998), 'Spaces of Poverty: the Geography of Social Change in Rural Egypt', in N.S. Hopkins and K. Westergaard (eds), *Directions of Change in Rural Egypt*, The American University in Cairo Press, Egypt, pp. 256-276.

Phillips, M., Fish, R. and Agg, J. (2001), 'Putting Together Ruralities: Towards a Symbolic Analysis of Rurality in the British Mass Media', *Journal of Rural Studies*, Vol. 17, pp. 1-27.

Philo, C. (2000), 'More Words, More Worlds: Reflections on the 'Cultural Turn' in Human Geography', in I. Cook, D. Crouch, S. Naylor and J. Ryan (eds), *Cultural Turns/Geographical Turns*, Pearson, Harlow, pp. 26-53.

Potter, C. and Lobley, M. (1996), 'The Farm Family Life Cycle, Succession Paths and Environmental Change in Britain's Countryside', *Journal of Agricultural Economics*, Vol. 47, pp.172-190.

Pratt, A.C. (1995), 'Putting Critical Realism to Work: the Practical Implications for Geographical Research', *Progress in Human Geography*, Vol. 19, pp. 61-74.

Roe, E. (1998), *Taking Complexity Seriously: Policy Analysis, Triangulation and Sustainable Development*, Kluwer Academic Publishers, USA.

Röling, N. (1994), 'Platforms for Decision-making About Ecosystems', in L.O., Stroosnijder, L., J. Bouma and H. van Keulen, (eds), *The Future of Land: Mobilising and Integrating Knowledge for Land Use Options*, John Wiley & Sons Ltd, Fresco, pp. 385-393.

Röling, N. (1997), 'The Soft Side of Land: Socio-economic Sustainability of Land Use Systems', *ITC Journal*, Vol. 3-4, pp. 248-262.

Said, E. (1995), *Orientalism*. Penguin Books, Harmondsworth (1st edn, 1978).

Sayer, A. (1992), *Method in Social Science: a Realist Approach*, Routledge, London.

Taylor, P.J. (1999), 'Places, Spaces and Macy's: Place-space Tensions in the Political Geography of Modernities', *Progress in Human Geography*, Vol. 23, pp. 7-26.

Thom, B.G. and Woolmington, E. (1988), 'The Integrative Power of Interdisciplinary Geography', *Interdisciplinary Science Review*, Vol. 13, pp. 52-63.

Tuan, Y.E. (1977), *Space and Place*, Arnold, London.

Turner, M.D. (1999), 'No Space for Participation: Pastoralist Narratives and the Etiology of Park-herder Conflict in Southeastern Niger', *Land Degradation and Development* Vol. 10, pp. 345-363.

Unwin, T. (2000), 'A Waste of Space? Towards a Critique of the Social Production of Space...', *Transactions, Institute of British Geographers*, Vol. 25, pp. 11-29.

Van Hoven (2001), 'Women at Work – Experiences and Identity in Rural East Germany', *Area*, Vol. 33, pp.38-46.

Vayda, A.P. (1983), 'Progressive Contextualisation: Methods for Research in Human Ecology', *Human Ecology*, Vol. 11, pp. 265-281.

Walmsley, D.J. and Lewis, G.J. (1984), *Human Geography: Behavioural Approaches*, Longman Group Limited, Singapore.

Weber, M. (1949), *The Methodology of Social Sciences*, The Free Press, New York.

Whatmore, S. (1991), 'Agricultural Geography', *Progress in Human Geography*, Vol. 15, pp. 303-310.

Whatmore, S. (1993), 'Agricultural Geography', *Progress in Human Geography*, Vol. 17, pp. 84-91.

Wilson, G.A. (1996), 'Farmer Environmental Attitudes and ESA Participation', *Geoforum*, Vol. 27, pp.115-131.

Winchester, H.P.M. (1999), 'Interviews and Questionnaires as Mixed Methods in Population Geography: the Case of Lone Fathers in Newcastle, Australia', *Professional Geographer*, Vol. 51, pp. 60-67.

Winter, M. (2000), 'Strong Policy or Weak Policy? The Environmental Impact of the 1992 Reforms to the CAP Arable Regime in Great Britain', *Journal of Rural Studies*, Vol. 16, pp.47-59.

Yeung, H.W. (1997), 'Critical Realism and Realist Research in Human Geography: a Method or a Philosophy in Search of a Method?', *Progress in Human Geography*, Vol. 21, pp. 51-74.

Yeung, H.W. (2000), 'Practicing New Economic Geographies? Some Methodological Considerations'. Paper presented at the RGS-IBG annual conference, Brighton, University of Sussex, 4-7 January 2000.

PART II
RURAL SOCIETIES: INCLUSIONS
AND EXCLUSIONS

Chapter 6

Politics and Protest in the Contemporary Countryside

Michael Woods

22 September 2002

In late September 2002, the centre of London was occupied on successive weekends by two very different demonstrations. On 28 September, up to a quarter of a million people took to the streets to protest at the prospect of military action against Iraq. It was a political protest in the classic mould, organized by left-wing campaign groups and parties and standing in the long tradition of anti-war and anti-nuclear mass rallies. Remarkably, however, the demonstration was overshadowed, at least in numerical terms, by the gathering six days earlier of over 400,000 protesters for the 'Liberty and Livelihood March', organized by the Countryside Alliance.

For many participants, the Liberty and Livelihood March was their first political demonstration. Others were veterans of the Countryside Alliance's two previous London rallies, the Countryside Rally in 1997 and the Countryside March in 1998, but few could claim a history of political activism stretching back more than five years – unlike many of the hardened anti-militarism campaigners treading the same route a week later. Many had travelled to London on chartered coaches and trains from all regions of Great Britain. Some marched as families, others in groups as young farmers' clubs or hunt supporters' clubs. They wore Barbour jackets and sweatshirts, and brandished placards supplied by *Farmers Weekly* and the *Daily Mail*.

A few wore the yellow t-shirts of Farmers For Action, a radical group engaged in direct action including the notorious fuel depot blockades of September 2000. Equally present, if more anonymously, were supporters and sympathizers of the Countryside Action Network and the Real Countryside Alliance, two hardline pro-hunting groups also practising direct action and, in the latter's case, advocating civil disobedience. More generally, a current of latent militancy bubbled tantalisingly beneath the surface of the march – sweatshirts emblazoned with the slogan, 'Born to Hunt, Forced to March, Ready to Fight', and banners starkly warning: 'This is our last peaceful march'.

Rural Protests

The three mass demonstrations organized by the Countryside Alliance have been the most visible manifestations of an extraordinary political mobilization of rural people that has taken place in Britain since the mid 1990s (Figure 6.1). Only a decade ago most of the British public, media, political class and even rural studies literature still held to the discourse, carefully crafted in the 1930s, that the countryside was an 'apolitical space' (see Woods, 1997). A newspaper article in 1992 which reported on a particularly turbulent parish council election opened by suggesting that a Somerset village 'seems an unlikely setting for infighting you would expect in Tower Hamlets or some other troubled corner of metropolitan Britain' (Dunn, 1991, p. 3). Similarly, a 1990 review by Wyn Grant of literature on British rural politics commented that, 'Britain is a highly urbanized society, and rural issues are of marginal importance in national politics' (Grant, 1990, p. 286). Neither statement would seem possible today, when local conflicts are an accepted part of rural life and the politics of agriculture, hunting, housing development, access to land and rural services have claimed considerable space in the national media, and the attention of mainstream politicians.

This transformation has been structured in Britain around a number of key issues and watershed events. The election of a Labour government in 1997 with a manifesto commitment to ban hunting, for example, provided the stimulus for the formation of the Countryside Alliance and the organization of the first Countryside Rally. However, it is unlikely that the Countryside Alliance could have maintained the momentum that it has, or rural politics achieved such prominence, had not the first hunting debate been quickly followed by a crisis in agricultural income, concerns over rural services and housing development, anger over fuel taxes, and, finally, a foot and mouth epidemic, each of which fuelled a popular notion of a countryside under siege. At the same time, it should also be noted that rural issues were beginning to intrude on the national political agenda before the 1997 election, notably over the handling of BSE and targets for housing development (see Woods, 1998a, 1998b). Such political contingencies have therefore influenced the nature and form of the new rural politics, but are not its cause. For that we must look to the processes of social and economic restructuring that transformed rural space and rural society in the late twentieth century, and their impact on political discourses of the rural and on local and national power and policy-making structures.

Indeed, common experiences of rural restructuring have produced a similar resurgence of rural politics in several countries other than Britain, albeit shaped by local circumstances. In France, for example, where rural society has always been politicized, and where there is a long tradition of farmers' protests (Naylor, 1994), the last decade has witnessed the emergence of a new strand of rural protests, including direct action against global corporations such as McDonalds, and conflicts over issues such as hunting, quarrying and road-building. In Germany, food safety scares and scandals, for instance over BSE, have prompted a radical reform of agricultural policy enthusiastically driven by a Green Party Agriculture Minister drawing on a very different discourse of rurality than

Figure 6.1   Some of the **400,000** participants in the Countryside Alliance's Liberty and Livelihood March through London, September 2002

that of the established agrarian elite. Meanwhile, in the United States, the 2001/2 review of the Farm Bill provided the focus for an extensive debate between politicians, lobbyists and pressure groups about the future of rural America. Whilst these machinations did not produce the same level of public demonstration as at the height of the American Agricultural Movement in the 1970s (Stock, 1996), the ultimate failure of the new bill to introduce significant reforms may lead to a more militant mobilization. Groups representing the remaining minority of 'family-farmers' are becoming more radical in their rhetoric and right-wing militia groups have drawn on feelings of rural dislocation (Dyer, 1998; Stock, 1996).

Notions of the 'beleaguerment' and 'powerlessness' of rural populations in an urban-dominated modern world are important rhetorical devices for rural campaigners and have helped to mobilize rural resistance to a perceived urban threat. Ironically, however, it can be argued that the rural lobby has achieved a political significance above that which its numerical strength would warrant. In June 1999, after a pro-hunting party, *Chasse, Pêche, Nature et Tradition* (CPNT), had polled an unexpected seven per cent of the vote in European Parliamentary Elections in France, the *Ouest France* newspaper published a cartoon showing the then Prime Minister, Lionel Jospin, amorously courting hunting-gear-clad CPNT supporters in his office. Rural protests have at times provoked a loss of nerve by the British government, with vacillation over action on issues such as hunting and the handling of the foot and mouth epidemic. Whilst the Countryside Alliance may ultimately fail to prevent a ban on hunting, and may not be successful in persuading the government to adopt its agenda, its prominence has forced the Labour government to develop a far more comprehensive and more strident rural policy than may otherwise have been the case.

In these ways, the recent plethora of rural protests and campaigns can be regarded as the most significant new social movement to emerge since the rise of environmentalism in the 1980s. Yet, the speed of the bandwagon of rural protest has been such that it remains virtually untouched by academic research. This chapter begins to redress that neglect by exploring the emergence and organization of the rural movement, illustrated through three brief case studies, of the Countryside Alliance and Farmers for Action in Britain and the *Confédération paysanne* in France.

The Mobilization of the Rural Movement

Rural protests are not, of course, a new phenomenon. Despite the prevalence of twentieth-century notions of the countryside as an apolitical space, history is littered with records of peasants' revolts and rural rebellions. Even for today's rural campaigners, legends of the Diggers of seventeenth century England (Parker, 2002), the *jacqueries* of fourteenth century France (Naylor, 1994), or the rural radicals in the early United States (Stock, 1996), have resonance, allowing contemporary protest to be positioned in a long historical tradition. Yet, such historical examples belong to an era of pre-modern protest, in which a disenfranchized peasantry employed the only political tool at their disposal to challenge their own exploitation. They were essentially intra-rural class conflicts – described as rural protests because of their rural setting, but not actually protests in which 'rurality' itself was the object of contestation.

The twentieth century introduced a new set of political dynamics, yet the 'rural' essentially remained the context for political activity rather than its object. As the food-security interests of the State became aligned with the private economic interests of an ascendant class of capitalist farmers, so rural policy became acquiescent to the concerns of the dominant elite in rural society and the necessity of public protest diminished. Agricultural unions were drawn into corporatist or policy community policy-making arrangements, creating a private, backstage, route for the representation of rural interests (Browne, 2001; Grant, 2000; Winter, 1996). Where demonstrations and protests still occurred, they tended to be associated with particular agricultural fractions whose practices were out of kilter with the demands of agrarian capitalism – notably smallholders in France and the United States (Naylor, 1994; Stock, 1996). However, again, these were *agricultural* protests as opposed to *rural* protests in that the core concern was with the management of the agricultural sector. As such, they conformed to a modernist mode of politics, driven by sectoral economic interests and fought broadly along class lines.

The new wave of rural protests, in contrast, are of a different order. In what may be characterized as a shift from *rural politics* to a *politics of the rural*, the very definition and meaning of 'rurality' has become the central object of conflict. Thus, there are demands to protect agriculture because of the importance of farming to the maintenance of rural social structures and rural landscapes; hunting is defended as a central facet of a 'rural way of life'; the closure of village shops and schools is resisted because they are pivots of rural communities; and new housing developments and new roads are opposed because of the damage to the rural environment and/or the potential influx of 'urban values'. As protests have become oriented around cultural signifiers of rurality, so they also have begun to adopt the tactics of what Routledge (1997) labels 'the post-modern politics of resistance', which 'mounts symbolic challenges that are extensively media-ted in order to render power visible and negotiable, and to attract public attention...Eschewing the capture of state power, they nevertheless pose challenges to the state' (pp. 372-373).

The transition between styles of politics reflects the wider processes of social and economic restructuring that transformed rural areas in the last quarter of the twentieth century. Restructuring disrupted the imagined unity of a rural space built upon the primacy of agriculture. As the contribution of agriculture to the rural economy and labour market declined, and those of alternative economic practices expanded, so the meaning and regulation of rural space became fragmented:

> There is no longer one single space, but a multiplicity of social spaces for one and the same geographical area, each of them having its own logic, its own institutions, as well as its own network of actors – users, administrators, etc. – which are specific and not local (Mormont, 1990, p. 34).

As Mormont implies, restructuring has also drawn a new range of actors into rural society and politics. Some of these are investors in rural space – companies and entrepreneurs behind new business ventures, government agencies providing funds for rural development and subsidies for agriculture, and counterurbanizers, who have invested both financially and emotionally in a personal pursuit of the rural idyll. Others have remained external to the rural – tourists and day-trippers, supermarkets that have exerted increasing power over agricultural markets, animal welfare protestors and environmental campaigners, the media and marketing industries that have reproduced particular cultural representations of rurality, and, of course, government. Each actor promoted their own discursive understanding of the rural, in ways that were frequently incompatible. Thus, as Mormont again noted, a new focal point for political conflict emerged:

> The structural characteristics of the development of rural space hide a series of economic, social and political contradictions which make up the basis of 'rural' forms of opposition. Our analysis of these embryonic social movements will examine the setting up and development of new rural struggles whose particular nature, in our view, is no longer to focus on specific aspects of the situation of rural populations, but really to pose the problem of rural space. From now on, if what could be termed a rural question exists it no longer concerns issues of agriculture or of a particular aspect of living conditions in a rural environment, but questions concerning the specific functions of rural space and the type of development to encourage within it (Mormont, 1987, p. 562).

Conflicts of this type appeared first at a local scale, often produced by disagreements over the development or management of specific territories or places (see Lowe et al, 1986; Murdoch and Marsden, 1994; Spain, 1993; Woods, 1998a, 1998b). However, by the mid 1990s, such conflicts had begun to impinge on national politics, partly because conflicts could not be resolved locally (see Woods, 1998a, 1998b), but more importantly because diverse circumstances forced government and politicians to engage with issues such as agricultural reform, rural planning and hunting.

As such conflicts inevitably involved challenges to the *status quo* of rural society, so those sections of the rural population directly connected to traditional rural pursuits like farming and hunting, whose interests had been closely served by established discourses of rurality, came to feel they were under attack on multiple fronts. A sense of beleaguerment and isolation has thus been generated, in which a so-called 'indigenous' rural culture and way of life is perceived to be under threat (see also Cox and Winter, 1997). As the *Guardian* newspaper reported of one participant in the 1997 Countryside Rally, 'he sees country people as "Red Indians", indigenous people as threatened "as any poor bugger in the rainforest"' (Vidal, 1997, p. 3). This perception enables issues as diverse as agricultural incomes, the closure of village services, new housing developments, public access to land, and the prohibition of hunting to be linked in a single political struggle, and positions the 'urban' or 'urban values' as the enemy. Moreover, it places rural identity as the source of solidarity in the face of adversity. Yet, the very fact that core facets of rural life were considered to be threatened was read as a failure by the traditional representatives of rural society – the agricultural unions and the landowning elite who had formed the rural political class – to protect rural interests. Instead, 'rural' people looked to new organizations and new forms of protest.

The Rural Movement as a Social Movement

Contemporary rural protests cannot be researched using the conventional interpretative tools of rural political analysis – concepts such as policy communities and insider and outsider pressure groups. Rather, the protests exhibit many of the characteristics of a new social movement – hence the earlier claim that the rural movement could be regarded as the most significant new social movement to emerge since environmentalism. However, this identification of a 'rural social movement' is contentious. As is discussed at greater length elsewhere (Woods, 2003), the rural movement meets Diani's definition of a social movement as 'a network of informal interactions between a plurality of individuals, groups and/or organizations, engaged in political or cultural conflict on the basis of a shared collective identity' (1992, p. 13), as well as Offe's description of new social movements as characterized by 'an open, fluid, organization, an inclusive and non-ideological participation, and greater attention to social than to economic transformation' (della Porta and Diani, 1999, p. 12). Yet, questions are still raised about whether the diverse collection of rural protests and campaigns have sufficient coherence to qualify as a 'social movement'.

Certainly, they embrace a full range of political and ideological perspectives. Some of the groups involved, such as the *Confédération paysanne* and the US Rural Coalition, adhere to a broadly progressive, left-leaning political philosophy; some are identified with mainstream conservatism, including the Countryside Alliance in Britain; whilst a few are linked to the extreme-right, notably the American militia movement. More significantly, perhaps, they also represent a range of different, and sometimes contradictory, discourses of rurality. To categorize crudely, these can be distilled into three distinct 'ruralisms':

- *Reactive ruralism* - the mobilization of a self-defined 'traditional' rural population in defence of purportedly historic, natural and agrarian-centred 'rural way of life' in response to a perceived challenge from 'ill-informed' urban intervention. Examples include the pro-hunting Countryside Alliance and militant Farmers for Action movements in Britain, landowners' resistance to a right to roam, French farmers demonstrating against CAP reform, the pro-hunting CPNT political party in France, the alignment of disaffected American farmers with the far right militia or patriot movement, and the rural electoral support for the One Nation party in Australia.

- *Progressive ruralism* - action in opposition to modern farming practices and agricultural policy, the globalization of the food trade, land-ownership patterns, road developments or other activities which conflict with a discourse of a simple, close-to-nature, localized and self-sufficient rural society. Explicit ideological alliances are often made with other social movements, including notably environmental and anti-globalization protesters. The most prominent example is Bové's *Confédération paysanne* in France, both others include anti-road protesters in Britain and Germany, land rights campaigners and experimental communities in Britain, and the minority farmers movement in the United States.

- *Aspirational ruralism* - the mobilization of in-migrant and like-minded actors to defend their fiscal and emotional investment in rural localities by seeking to promote initiatives which further the realization of an imagined 'rural idyll', and resisting developments which threaten or distract from this imagined ideal rural. Examples include moves by in-migrants to oppose hunting in rural Britain (see Woods, 1998b), proactive intervention in the planning process to protect rural 'middle class space' (Murdoch and Marsden, 1994; Woods, 1998a), and the campaigning of British ex-patriots against the installation of overhead power cables in the Quercy in south-west France (Henley, 1997).

However, to dismiss rural campaigners as a social movement because of this apparent fragmentation would be to both overestimate the coherence of other social movements, and underestimate the degree of interaction between rural campaign groups. Unity of purpose is not a defining characteristic of social movements and many key writers in the field have emphasized that social movements are by nature segmented, policephalous or multi-centred, and reticular, with 'multiple links between autonomous cells forming an indistinctly-bounded network' (della Porta and Diani, 1999, p. 140; see also Gerlach, 1976, Offe ,1985). Even archetypal social movements such as the environmental movement or the gay movement are very loose collections of groups promoting different ideological positions, employing different tactics, working towards different ends, and – on occasion – being drawn into conflict with each other.

At the same time, there is an unappreciated degree of correspondence between different rural campaign groups. For example, in conventional political terms the Countryside Alliance and the *Confédération paysanne* appear to occupy opposite ends of the left-right spectrum. Yet both are concerned with resisting a perceived external threat to a rural way of life; both are pro-hunting, both support local food production and consumption, and both are anti-bureaucracy. Whilst conflicts can arise between groups advocating incompatible discourses of rurality, occasional alliances can also be forged. Protests against road-building and housing construction have united radical ecologists pursuing a progressive ruralism and middle class residents upholding an aspirational ruralism (see for example, Bryant, 1996; Merrick, 1996); the anti-fuel-tax protests in Britain in September 2000 were essentially an act of reactive ruralism, but drew support from car-using adherents of aspirational ruralism (Doherty et al, 2003); whilst the iconization of José Bové in France has been attributed in part to his ability to appeal to all three ruralisms (Viard 2000). Campaigns to defend rural services or for greater investment in rural areas can similarly produce coalitions of all three ruralisms.

Fainstein and Hirst (1995) describe social movements as 'forever in process of becoming' (p. 183), emphasising the dynamism and continuing evolution of social movements. The Deleuzian allusion is appropriate as it provides a means of breaking free from modernist perceptions of organizational forms, and supplies a new language that is more faithful in representing the complexities of the rural movement. Thus, rather than the restrictive forms of the hierarchy and the network, which imply coherence and direction, the rural movement more closely resembles the rhizome – decentred, polymorphous, resilient, growing from the bottom-up not the top-down. Rather than an organization constrained by the ordered conventions of striated space, the rural movement exists within the possibilities of smooth space (see Deleuze and Guattari, 1988). As such, employing a social movement perspective to examine contemporary rural protests can allow for the ambiguous coherence of the 'rural movement' to be accommodated. In recognising that social movements are decentred, fractious and dynamic, it becomes possible to conceive both of a singular 'rural *movement*', defined by the centrality of the notion of rurality to its rationalities of action, and of plural 'rural *movements*', structured around particular understandings of rurality. This distinction is essentially rhetorical, however, as in practice the strands of rural protest that might be identified as different 'rural movements' exhibit a fluidity of purpose, membership, ideology, strategy and allies, that renders any attempt to impose a formal segregation or classification impossible.

Three Case Studies

The complexities discussed above are evident in the three case studies that comprise the remainder of this article. The case studies represent points of mobilization of the rural movement – the Countryside Alliance in Britain; British farmers' protests that coalesced into Farmers for Action; and the *Confédération paysanne* in France. If the distinction between reactive, progressive and aspirational ruralism made earlier were to be followed, strong elements of reactive ruralism could be identified in the Countryside Alliance and Farmers for Action, and of progressive ruralism in the *Confédération paysanne*. However, all three cases defy easy categorization. All three have drawn support from advocates of an aspirational ruralism, and all three have employed elements of the aspirational discourse in their rhetoric. Moreover, there is substantial correspondence in the concerns, objectives and tactics of the *Confédération paysanne* and Farmers for Action, even if the leaders of the two organizations choose to position their campaigns differently in the wider context of conventional politics.

More significantly, the three examples illustrate some of the different patterns of engagement, mobilization and reasoning evident in the rural movement.

Different groups have mobilized at different times in response to different triggers, and have adopted different approaches to questions about organizational form, strategy and tactics. In highlighting these factors, the case studies seek to point towards issues of strategy, organizational dynamics, leadership and participation around which a future research agenda might be constructed.

Case 1 - The Countryside Alliance

The Countryside Alliance represents the classic example of an established 'insider' group adapting to new circumstances. Formed in 1997 as a merger of the British Field Sports Society (BFSS), the Countryside Movement and the Countryside Business Group, it has its roots firmly in the British pro-hunting lobby. Although hunting has been a contentious issue in Britain for over 70 years, the 'insider' status of the hunting lobby and its informal back-stage lobbying of political leaders had succeeded in resisting attempts to ban hunting with hounds despite popular support for such a move. By the 1990s, however, the continuing viability of this strategy was being brought into question. Opinion polls were pointing towards a Labour victory in the 1997 general election, with the prospect of both a manifesto commitment to a free vote on hunting, and an anti-hunting majority in the House of Commons. Pressure hence mounted for a change of tactics from two sources – hunting supporters in the business community led by Eric Bettleheim, an American-born entrepreneur, who wished to adopt a higher profile and more professionally organized approach, and grassroots supporters who were anxious to do something themselves. From the former came the impetus to pool resources and re-brand as the Countryside Alliance, from the latter came the idea for the experimentation with mass demonstration in the shape of the Countryside Rally:

> Nobody can say who first had the idea for a latter day peasants' revolt – for by the beginning of 1997 discontent has become so widespread that all across the country people were wondering how they might demonstrate their ever-increasing annoyance at the manner in which their jobs, traditions and very way of life were being threatened by the uncomprehending dogmatism of urban politicians. One potential rebel was Chipps Mann, wife of a Gloucestershire farmer, who, as she was driving home on a January evening abruptly announced to a friend, 'I'm going to walk from Cornwall to London (Hart-Davis, 1997) p. 2).

The selection of a march as a form of protest drew upon a strong and popularly-known tradition of protest marches in British history. This historical precedent was subsequently developed in the 'marketing' of the protest, playing an important role in consolidating a sense of identity for the protesters by establishing a set of historical references with which the contemporary protest could be framed. Thus, comparisons were drawn not just the with Peasants Revolt of 1381, as in the quote above, but also with the Jarrow March of 1936 and the 'Piccadilly Hunt' of 1949 against an earlier threat of legislation. The historical resonance was incorporated into the location of the Rally, at the Reformers' Tree in Hyde Park.

The translation of the idea into practice involved a refinement both for practical reasons and as a result of learning from the experience of others. Thus, discovering that it is illegal to march on Parliament whilst it is sitting, the proposed march became transmuted into a fixed rally in Hyde Park, which would also form the end of three long-distance marches from Cornwall, Wales and Scotland. Most significant, however, was the experience of key actors participating in an earlier protest against handgun legislation organized by the Sportsman's Association in November 1996:

> Among the crowd that day were Simon Clarke and his newly-recruited secretary, twenty-four year old Mary Eames. They enjoyed the march ... 'There was a buzz about the event, and at the BFSS we realized that was what we would have to create if we held a rally of our own.' Next day, however, he was dismayed to find that the demonstration received minimal press coverage: tiny, single-column reports in the Times and the Daily Telegraph and a little bit more in the Daily Mail (Hart-Davis, 1997, p. 4).

> Mann and Butler had also joined the Sportsman's rally and thought it brilliantly organized. When they looked back along Piccadilly and realized that half the thoroughfare was a solid mass of marchers, as far as they could see, they thought 'Fantastic! This must be bringing London to a halt.' Yet when they slipped out of Trafalgar Square with speeches still in progress, and went round one corner, they found everything perfectly normal. 'London didn't know what was happening,' Mann recalled afterwards. 'We saw that over 20,000 people could march through the streets, and London didn't notice.' They considered this, and the lack of publicity, a disaster, and felt that they themselves could not risk staging a rally if it was going to attract so little attention (Hart-Davis, 1997, p. 5).

These assessments of the Sportsmen's March led to a series of rationalizations which served to shape the eventual form of the Countryside Rally. First, that a rally would only be worthwhile if it attracted significant media attention. Second, that to achieve extensive media coverage – and be noticed by the people of London – it would have to attract a large number of participants. Third, that the hunting issue alone was too narrow an issue to motivate sufficient numbers, and that the primary objectives of the protest would only be achieved if it could succeed in mobilizing a wider coalition of rural people. In consequence, the BFSS re-branded itself as the Countryside Alliance, and the rally became a Countryside Rally, a protest about defending not just hunting, but a whole rural way of life.

The Countryside Rally achieved an immediate effect in effectively panicking the British government into refusing government time for a bill to ban hunting and shoring up its rural policy. However, at best the rally was judged to have delayed a ban rather than defeated one. Pressure was kept up by a second mass demonstration, the Countryside March in March 1998, which again bore the hallmarks of pragmatic rationalization by the organizers:

> [We] were all convinced another big show of strength would be needed before the Report Stage of the [hunting] bill.... Park authorities would not permit a rally in Hyde Park during the winter months, and standing around for hours on wet or icy

ground would be too unpleasant to contemplate anyway. A march was the obvious answer (George, 1999, p. 140).

After the March, however, the Countryside Alliance was faced by the challenge of maintaining momentum over the longer term. The Alliance, which had been put together very hurriedly in 1997 as a genuine coalition, was formally constituted as a single organization, with a professional secretariat headed by Edward Duke, a Yorkshire businessman. Duke and his allies approached the problem of momentum from a business perspective, aiming to expand the 'market' of the organization by seeking to appeal to a wider range of rural interests. This strategy, however, was an anathema to many of the pro-hunting activists who believed that the Alliance should represent a particular traditional type of rurality. As the organization's press officer, who resigned over the dispute, remarked:

> The decision to allow ramblers and canoeists to join annoyed many of the members – ramblers are only of use when we run out of pheasants to shoot, and canoeists scare off the fish. The Countryside Alliance can't be all things for all people (Janet George, quoted in Carter, 1998, p. 3).

Following a further change of chief executive, the Countryside Alliance adopted a *modus operandi* that mimics that of large environmental groups such as Greenpeace and Friends of the Earth in combining occasional protests with more mainstream lobbying and policy work. Between the Countryside March in 1998 and the Liberty and Livelihood March in 2002 (Figure 6.2), the Alliance concentrated on smaller-scale events, including regional rallies and marches, protests at party conferences and 'country comes to town' events in county and market towns, as well as letter-writing and petition campaigns. It has developed a sophisticated campaigning structure with regional campaign officers and regular paper and e-mail newsletters. At the same time, it has produced its own policy handbook and manifesto, made representations to government inquiries and run promotional campaigns for local foods and farmers markets. By 2001 its membership had expanded to over 90,000.

Indeed, the Countryside Alliance has gone further than the environmental organizations by also seeking to act as a social organizer for rural communities, promoting and arranging social events such as fairs and fundraising hog-roasts. In some ways it has moved to fill a perceived vacuum of rural leadership, and in doing so has effectively 'naturalized' and 'depoliticized' its role in rural society, to the extent that it has managed to get its information points into village post offices and its notices into parish magazines. In September 2002, some church services were even relocated to the coaches en route to the Liberty and Livelihood March.

In these ways the Countryside Alliance has probably guaranteed its longevity even beyond a ban on hunting. However, it has also exposed itself to criticisms that it has neglected the core issue of hunting. Dissatisfaction with the Countryside Alliance has prompted the formation of at least two breakaway

Figure 6.2 A pro-hunting supporter on the Liberty and Livelihood March indicates the depth of their feeling

factions, each engaged in more direct action tactics. The Countryside Action Network, founded by Janet George, has organized road-blocks; whilst the Real Countryside Alliance, which has a looser structure and no formal membership, but has been associated with Edward Duke, has been involved in fly-posting MPs offices and signposts. Though such actions have been moderate thus far, the threat of more militant and disruptive actions has been raised by some activists (Walton, 2002), encouraging a further evolution in the Countryside Alliance's position, with its chairman drawing on Henry David Thoreau to suggest that civil disobedience can be justifiable (Jackson, 2002).

Case 2 - Farmers for Action

Farmers for Action also came into existence in response to the declining ability of established insider rural groups to deliver – but in a very different way to the Countryside Alliance. It was born from the desperation of farmers experiencing dramatic falls in their income resulting in part from a strong pound which made imported food cheap, and frustration at the apparent impotence of the established farming unions to change government policy on the continuing EU ban on the export of British beef imposed during the BSE crisis.

    The initiative came from North Wales farmers talking at livestock marts and other events and turned on a spontaneous decision of 400 demonstrators to go to Holyhead docks on Anglesey with the intention of *reasoning* with lorry drivers importing Irish beef and persuading them to turn back. The initial protest on 30 November 1997 achieved some success in gaining media coverage and as such fuelled the motivation of participants in extending the action, which became repeated nightly. As the protest continued it also matured – becoming more planned and organized, more media savvy, and with clear leaders emerging. The Anglesey protests also set a precedent for farmers elsewhere, with local groups holding meetings and organizing their own protests, sometimes imitating the North Wales tactics, sometimes innovating their own. Within a week pickets had spread to docks at Fishguard, Swansea, Pembroke Dock, Dover, Plymouth, Portsmouth, Heysham and Stranraer and to the Channel Tunnel, and had been accompanied by demonstrations at supermarkets and in town centres in Abergavenny, Cardigan, Cardiff, Newtown and Haverfordwest. The participants in these local protests began to form a loose confederation, with leaders communicating by phone to co-ordinate action.

    At the same time however, the protests were marked by hesitancy and reluctance. The farmers' conservative political socialization became apparent in their rejection of comparisons to industrial disputes and their eagerness to be seen as 'law-abiding'. As such the docks protests came to an abrupt end in late January 1998, when more than 500 farmers at Holyhead clashed with over 200 police in a confrontation in which baton charges and CS gas were deployed.

    The protests had, however, created a movement for direct action in British agricultural politics. Over the course of the next two years a rolling series of demonstrations were held at supermarkets, distribution depots, ports, creameries and political meetings as the impact of recession spread through agricultural sectors. At the same time, the organization of these protests became more co-ordinated and more formalized with the establishment of an umbrella group,

Farmers for Action, at a meeting of protest leaders from 16 regions held at a motorway service station in Spring 2000.

Yet, as the novelty of the farmers' protests wore off, so their ability to capture media attention faded, whilst the ephemeral and largely peaceful nature of the protests rarely caused any significant problems for those being targeted. Here again, the political naïvety and hesitancy of the protesters was evident, effectively compromising the objectives of the action. Thus for example, in June 2000, Farmers for Action members from Wales, the North West and Midlands congregated for a mass demonstration in Manchester against the high price of fuel. The demonstration was organized with the co-operation of the police and routed to minimize disruption to city centre traffic and shoppers. In consequence it failed to gain any news media coverage. A BBC Wales documentary which followed two of the leaders organizing the event – David Handley and Brynle Williams – gave voice to their frustrations at the lack of media attention, and, by implication, at the limitations of their strategy. Towards the end of the film, Brynle Williams is shown reluctantly expressing the opinion that the only way forward was to 'copy the French' in adopting more militant blockades.

The escalation of action came in September 2000. During the summer, FFA had been associated with a 'Dump the Pump' campaign which attempted with little success to encourage motorists to boycott petrol stations on specific days in protest at the high fuel prices. Fuel tax had become the key issue as for many farmers increasing fuel prices marked a further erosion of their finances. In August 2000, blockades of ports, oil refineries and depots by French farmers, hauliers, taxi drivers and fishermen – organized and co-ordinated by mainstream trades unions – succeeded in winning concessions from the French government (Mitchell and Dolun, 2001). This achievement was not lost on the British protesters:

> We looked at the French and we thought enough is enough. We just cannot survive. We have got to get the price of fuel down (Paul Ashley, Farmers for Action, quoted in *The Guardian*, 8 September 2000).

Whilst the stimulus came from France (see also Doherty et al, 2003), the ignition again came from farmers in North Wales, many of whom had been involved in the earlier Holyhead blockade. A hurriedly convened meeting of over 160 farmers at St Asaph on 7th September resolved to organize an immediate blockade of the Shell oil depot at Stanlow in Cheshire. As Doherty et al observe,

> The decision to mount this protest was essentially spontaneous. By this we do not mean that it was an irrational or instinctual action, but rather an on the spot decision, the consequences of which were unanticipated. Current social movement theory tends to over-emphasize the strategic aspects of action at the expense of spontaneity. Some who went to the meeting in St Asaph were prepared to take some kind of protest action, but the decisions to act immediately and to target the fuel depot were unplanned and unanticipated (Doherty et al, 2003, p. 8).

Again the protests spread to other depots and refineries around the country, transmitted through farmers' private social networks but also this time the loose organizational structure of FFA which enabled rapid movement and the presentation of a more united and coherent front. Internally, experimental forms of decision making were adopted. Early morning 'board meetings' were held at the blockades to decide strategy, with the various local protest leaders communicating by mobile telephone and with Brynle Williams and David Handley emerging as national spokespersons. The farmers were also joined by hauliers, increasing their capacity to act.

Most significantly, however, the fuel protests stepped over the line from demonstration to disruption and did so with a commodity vital to modern life. Aided by the effects of panic-buying the protesters were able to seriously impact on the everyday life of British citizens, from whom they nonetheless appeared to gain widespread sympathy. The populism of the cause, the suffocating effect of the blockade and the continuity provided to an already running news story led to widespread exposure in the media, which subsequently became a major driver of events in its own right, successfully turning the pressure on to the government which suffered a major slump in its ratings as it appeared to be caught by surprise. At the same time, the protests gained effect largely through the power of perception, with little actual physical intimidation or halting of vehicles – the role of the oil companies and of tanker drivers remaining enigmatic as deliveries were in effect voluntarily ceased.

Yet, as the fuel protests gained strength, so the political frailties of Farmers for Action and their allies became apparent. There was no clear agreement on desired outcomes and no clear endgame strategy. As the government wobbled, the protesters appeared unable to seize the initiative and drive home their advantage. Differences emerged between farmers and the more militant hauliers, as well as between the leaders of FFA, notably Brynle Williams and David Handley, about how far the protesters would be prepared to go. Underlying these tensions remained both the political inexperience of the demonstrators, and their inherent conservatism. Williams had, according to his own description of events, set out a moderate course at the beginning of the protests, telling demonstrators before the blockade of Stanlow:

> You have to remember what you are getting yourself into: you are not challenging a commercial company. You are challenging the Government. Be under no apprehension. This is a very, very serious thing that you do....... Emergency services get total priority. You don't hinder them in any shape or form. And the first senseless act of violence and that's it. It's over. You are all businessmen in your own right and hopefully gentlemen too, and will act accordingly (Brynle Williams, quoted in Treneman, 2000, p. 4).

As the effects of the protests grew, this wariness mushroomed into unease not that the protests may fail, but that they may achieve too much. It was thus as reluctant revolutionaries that the protest leaders decided to prematurely end the blockades in favour of setting a sixty day ultimatum to the government to cut fuel tax. As Williams again explains,

It was in the interest of common sense and decency. We weren't there to topple a government. We were there to demonstrate our problems. And we walked away. We walked away as businessmen. Not as bloody hooligans. Not as vandals. Not as anarchists. Not as leftists, rightists, centrists, whatever (Brynle Williams, quoted in Treneman, 2000, p. 4).

However, in capitulating when they did the fuel protesters wasted their main weapon, that of momentum. The government allowed the sixty day ultimatum to pass without meeting the protesters' demands and did not suffer any sanction. Attempts to revive the protest - led mainly by hauliers - failed to affect oil supply or receive public backing. Farmers for Action as an organization, though, remains significant having achieved some sort of victory over the established farming unions in its ability to draw attention to its cause, thus positioning itself as a focal point for future militant direct action.

Case 3 – Confédération paysanne

The French *Confédération paysanne* differs from the two British cases above in its ideological rationale, in the sophistication of its actions, and in the political socialization of its followers. The *Confédération paysanne* builds on an established history of agricultural militancy and direct action in France often undertaken with at least the tacit condonement of the 'all-powerful' FNSEA farming union. More generally, however, the FNSEA had operated through corporatist and policy community arrangements in which the union itself had acted as an agent of productivist policy (see Boussard, 1990). Whilst this benefited larger farmers, many smaller farmers felt disadvantaged by productivism and free trade, and came into confrontation with the FNSEA:

> I have a fundamental disagreement with the FNSEA, which defends the logic of productivism. For them, France should feed the world and to realize this ambition, it needs to conquer the markets. That conception is the antipode of ours, of the *Confédération paysanne* (José Bové, in Ariès and Terras, 2000, p. 29 (my translation)).

Tensions arose in the late 1960s and 1970s between the FNSEA and regional farmers' movements in Brittany and the Loire region, and eventually led to the formation of *Confédération nationale des syndicats de travailleurs paysans* (CNSTP) (National Confederation of Unions of Rural Workers) in 1981 (Bové and Dufour, 2001). Unlike Britain, French farmers have not traditionally been homogeneously associated with conservative politics, and the militant farmers in the breakaway unions represented a strain of left-wing agrarian politics. As such they formed links with left-wing ecological campaigners such as José Bové, who had first come to the rural Larzac region as an anti-militarism protester working with local farmers in opposing the establishment of a military base.

In 1987 the splintered CNSTP re-grouped as the *Confédération paysanne* under the leadership of Bové. Initially its primary objective was to challenge the productivist hegemony by competing with the FNSEA in representing farmers' interests. As such, in 1989 it obtained 18 per cent of the vote in elections to the regional Chambers of Agriculture which are charged with administering aspects of agricultural policy, placing the system under pressure when the FNSEA refused to sit on bodies with it (Naylor, 1994). The *Confédération paysanne* subsequently emerged as a leading agitator in farmers' protests against the 1992 GATT agreement on agricultural trade, participating in demonstrations at McDonalds restaurants and Coca Cola bottling plants, the use of political graffiti, and a blockade of Paris in alliance with the CNJA (Young Farmers) and *Co-ordination rurale* in September 1993 (Naylor, 1994). By the end of the decade the *Confédération paysanne* had consolidated its position, taking 19 per cent of the vote in the 1999 Chambers of Agriculture elections, and claiming over 45,000 members.

Significantly, one of the grounds on which the *Confédération paysanne* differentiated itself from the FNSEA was by emphasizing that it had a duty to wider society beyond individual farmers as traditional agriculture was central to rural life:

> Our job is not only to produce; we live on the land, we look after it and we are part of a rural social network (François Dufour, General Secretary of the *Confédération paysanne*, in Bové and Dufour, 2001, p. 48).

As well as the clear differences in organization and pedigree, the *Confédération paysanne* is also distinguished from militant British rural groups in three respects. First, it's actions are underpinned by an unashamedly left-wing political ideology, marking both the greater consciousness of political philosophy in France and the influence on the *Confédération paysanne* of intellectuals such as Bové. Bové is keen in interviews and writing to provide an intellectual justification of the group's actions, frequently quoting Marxist, anarchist and syndicalist thinkers, and explicitly positioning the *Confédération paysanne* as part of both the workers' struggle and resistance to globalization, making reference to groups such as Brazilian peasant farmers with which he identifies (see Ariès and Terras, 2000). Second, the strategies of the *Confédération paysanne* are shaped by the political experience of its leaders, including both those who have participated in sporadic farmers' protests since the 1950s, and Bové's own experience of anti-military protests and involvement with Greenpeace's operation to sabotage French nuclear testing in the Pacific in 1995. Third, this political experience has developed a highly attuned understanding of the media and of the importance of *symbolic actions* in contemporary protest. Much of this has focused on Bové himself, who has played up his characterization as '*une sorte d'Astérix des temps modernes*' (Ariès and Terras, 2000, p. 8).

The melding together of these elements can be seen in the *Confédération paysanne's* high-profile campaign against '*malbouffe*', or 'rotten food'. The selection of the campaign focus is itself significant – *malbouffe* symbolizing the threat posed to traditional rural values by modern, globalized culture, and implicitly linking the self-interests of small farmers with wider environmental and consumerist agendas. The campaign was advanced most notably by the occupation and 'dismantling' of the McDonalds restaurant in Millau on 12 August 1999. The motivation for the action was the decision by the United States to impose a surcharge on imports of Roquefort cheese as part of a trade war with the European Union, but the selection of McDonalds as a symbol of American culture corresponded with a wider set of objectives:

> We thought McDonalds appropriate for several reasons: the type of food at McDo, which is industrial food requiring industrial agriculture (meat as cheap as possible, one type of potato for all McDonalds worldwide, and three or four varieties of salad). Everything is standardized. It is a multinational firm with a wish of hegemony. These elements show well that its is a target which corresponds to opposition to globalization. The decision was to dismantle the McDo under construction in Millau (José Bové in Ariès and Terras, 2000, p. 74 (my translation)).

The targeting of McDonalds also resonated with demonstrations at McDonalds restaurants as part of the anti-GATT protests in the early 1990s. But whilst the earlier protests had involved placard waving and US flag burning (Naylor, 1994), the Millau action went further, physically dismantling the building. This act of 'criminal damage' was deliberate, calculated and publicized:

> Three days before the action, we announced it to the press and appealed for a public demonstration at McDonalds. I had first contact with the police and established contact with the *sous-préfecture* [local government] (Bové in Ariès and Terras, 2000, p. 74 (my translation)).

As such, the point of the action was not so much the damage sustained by the restaurant, but rather the securing of arrests which allowed the campaign to move to its climax. Almost a year later, in June 2000, Bové and his co-defendants appeared on trial in Millau and proceeded to exploit the platform it provided to their own ends. Bové called in his defence a series of witnesses including academics, tribal leaders and peasant farmers from three continents who provided 'evidence' of the effects of globalization and inverted the trial into a trial of McDonalds. With an estimated 100,000 well-wishers attracted to a concurrent 'anti-globalization festival' in the town, with concerts by big name French pop stars and addresses from actors and environmental campaigners, the trial became a means by which the *'malbouffe'* campaign was projected to a global audience (Bové and Dufour, 2001). Bové was nevertheless convicted and imprisoned. In an act rich in symbolism he made the journey to jail by tractor.

Conclusion

The three case studies discussed above illustrate both the contingencies and the commonalities of contemporary rural protests. In each of the three cases, the narrative has been shaped by particular decisions that have been taken in response to specific circumstances. These include decisions about who to target, about which tactics to adopt, and about how far protesters are prepared to go. Some of these issues have reflected the significance of national context, which, along with the sub-national regional context, is important in influencing the political socialization of participants, and in determining the institutions with which they must engage. At the same time, the three groups have all emerged in response to broad processes of rural restructuring, and all share concerns with the meaning or regulation of rurality and perceived threats to rural life, culture, landscape or identity – even if their understanding of 'rural' differs.

It was noted earlier that conventional frameworks for the analysis of rural politics, including concepts such as policy communities and insider and outsider pressure groups, are not sufficient for research on the new politics of the rural. Models of policy communities and policy networks can still help to interpret the interface between rural interest groups and the state; the categories of insider and outsider groups can assist in illuminating the trajectory followed by some groups such as the Countryside Alliance; but neither approach provides guidance on issues like the mobilization of grassroots protesters, the motivations for participation, the development of organizational form and the selection of tactics. The social network approach, as advanced in this article, offers one alternative framework that could accommodate this broader research agenda. Its attraction lies in the emphasis placed on issues of identity, mobilization, organizational form and strategy, as well as its understanding of social movements as segmented, multi-centred, reticular and fluid entities (see also Woods, 2003). However, the social movement approach does not provide a comprehensive, totalising theory for rural political analysis, rather it is one interpretative tool that might be drawn upon alongside others, including theories of citizenship, participation, identity politics and the politics of resistance. Neither should it be anticipated that the social movement perspective will be a 'perfect fit' when applied to the politics of the rural. There are serious questions that need to be explored and debated further, including whether the plethora of rural protests are most accurately represented as a single 'rural movement', or as an 'emergent rural movement', or as a collection of discrete 'rural movements'; whether it is possible to make transnational connections between rural protests, or whether they are resolutely rooted in national contexts; and the extent to which the contestation of 'rural identity' can be positioned as a primary motivation for mobilization, as opposed to other material and ideological factors.

The exploration of these issues forms one potential avenue for further research. Additionally, there are a large number of foci for empirical research that are highlighted by the social movement perspective, but whose study does not necessarily require its application – of which six can be flagged up here. Firstly, studies are needed to examine the conditions of emergence of rural protest campaigns, identifying the trigger events and understanding the micro-processes by which individual discontent is translated into collective action. Secondly, research is required on the participants in rural protests, examining their backgrounds and characteristics, their motivations for involvement, and their perceptions of rurality and the connections made between their 'rural identity' and their political action. Thirdly, the organizational form and dynamics of the rural movement and its component groups needs to be explored, investigating for example, the selection of particular structures, the significance of individual leaders, the nature and practice of leadership, the mechanisms for decision-making, the points of internal tension, and the relationships between different rural campaign groups. Fourthly, work is required on the selection of campaign strategies and tactics, including the rationalization of different modes of engagement from electoral participation to direct action. Fifthly, analysis could be undertaken of the rhetoric, language and representational strategies employed by the rural movement and the ways in which its actions and objectives are made meaningful both to its participants and to the wider public. Finally, the effectiveness of the rural movement might be evaluated, considering its impact on the policy process and on public opinion, and the implications for broader politics.

Rural protests and campaigns are likely to remain a feature of the political landscape in the early twentieth century. The trend of rural restructuring is continuing to disrupt established discursive constructions and expectations of the rural, heightening a perception of beleaguerment amongst those who chose to identify themselves as 'rural'; whilst some major policy flash-points, such as the reform of the EU Common Agricultural Policy, are only beginning to take shape. The emergent 'politics of the rural' will have implications across rural culture and society – producing new structures of social organization and leadership; politicising and contesting cultural events and traditions; and shaping the future development of rural policy. As such, the analysis of the 'rural movement' of campaigning and protest groups engaged in perceived 'defence' of the rural (and of those organizations and institutions against which they direct their campaigns) has much to contribute to our understanding of the contemporary countryside and to the development of new agendas for rural research.

References

Ariès, P and Terras, C. (2000), *José Bové: la Révolte d'un Paysan*, Éditions Golais, Villeurbanne, France.

Boussard, I. (1990), 'French Political Science and Rural Problems', in P. Lowe and M. Bodiguel (eds), *Rural Studies in Britain and France*, Belhaven, London, pp. 269-85.

Bové, J. and Dufour, F. (2001), *The World Is Not For Sale*, Verso, London and New York.

Browne, W.P. (2001), *The Failure of National Rural Policy*, Georgetown University Press, Washington D.C.

Bryant, B. (1996), *Twyford Down: Roads, Campaigning and Environmental Law*, Spon, London.

Carter, H. (1998), 'Morale Dilemmas', *The Guardian, Society Supplement*, 12 August, pp. 2-3.

Cox, G. and Winter, M. (1997), 'The Beleaguered "Other": Huntsmen in the Countryside', in P. Milbourne (ed), *Revealing Rural 'Others': Diverse Voices in the Countryside*, Countryside and Community Press, Cheltenham, pp. 75-89.

Deleuze, G., and Guattari, F. (1988), *A Thousand Plateaus*, Athlone Press, London.

della Porta, D. and Diani, M. (1999), *Social Movements: an Introduction*, Blackwell, Oxford.

Diani, M. (1992), 'The Concept of Social Movement', *The Sociological Review*, vol 40, pp. 1-25.

Doherty, B., Paterson, M., Plows, A. and Wall, D. (2003), 'Explaining the fuel protests', *British Journal of Politics and International Relations*, Vol. 5, pp. 1-23.

Dunn, P. (1991), 'Villagers Bugged by Parish-pump "Plot"', *The Independent*, 20 April, p. 3.

Dyer, J. (1998), *Harvest of Rage*, Westview, Boulder, CO.

Fainstein, S. S. and Hirst, C. (1995), 'Urban Social Movements', in D. Judge, G. Stoker and H. Wolman (eds), *Theories of Urban Politics*, Sage, London, pp 181-204.

George, J. (1999), *A Rural Uprising? The Battle to Save Hunting with Hounds*. Allen & Unwin, London.

Gerlach, L. (1976), 'La Struttura dei Nuovi Movimenti di Rivolta', in A. Melucci (ed), *Movimenti di Rivolta*, Etas, Milan., pp 218-232.

Grant, W. (1990), 'Rural politics in Britain', in P. Lowe and M. Bodiguel (eds), *Rural Studies in Britain and France*, Belhaven, London, pp. 286-98.

Grant, W. (2000), *Pressure Groups and British Politics*, Macmillan, Basingstoke.

Hart Davis, D. (1997), *When The Country Went To Town*, Excellent Press, Ludlow, UK.

Henley J. (1997), 'Militant Brits Lead Fight to Save a French Haven', *The Guardian*, 5 November.

Jackson, J. (2002), 'We Would Support the Lawbreakers', *The Guardian*, 23 October, p. 20.

Lowe, P., Cox, G., MacEwan, M., O'Riordan, T. and Winter, M. (1986), *Countryside Conflicts*, Gower, Aldershot, UK.

Merrick (1996), *Battle for the Trees*, Godhaven Ink, Leeds.

Mitchell, J. V. and Dolun, M. (2001), *The Fuel Tax Protests in Europe, 2000-2001*, Royal Institute of International Affairs, London.

Mormont, M. (1987), 'The emergence of rural struggles and their ideological effects', *International Journal of Urban and Regional Research*, Vol. 7, pp. 559-78.

Mormont, M. (1990), '"What is Rural?" or "How to be Rural": Towards a Sociology of the Rural', in T. Marsden, P. Lowe and S. Whatmore (eds), *Rural Restructuring*, David Fulton, London, pp. 21-44.

Murdoch, J. and Marsden, T. (1994), *Reconstituting Rurality*, UCL Press, London.

Naylor, E. (1994), 'Unionism, Peasant Protest and the Reform of French Agriculture', *Journal of Rural Studies*, Vol. 10, pp. 263-73.

Offe, C. (1985), 'New Social Movements: Changing Boundaries of the Political', *Social Research*, Vol. 52, pp. 817-68.

Parker, G. (2002), *Citizenships, Contingency and the Countryside*, Routledge, London.

Routledge, P. (1997), 'The Imagineering of Resistance: Pollock Free State and the Practice of Postmodern Politics', *Transactions, Institute of British Geographers*, Vol. 22, pp. 359-76.

Spain, D. (1993), 'Been-heres Versus Come-heres: Negotiating Conflicting Community Identities', *Journal of the American Planning Association*, Vol. 59, pp. 156-171.

Stock, C. M. (1996), *Rural Radicals: Righteous Rage in the American Grain*, Cornell University Press, Ithaca.

Treneman, A. (2000), 'The Maverick who Rattled Tony Blair', *The Times, section 2*, 26 September, pp. 3-5.

Viard, J. (2000), 'José Bové, Pont Entre le Rural et l'Urbain', *Libération*, 30 June, p. 5.

Vidal, J. (1997), 'Nobody Likes a Fox More Than Me...', *The Guardian*, 5 July, p. 3.

Walton, E. (2002), 'They Can't Lock Us All Up, Can They?', *The Field*, August, pp. 48-52.

Winter, M. (1996), *Rural Politics*, Routledge, London.

Woods, M. (1997), 'Discourses of Power and Rurality', *Political Geography*, Vol. 16, pp. 453-478.

Woods, M. (1998a), 'Advocating rurality? The Repositioning of Rural Local Government', *Journal of Rural Studies*, Vol. 14, pp. 13-26.

Woods, M. (1998b), 'Researching Rural Conflicts: Hunting, Local Politics and Actor-networks', *Journal of Rural Studies*, Vol. 14, pp. 321-40.

Woods, M. (2003), 'Deconstructing Rural Protest: the Emergence of a New Social Movement', *Journal of Rural Studies*, Vol. 19, pp. 309-325.

Chapter 7

Geographies of Invisibility: The 'Hidden' Lives of Rural Lone Parents

Annie Hughes

Introduction

Increasingly academic and policy research has focused on the changing nature and extent of lone parenthood in the UK. This is unsurprising given that the rising rate of lone parenthood in the UK is one of the most visible outcomes of changing patterns of family formation and dissolution over the past thirty years (Millar, 2000). There are now an estimated 1.75 million lone parent families in the UK, a trebling since 1971 (NCOPF, 2003). Lone parent households represent one in four families with dependent children (General Household Survey, 2000), or three million children nationwide (Gingerbread, 2000). It has been estimated that one-third to one half of children will spend some time in a one-parent family during their childhood.

There is a growing body of literature which focuses on the lives of lone parents, or more particularly lone mothers[1] (Silva, 1996; Ford and Millar, 1998; Kiernan et al, 1998; Rowlington and McKay, 2002). Attention has been focused on three key areas; the (under)employment of lone parents (Bryson et al, 1997; Finlayson and Marsh, 1998; Ford et al, 1998; Marsh et al, 1997), the link between lone parenthood, poverty and social exclusion (Millar, 1992; Rowlingson and McKay, 2002) and State Social Policies as related to lone parents (Evason and Robinson, 1998; Millar, 2000; Rake, 2001; van Drenth et al, 1999). This literature acknowledges the diversity and dynamics within the category lone parent, uncovering the multi-faceted nature of this phenomenon.

Although these studies have contributed substantially to our knowledge of the lives and difficulties of lone parents, they can be criticized for their failure to explore the *spatiality* of lone parenthood. Existing work which has taken a geographical perspective has primarily focused on spatial differences in lone parent experience at the national scale, considering particularly cross-national comparisons in employment-related and welfare policies (see for example Millar and Rowlingson, 2001). However, at a more local level, the *contexts* within which lone parents make decisions about their lives have been largely ignored. This omission has led Simon Duncan and Rosalind Edwards, in their book *Lone Mothers, Paid Work and Gendered Moral Rationalities*, to identify what they term the 'spatial deficit' of research into lone parenthood. They argue that previous work has tended to 'downplay...the differences...in the contexts in which (lone parents) live' (p. 8). They point to the need to redress the imbalance and provide a sense of 'place' in research on lone motherhood (p. 12).

While there is a dearth of research on the geographies of lone parents (see Duncan and Edwards, 1997, 1999; Speak, 2000; Winchester, 1990, for notable exceptions), there is even less work on the manifestations of *rural* lone parenthood. Academics who have detailed the growth of lone parenthood and the experience of lone parents have based their research and writings on urban locations. Similarly, voluntary organizations such as the One Parent Families (formerly known as the National Council for One Parent Families) and Gingerbread have remained relatively silent on the particular difficulties experienced by rural lone parents. It is clear that within political, policy and media discourses the lone parent 'problem' has been firmly set within an urban landscape. There is little doubt that lone parenthood has been linked with symbols of urban spaces, particularly the rundown council estate, criminality and the urban underclass. It is argued here that by setting the lone parent problem in an urban landscape, other spaces are excluded from consideration. In this way, spaces such as the village, the farm and the country cottage are precluded and lone parents and their particular needs become invisible (or at least less visible) in rural spaces.

There are several obvious reasons for an urban emphasis to lone parent research. The limited work which has explored the spatial distribution of lone parent families has pointed to the fact that they are concentrated, in the main, in inner London, large cities such as Birmingham, Manchester and Liverpool and in depressed, industrial, mainly Northern towns such as Middlesborough, Sunderland and Hartlepool (Forrest and Gordon, 1993; Haskey, 1994). Gordon and Forrest (1995) state quite categorically 'the message is clear – lone parents and cohabitee families live in cities. The traditional family survives most in remoter rural areas and in the commuter belt' (p. 47). Nevertheless, the proportion of lone parent households is rising in *all* areas. Rural areas are no exception. Recent trends from the 2001 UK population census show that 4.8 per cent of all rural households are headed by a lone parent with dependent children, compared with 6.4 per cent of all households nation-wide. There are now over two hundred and eighty thousand lone parent households (with dependent children) in rural districts. In fact, rates of lone parenthood are rising slightly faster in rural districts than compared to the national average; 120 per cent in rural areas since 1991 compared to 89 per cent nation-wide.

Notwithstanding these statistics, this chapter contends that the growth in lone parenthood has been and continues to be viewed as an urban concern. It draws on the conceptual ideas adopted by Paul Cloke, Paul Milbourne and Rebecca Widdowfield in their work on rural homelessness regarding the 'out-of-placeness' of social problems in the countryside (Cloke et al, 2000a, 2000b, 2001). Cloke and his co-workers employ the concept of socio-spatial dialectics to argue that

> certain spatial arrangements and cultural forms can, together, produce some accepted codes of what are, or are not, acceptable practices in particular places. (Cloke et al, 2000a, p. 725)

In so doing, they point to the 'non-coupling' of the socio-spatial arena of 'the rural' and the social problem of homelessness, arguing that failure to 'couple' together these two discursive constructs has led to rural homelessness being denied as a set of 'recognizable and acceptable problems which demands political and policy responses' (Cloke et al, 2000a, p. 715). In this chapter, three lines of argument are pursued to explain why lone parenthood has been, and continues to be, viewed as an urban issue. The first involves the ways in which the discursive concept of 'the lone parent' has been constructed in recent political and popular discourses. In particular, it is argued that lone parenthood has been constructed as problematic at every level. In particular, it has become directly associated with poverty, welfare dependency, delinquency and bad mothering. Furthermore, debate has embodied spatialized representations of lone parenthood which set this 'problem' in an urban landscape. This is due to a great extent to the links drawn between the growth in lone parenthood and the development of the so-called 'urban underclass'. Secondly, the chapter recognizes how socio-cultural constructs of rural life render lone parenthood less visible than in urban areas by embodying a construction of rurality which draws on bucolic and problem free lifestyles where traditional values are maintained and respected. Rurality has come to be viewed as a natural space of home and community, at a distance (both physical and metaphorical) from the immoral and deviant spaces of the city. The problem of the delinquent, immoral and welfare-dependent lone parent sits uneasily with the bucolic notions of rural idyll. As such, this chapter draws on recent discussions in rural studies concerning the ways in which popular ideals of rurality can, in fact, conceal social issues such as poverty, social exclusion, and homelessness in the countryside (Cloke and Little, 1997; Milbourne, 1997). Finally, it is argued that certain characteristics of rural areas make it less easy for a lone parent to move to, or remain in, rural locales thereby reducing the absolute numbers of lone parents in rural areas. This is due to the lack of social housing, the lack of conveniently located support services such as childcare facilities and social care services, and the lack of flexible and convenient employment. The chapter concludes by pointing to the need to 're-couple', in the words of Cloke et al (2000a), lone parenthood with the discursive category rural. In this way, I hope to highlight the need for social policies which take account of changing family forms, particularly familial dissolution, in rural localities and present an understanding of rurality which steps beyond bucolic notions of rural life. Furthermore, and in so doing, I wish to expose the ways in which the lived realities of lone parents are not only highly dependent on their social circumstances, but also on their spatial contexts.

Evidence is drawn from a pilot study undertaken in the summer of 2001 in which thirty-one lone parents from villages in North Norfolk were interviewed. The interviews were conducted in the lone parents' own homes and lasted between one and three hours. Interviewees were contacted after they completed questionnaires distributed by local schools on the researcher's behalf. These interviews provide detailed experiential data and uncover the differing contexts within which rural lone parents live their lives. Firstly, however, I wish to turn attention to the ways in which the social issue of lone parenthood has been constructed in Britain through the 1990s and into the twenty first century.

Conceptualizations of Lone Parenthood in Britain

Motherhood is socially constructed not biologically inscribed. The existence of historical, social and geographical variations in what it is to be a mother confirms this. However, as Glenn et al (1994) note, one very particular construction of motherhood has come to dominate contemporary discourse.

> Mothering is a contested terrain in the 1990s. In fact, it has always been contested terrain. However, a particular definition of mothering has so dominated popular media representations, academic discourse, and political and legal doctrine that the existence of alternative beliefs and practices...has gone unnoticed. (pp. 2-3)

This particular definition of motherhood which has been constructed as universal is epitomized by the idealized model of motherhood, derived from white, middle class English society. In this model, good mothering involves the biological mother bringing up her children during their formative years with the support (both financial and emotional) of the children's biological father. Paralleling the construction of 'good mothering' has been the normalization of the nuclear family as the most appropriate place to bring up children. While Chambers (2001) points out that currently people are living in complex configurations of living arrangements, it is clear that 'the modern nuclear family *does* exist and is flourishing as an ideal: as a symbol, discourse and powerful myth within the collective imagination' (p.1, original emphasis).

Undoubtedly, as we move further into the twenty-first century there has been an increased entrenchment of ideological positions in British political discourses about what families should be like and should do. Both Labour and Conservative governments have spoken in favour of supporting 'the family'[2]. In addition, 'the family' has been heralded as congruent with society's social and moral health, being both the source and the saviour of society's ills (Chambers, 2001). However, it is clear that the 'family' to which politicians refer is the 'traditional' nuclear family - two married parents and their children. The nuclear family is the only version of 'the family' that the government will support and actively reproduce. Certainly, recent government policy has moved away from giving additional support to lone parents, removing benefits such as the lone parent premium on new child benefit claims (Smith, 1999). Tony Blair, for example, sets up a dichotomy where 'bad families' threaten 'good families' and where good families must be protected to 'gain strength to save the nation' (Chalmers, 2001, p. 8). To be clear then, the dominance of the nuclear family form is strengthening in political circles, and forms the 'real' family in British society and politics (Reinhold, 1994).

It follows then that like motherhood and 'the family', lone motherhood is also a discursive concept. As Wallbank (2001) suggests, 'the [re]constitution of the lone mother occurs in a kaleidoscopic network of legal, academic, media and state discourses of motherhood and "the family"' (p. 17). This is a perpetual and dynamic process with the conceptualization of lone parenthood continually being constituted, reproduced, changed and contested. For example, Song (1996) explores the conceptualizations of lone parents from the 1960s to the present day, pointing to the changing construction of the social and moral 'problem' of lone parenthood (see Brush, 1997, for a US case-study). Specifically, she points to the changing public and political concerns central to the conceptualization of lone parenthood, from illegitimacy in the 1960s, through a concern for the poverty of lone parents in the 1970s, to concerns about 'dependency culture', family breakdown and criminality in the 1990s. As such lone parenthood is not a neutral category, but a construction suffused with political and moral evaluations (Silva, 1996).

There is an increasing body of literature which explores the *sociality* of lone parenthood and uncovers how lone parents are constituted in and by complex networks of power relations (Wallbank, 2001). This literature has pointed to the fact that in contemporary media, policy and political discourse, the lone parent family has been constructed in opposition to the 'real' family form discussed above. If the nuclear family represents the 'good' and 'proper' family then the lone parent family is the quintessentially 'bad family'. The dramatic increase in lone parent families is viewed as concomitant with the disintegration of the traditional family form and the breakdown of society's moral framework. In other words, lone parents have been held responsible for the decline in the traditional nuclear family. As Wallbank (2001) notes, lone mothers have been seen as 'engineering the expulsion of fathers from the home for the sake of their independence' (p. 3). Lone parents or 'single mums' as they are more commonly known have not only been vilified as a social and moral threat to society, but also as a financial burden. As Song points out:

> (t)he current emphasis...resuscitates moral discourses around putatively feckless and irresponsible 'single mums' and 'absent' fathers, who are contributing not only to family breakdown, but also to taxpayers' burdens. (1996, p. 392)

Rather than framing the poverty experienced by lone parent families in terms of structural causes such as inadequate childcare and low wages, single mothers have been depicted as being responsible for their own situation (Atkinson et al, 1998). The popular caricature of a lone parent constructed throughout the 1990s is one of a feckless teenager, wedded-to-welfare, intentionally getting pregnant to jump the housing queue. This representation of the single parent was actively reproduced through the popular media, political rhetoric and government policy throughout the 1990s and into the twenty first century. Political discourses, particularly but not exclusively New Right Policies[3] have been particularly vitriolic in their treatment of lone motherhood. For example, increasingly during the 1990s, Tory strategists adopted the ideologies of American New Right commentators concerning the growth of an underclass defined by their behaviour (Murray, 1990). This so-called 'underclass theory' posited that a new class of people was developing who had very little stake in the social order, were welfare dependant and were the major source of crime, deviancy and social breakdown. Proponents of such a thesis, such as the American social scientist and writer Charles Murray, pointed to the rise of 'illegitimacy' as 'the best predicator of an underclass in the making' (Murray, 1990, p. 4). Although Murray recognizes that not all single parents are in equal circumstances, and as such not all are responsible for the growth in the underclass, he suggests that *all* single-parenthood is more problematic than nuclear families;

> single-parenthood is a problem for communities...illegitimacy is the most worrisome aspect of single-parenthood (Murray, 1990, p. 11).

In this way, lone mothers have been constructed as active agents in the creation of an under-class (Murray, 1990, 1994). Single parents are condemned and demonized as undeserving 'scroungers' responsible for the breakdown of traditional family values, the increase in crime and the 'benefits crisis' (Atkinson, 1998). By portraying single parents as promiscuous, uneducated and poor, neglectful mothers, the underlying structures which perpetuate poverty could be overlooked and their poverty presented as an individual problem rather than a societal issue rooted in economic and social inequality (Bullock et al, 2001). Such a thesis has become ubiquitous and frequently espoused by the popular media. Both broad sheets and the tabloid press have run stories about the 'moral panic' ensuing from the rapid increase in single parent families.

> The creation of an urban underclass, on the margins of society, but doing great damage to itself and the rest of us, is directly linked to the rapid rise in illegitimacy (Editorial, Sunday Times, 28 February 1993).

As the 1990s have given way to a new Millennium with a new political orthodoxy, the political response to lone parents has remained negative. While playing down the New Right's view that lone parents are a social threat, the New Labour construction remains aligned to the dependent nature of this group and continues, to some extent, to purport its character as an emerging underclass (Smith, 1999).

As in the US, there is little doubt that the underclass in the UK is perceived to be an urban problem[4]. Murray (1994) talks about the increasingly spatial segregation between two-parent working class families and the (one-parent) under-class. The underclass, he argues, will become more concentrated in council-owned sink estates in or on the edge of cities and large towns. He continues:

> Two-parent working-class families will increasingly leave council-housing, and council housing will be the place where the underclass congregates (a process that is already well-advanced in many cities)...The rich will tend to seek areas that are not only physically distant from *the inner-city*, but defensible. (1994, p. 115, emphasis added)

By defining lone parents as the perpetrators of, or at the very least contributors to, the growing underclass, lone parenthood has been linked to symbols of urban spatiality and firmly set in an urban landscape. Central to the 'urban spatialization' of the lone parent 'problem' is the perceived causal link between the provision of council housing and increasing levels of lone parenthood.

Comments espoused by political commentators and the press clearly give the impression that the spatiality of lone parenthood is encompassed in urban estates. For example, Conservative MP Gerry Malone was quoted as saying 'in some urban estates the sighting of a father is as common as spotting the dodo' (The Guardian, 5 May 1993, p12). Similarly, John Redwood, in the summer of 1993, arguably the first politician to quite literally 'place' the 'problem' of lone parents on the political agenda, used the example of the St Mellons Estate in Cardiff to substantiate his claims concerning the menace of lone parenthood and the growth of a welfare dependent urban under-class. He argued that over 50 per cent of the households on this urban estate were headed by lone parents. The perceived causal link between the increasing number of lone parents and the provision of council housing was constructed in such a forceful way as to incite the Housing Minister, Sir George Young, to review the obligation on local authorities to give young mothers council accommodation as a solution to the lone parent problem[5]. The popular media also reinforced these dominant constructions of lone parenthood and their associated urban spatialities. Intuitive connections between lone parenthood and urban spaces have been reinforced in newspaper articles and television and radio programmes. Whilst lobby groups working on behalf on lone parents, such as Gingerbread and One Parent Families, have worked exceedingly hard to dispel the stereotypical images of lone parents, they have systematically failed to challenge the 'urban myth' of lone parenthood. Indeed, in interviews with these lobbying organizations in 2001 none were addressing the particular concerns of their rural membership.

Lone parenthood, then, has been strongly associated with the under-class. Single parents as the active perpetrators of the under-class phenomenon have become associated with urban areas and carry with them 'symbols of urban spatiality', particularly the rundown council estate, criminality and violence. Lone parenthood has been normalized as an urban problem and intuitive connections have been formalized in the minds of many people which place lone parent families in urban settings. Moreover, as Cloke et al (2000) argue in the context of homelessness, setting a problem in a particular type of landscape 'serve(s), by implication, to exclude other spaces from consideration' (p. 715). As such, the association between lone parenthood and rurality becomes 'unnatural' and violates the popular representation of this social group. In the next section, I argue that it is not only the popular conceptualizations of lone parenthood which act to render rural lone parenthood less visible, but also socio-cultural constructions of rurality which contribute to the assumption that lone parenthood is an urban phenomenon.

Socio-Constructs of Rural Life

Rural geographers' recent concerns with socio-cultural issues have heralded a focus on two key issues. The first has emphasized how some individuals and groups have been marginalized in academic and policy writings on the rural, drawing attention to a range of hidden others in rural areas (Cloke and Little, 1997; Milbourne, 1997). The second has begun to unpack constructions of rurality, and investigate how the key meanings attached to the rural are capable of concealing and/or excluding so-called 'others' in rural society (Cloke, 1997). Research has pointed to the ways in which representations of rurality serve as 'devices of exclusion and marginalization' obscuring the experiences of social groups with less idyllic experiences of the countryside (Cloke and Little, 1997). Whilst the emphasis on hidden others has produced some excellent research on the experiences of some social groups whose lives have been neglected within mainstream rural studies, this work has failed to explore the experiences of lone parents.

Recent research has pointed to the existence of dominant socio-cultural constructions of rurality associating rural England, at some level, with notions of idyll. While it is clear that a single construction of 'rural as idyll' is problematic, there are a number of key meanings that have become associated with rurality (Cloke and Milbourne, 1992). These include bucolic and problem free lifestyles where traditional values are maintained and respected. Rurality has come to be viewed as a natural space of home and community, at a distance (both physically and metaphorically) from the immoral and deviant city spaces. Indeed, there is an increasing body of research which points to the centrality of sets of discourses and practices concerned with domestic, familial and community relations to representations of rurality (Bell, 1994; Short, 1991). 'The family' has been constructed as particularly central to rural ideology and people's everyday understanding or imagined geographies of rurality (Bell, 1994; Hughes, 1997; Little, 1986, 1987; Little and Austin, 1996). However, the family of rural ideology is the nuclear family with a married couple bringing up their dependent children under the same roof[6]. There is very little room within these pastoral idylls for other, less 'traditional', family forms such as one-parent families or lesbian and gay families. Indeed, Murdoch and Day (1998) argue that this particular version of the rural is currently being entrenched as the rural is being 'recast as social actors seek out new forms of natural and communal stability within the context of increasingly globalized forces and relations.' (p187).

It has been extensively argued that these socio-cultural constructs of rurality exert a pervasive influence over the thoughts, practices and actions of urban and rural dwellers alike. They act as social codes which steer ideas of appropriate behaviour and acceptable practices in particular places. They guide what is deemed to be 'in place' or 'out of place' and become powerful mechanisms in the spatial differentiation of social issues. Recent commentators have argued that constructions of rurality which associate rural England with some form of pastoral idyll have led urbanites and rural dwellers to deny the existence of social issues such as poverty and homelessness in the countryside (see also Cloke, 1995; Woodward, 1996). Similar arguments can be used here to argue that certain symbols of the rural idyll, most notably the centrality of the family, close community and problem-free lifestyles, lead to the disconnection between rurality and the social problem of lone parenthood. This is particularly the case given that, as discussed above, lone parent families have been linked to poverty, deprivation and societal breakdown; problems and practices which fit uneasily with dominant meanings of rural places. These socio-cultural barriers, then, negate the possibility of the 'lone parent problem' existing in rural areas. This research points to the ways in which rural lone parents (and rural dwellers) denounce the existence of the 'lone parent problem'. They do so not by denying the existence of lone parent households in their communities, but by adopting strategies to distance themselves from the cohort of undesirable lone parent families.

It is clear from this research that some rural dwellers did not readily accept that there was any significant level of family dissolution, and as such lone parenthood, in their rural locality. For example, Hazel (all names have been changed), a Parish Councillor, distanced her community from such 'problems'.

> We don't have many problems like that around here. People try harder and work together more [and]....there are fewer temptations....that is more of a problem in the more urban areas.

Where family breakdown is accepted as part of life in rural areas, there was a clear sense that it was less 'problematic'. Samantha, an active community worker and herself a lone parent, comments:

> I think that this is such a nice environment and people round here are so caring that if anybody was really in the doldrums you would here about it. Somebody would say did you know that so-and-so was going through a rough patch....It is a fabulous place...I hope there is nobody I know who is really struggling and hasn't been able to say something. It is a very caring community being a small village.

Samantha is clearly drawing on romanticized notions of the 'rural' and 'placing' the notion of care in village communities. The implication of this being that any lone parents who do live in rural communities will be looked after and therefore deemed less of a problem.

Both rural lone parents and rural dwellers alike drew on the categories of 'urban' and 'rural' as meaningful in accounting for the different experiences of lone parents in these different spatial arenas. Moreover these categories appropriated added significance when the interviewees talked about the *nature* of lone parents in urban and rural areas. The distinctions between lone parents in urban areas and rural areas were clearly value laden, with urban lone parents being considered to be more undeserving, irresponsible and immoral than rural lone parents. Lone parents themselves actively distanced themselves from 'undesirable' lone parent families, labelling their situation as different to that of other lone parents. For example, Rebecca states 'the young girl getting pregnant to get a council house is more of a city thing. That doesn't happen here.' Sandra also alludes to this in her interview stating:

> When I lived in Norwich you would see teenage girls with pushchairs and I would think what an earth possessed you. They don't seem to think it through at all...The children get pushed from one grandparent to another and they don't have a good life...but that is in cities...the undesirables.

Sandra is clearly differentiating between herself and the 'otherness' of the undesirable lone parent which she locates firmly in the urban arena. Like Sandra, Ruth talked candidly about the different experiences of lone parents in rural and urban areas.

It is clear that rural lone parents' behaviour in and experiences of rural life were informed at some level by socio-cultural values attributed to rural life. The general consensus amongst the rural lone parents interviewed was that they were respected and treated fairly by their neighbours. Lone parents talked about the support that they received from the local community. Jenny comments 'I don't think that I am treated any differently...I get involved in a lot of things...People are good to me when I can't go to meetings and things cos I haven't got a babysitter or whatever.' There was a clear sense amongst the lone parents interviewed that their legitimacy in the village centred around their role as parent and that being partner-less made very little difference. Indeed, rural lone parents seemed to actively distance themselves from any notion that they were in any way different from other village parents. For example, Samantha comments:

> I have never really sat back and thought oh yes, I am a single parent. It is not really anything that has occurred to me. I guess the person you think of is the struggling mother with children...getting by on what they are given by the government each week. I am not sure that I am aware of that many people like that around here, and most of them are couples. I don't think that single parents are a in worse position than normal parents in this area.

The rural lone parents, then, felt that they were perceived to be 'less problematic' by their own communities than in urban areas. This was, in part, attributed to the fact that there were simply fewer of them. For example, Nancy states 'I think that it is worse in an urban area. Because there are more people in urban areas therefore there are more single parents and they are used as scapegoats.' Sandra also commented 'in urban areas there are often, usually council, areas where teenage single mums are numerous. Because of their concentration they give a bad impression.' Similarly, Jessica held this opinion arguing 'there are fewer groups of lone parents in rural areas to stereotype.'

However, while there are fewer lone parents in rural areas compared to urban areas, and as such they may be considered less visible, the social surveillance which ensues in settlements of smaller size inevitably shines a very public light on the experiences of lone parents in rural communities. However, it would seem from the viewpoints and opinions expressed during the interviews that this visibility experienced by lone parents, in fact contributes to their *in*visibility. Living in a small community acts to expose the realities of lone parenthood and villagers are made aware of their particular circumstances. As such, there is less room for stereotyping and prejudice. For example, Margaret felt that the visibility of being a lone parent in a rural area had worked to her advantage. When asked if she had felt any stigma from the local community she commented:

Everyone knew we were married here together and everyone knew that Colin had walked off and left me so they were very supportive and very kind. The men offered to cut the grass and someone came to mend the car so they have all done that sort of thing for me. Generally there is a stigma attached to being a single mum....The community steps in here and people see your right. I have not experienced any stigma for me or any of the few single mums that we have got at school. They are all treated as part of the community.

The stereotype of the lone parent as a feckless, welfare dependant, morally corrupt and inept parent (or more accurately, mother) are challenged and rural lone parents are distanced, in the minds of rural dwellers, from this 'problematic' social group. Interestingly, Jane talked from personal experience having become a lone parent twice, once in an urban location and then later in the Norfolk village where she now resides. She argues

It is a nice village. I don't feel judged and that is important. I don't feel alienated. I did the first time I was a lone parent [in an urban area]. I was made to feel very judged and labelled. One minute I was that nice Mrs so and so ...I do find people here are nosy but they are not invasive and they are friendly.... My social standing fell in (the urban location). I felt 'cheap'.

Rural dwellers and rural lone parents alike, then, have acted to deny the existence of 'lone parents' in rural areas. They have done so not by actively excluding the lone parent families in their communities, but rather by drawing on a conceptualization of lone parenthood which not only sits very uneasily with the bucolic constructions of rural life, but also which fails to reflect the lived realities of most lone parents living in Britain today. Here, I am talking about a conceptualization of lone parents which centres on aspects of welfare dependency, poverty, immorality and neglectful parenting. Rural dwellers have rendered rural lone parents invisible by integrating them into the rural community undermining the differences between their experiences and those of other members of the rural community. Lone parenthood is constructed by rural dwellers and rural lone parents themselves as an urban problem, with little resemblance to their situation.

So far this chapter has argued that the discursive categories 'lone parent' and 'the rural' have been constructed in such a way as to obscure rural lone parenthood as a recognizable problem. However, I also want to argue in this chapter that the structural characteristics of rural areas make it more difficult for rural lone parents to remain in, or to move to, rural areas thereby reducing their absolute numbers in rural locations, and thus reducing their visibility.

Structural Characteristics in Rural Areas

Generally, when parents (usually mothers) find themselves the sole carer for their children they experience a significant drop of income. For example, using the British Household Panel Survey (BHPS), Jarvis and Jenkins (1998) estimate that mothers and children suffer substantial declines in real net income after a marital split, with a median of 17 per cent decrease. In addition, post-separation, mothers become more reliant on state benefits and less reliant on their own (as well as their partners) income from paid employment. The reduction of income is particularly acute after marital dissolution given that married mothers are, in general, more financially dependent on their spouses than non-married or never-partnered women. This has particular significance for lone mothers in rural areas as the preliminary evidence from this research suggests that rural lone parents are more likely to have been married than lone parents residing in more urban locations. From this pilot study, 61 per cent of the lone parents were divorced, separated or widowed from a married partner. Only six per cent were never-partnered (compared to a national average of 15 per cent) and a further three per cent separated from a partner they were never married to.

Of course, these changes in income affect lone parents wherever they live, whether in metropolitan, urban or rural areas. However, the most visible outcome of a reduced household income is the requirement to move into cheaper, more often than not social, housing. In over half of cases in the UK (58 per cent), families are forced to move when they become the sole carers for their children, most commonly into local authority housing (OPF, 2003). Two-thirds of lone parents live in rented accommodation, often social housing, compared with only one-fifth of other families (ONS, 2002). The lack of publicly rented accommodation in many rural areas[7] necessitates lone parent families to move to nearby towns to secure accommodation for their families. This is further compounded by the fact that the rural housing market is dominated to a greater extent by owner-occupation (76 per cent of properties, compared to 69 per cent in England as a whole). As Table 7.1 highlights, only 34 per cent of lone parent families own their own home, compared to a staggering 79 per cent of other families.

Table 7.1 Housing tenure: profile by family type

Tenure	One Parent Families (%)	Other Families (%)
Owner-occupied, owned outright or with mortgage	34	79
Rented, social housing	52	15
Rented, private	15	6

Source: OPF, 2003

In addition to the dearth of social housing, rural areas also lack other service provision required by lone parents. Key difficulties centred around transport, particularly the lack of (convenient) public transport and the cost of private transport. The Countryside Agency estimates that 29 per cent of all rural settlements have no scheduled bus service at all. This lack of public transport means that lone parents, like other parents, need to rely on private transport to ferry their children to activities as well as to get them to and from work if applicable[8].

The associated costs of purchasing and maintaining a private car cut deep into the purse of rural lone parents. For example Jennifer argues 'a car is not a luxury, but it is not easy to budget for'. Margaret reiterates this by stating 'transport to and from children's activities means a car and the expense of running it is a necessity.' Several lone parents talked about their plans to move out of their local area due to transport difficulties. Indeed, Liz had been forced to move out of the small village she was living in prior to becoming a lone parent due to the fact that she couldn't drive and could not afford driving lessons or indeed a private car.

Lone parents in the survey not only discussed the day to day expense of living in the countryside, but also the difficulty of 'packing in' the complex daily tasks with the inflexibility of rural services. For example, Chris was planning to move into the nearby town not only because she found it expensive but also because she found that rural living was not conducive to being a working parent.

> I think that it is a lot more stressful... at the moment I spend £50 a week on petrol. I have never had so many repairs on my car...The out-goings on the car and fuel is enough to cause stress levels....[But] the hardest thing about being rural is [in]convenience. I used to be able to go to Tesco's at ten o'clock at night because we were 24 hour there. Now just to find a shop open on a Sunday is hard. Being a working parent it is difficult trying to get everything done. Banks everything....To go shopping is a three hour trip.... It is not convenient to be a working lone parent here.

In addition to rural areas being problematic residential locations for lone parents in terms of the organization of children's day-to-day activities, there are also severe limitations for lone parents wishing to return to work. This will only become more salient given that welfare systems supporting lone parents are increasingly focused on paid work (Millar, 2000). Significant shifts in government thinking have rejected the idea that lone parents have a right to be *acceptably* outside the labour market. Paid work is now viewed as the route out of poverty for lone parents and the government is committed to 'making work pay', especially for those in the lowest paid sectors[9]. More fundamentally, the emphasis on paid work is constructed as the rational and moral choice to be followed by all people and households regardless of their caring responsibilities in the private sphere (Rodriguez Sumaza, 2001). Such a policy stance begs questions about the well-being of lone parents who live in rural areas where paid employment is difficult to secure or where labour market participation is difficult due to external factors such as a lack of childcare facilities or transport. Many rural economies are characterized by low wages and lack of employment opportunities. This is particularly the case if an individual's employment choices are restricted due to a desire to work flexible hours, difficulties with transport or childcare requirements. For example, Sandra is a trained teaching assistant but can only accept work in the small rural town where she lives because she cannot afford a car. She continues 'public transport is hopeless for work. It's very much a case that I can't get a job because I can't get a car. I can't get a car because I haven't got a job. So you go round in circles.' Sylvia comments 'you can't just walk down the road and get a job or get the local bus to a job...this is a small village, there are only four or five employers.' Employment that is available is generally on the level of the minimum wage. When Dawn was asked if she planned to return to work after the children's holidays she replied 'there are cleaning jobs, the very rare shop job, care assistant....there is not the choice. You just don't have the choice. And the wages are just so bad. Most of them are on minimum wage. You are very lucky to get over four pounds an hour which when you have got responsibilities you just can't do it'.

Rural areas do not only present problems for lone parents who are looking for unskilled or semi-skilled labour, but also for middle-class professional lone parents. Generally, it is difficult to secure well-paid employment in or indeed close to the villages in which they reside. As such, lone parents are having to juggle their child care responsibilities and employment responsibilities with long commutes. For example, Jane is trying to hold down a senior nursing position, as well as looking after her two daughters aged ten and fourteen. Her job is one hour away and she has to leave at just after seven o'clock in the morning to arrive at her place of work on-time. During her interview, she talked about the stresses and strains of everyday life.

I don't like it. I hate it. We have had some terrible times lately. I have a mobile phone with me all the time so they have contact with me all the time. The school has my number as well. But I have been having calls from Josie [name changed] in the morning, panicking...I should be there to get her off in the morning, but I can't...I feel dreadful.

She continues:

I am dreading getting a call from school saying would you come and get your daughter [if she is ill]. I couldn't do anything about it because if you are nursing you can't just down tools and say sorry, my child is ill. That terrifies me because I still haven't got an emergency plan. I am one hour away from them. At the moment I am just taking it day to day and hoping that they are not ill but if they are God knows. I won't know what to do.

To summarize then, the nature of rural areas – geographically, economically and socially – can make the lives of rural lone parents more difficult than their urban counterparts. This will become an increasing problem as lone parents are encouraged to gain paid employment outside of the home. The research shows quite clearly that the lack of services, transport and employment are already pushing rural lone parents into more urban locations reducing their numbers in rural villages. Moreover, this is a problem not only for working class lone parents, but also relatively well-off middle class lone parents with professional careers. Their complex daily routines are made increasingly difficult by, for example, commuting times as well as the rural childcare deficit and the lack of convenient social and retail services.

Conclusions

In this chapter, I have tried to present arguments to explain the intuitive connections between lone parenthood and *urban* areas in the popular imagination. Three lines of argument have been pursued to explain why this is the case. The first relates to the ways in which the category 'lone parent' has been discursively constructed in recent political and popular discourses. In particular, lone parenthood has become associated with poverty, welfare dependency, delinquency and criminality through claims that familial breakdown, and the correspondent single parent families, have become active agents in the creation of the new 'urban' underclass. Single parents as the active perpetrators of the under-class phenomenon have become associated with urban areas and carry with them 'symbols of urban spatiality'. In this way, lone parenthood has been normalized as an urban problem and intuitive connections have been formalized in the minds of people which 'place' lone parent families firmly within an urban arena. Moreover, as Cloke et al (2000a, p. 715) argue in the context of homelessness, setting a problem in a particular type of landscape 'serve(s), by implication, to exclude other spaces from consideration.' As such, the association between lone parenthood and rurality becomes 'unnatural' and violates the popular representation of this social group. The second set of reasons which have combined to simultaneously strengthen the connections between urbanity and lone parent families while obscuring their associations with rural localities, centres around the nature and intensity of the socio-cultural constructions of rural life. In particular, it has been argued that rurality, in England at least, has become associated with notions of idyll embodied in strong familial and community relations and problem-free lifestyles. These symbols of rural idyll lead to the disconnection between rural areas and the social 'problem' of lone parenthood. This is particularly so given that lone parent families have been linked to poverty, societal breakdown and moral panic; problems and practices which fit uneasily with the rural in the popular imagination. Moreover, the chapter points to the ways in which these socio-cultural constructs exert a pervasive influence over the thoughts, practices and actions of urban and rural dwellers steering rural lone parents and rural dwellers more generally, to distance themselves and their communities from this difficult social issue. Finally, in this chapter, I have argued that the structural characteristics of rural areas can work against lone parents in a variety of circumstances making it almost impossible for many to remain in rural locales.

My concerns in writing this chapter, and the aims of the pilot research more generally, have been two-fold. The first and foremost objective is simply to acknowledge and explain the particular obstacles which clearly lie ahead in recognising the existence of lone parents and other non-traditional family structures in the countryside. As such, the intention of the chapter is to add to the expanding literature in rural studies focusing on socio-cultural concerns. The chapter not only explores the lives and experiences of yet another 'hidden other' in rural society, but also contributes to the understanding of how the 'othering' process works in practice. However, the second objective is rather less 'discipline specific'. In fact, it calls for an inter-disciplinary dialogue which recognizes the salience of acknowledging and subsequently taking account of the *spatiality* of lone parenthood. Lone parents exist within, but also beyond, the limits of the city, they exist in the inner-city, in the suburbs, in small market towns, in coastal resorts, in new towns, in old towns and, as this chapter contends, in villages, farms and hamlets. They exist, and although many may share characteristics with one another, they also face unique experiences and challenges rooted directly in their geographical embeddedness in these specific spatial locations. I would argue that geographers have much to offer the often taxonomic and descriptive nature of contemporary research into lone parenthood. This is not simply in terms of mapping the whereabouts of lone parents (see Haskey, 1994; Forrest and Gordon, 1993; Bradshaw et al, 1996), but rather in terms of exploring lone parents' lives and decisions in the context of spatialized (and gendered) divisions of labour, regional and local cultures and neighbourhood social networks. It is crucial, then, that research which aims to explore the lives and experiences of lone parents begins to take account of the local geographical contexts in which lone parents live and, in so doing, challenge policy which fails to acknowledge the nuances of experience amongst lone parent households, both socially and spatially.

Notes

1. 90 per cent of lone parents are women. In this chapter, I use the terms lone 'parent' and lone 'mother' interchangeably.
2. In her memoirs Margaret Thatcher revealed her belief that policies must be directed at strengthening the traditional family and Major's 'Back to Basics' reflected support for family values. Tony Blair in one of his first interviews as leader of the Labour party spoke out in favour of the family (Chambers, 2001; Jones and Millar, 1996).
3. The denouncing of lone mothers was a theme not only of the former Tory governments' 'Back to Basics' policy, but also features in Labour's Green Paper *Supporting Families* (Chambers, 2001).
4. The underclass thesis developed by American commentators such as Murray has been closely related to the entrenchment of the inner-city ghetto where sectors of the urban population experience residential isolation in neighbourhoods with high concentrations of poverty and unemployment (see for example Wilson, 1993; Jencks and Peterson, 1991).
5. This said, there is, in fact, little credible evidence to suggest that lone mothers get pregnant to jump the housing queue (Institute of Housing, 1993). It would seem unlikely given the poor quality of much council accommodation, particularly on urban sink estates.
6. Although note that it may extend, for some mainly working class women who have lived all or a large part of their lives in the same village, to the extended family of parents and siblings (see Little and Austin 1996).
7. Statistics from the DETR Survey of Housing in 1998/1999 show that only nine per cent of the existing housing stock in rural areas is occupied by council tenants compared to 16 per cent in England as a whole. Compare this to owner-occupation where 76 per cent of properties in rural areas are owned, compared to only 69 per cent in England as a whole.
8. Twenty-five per cent of lone parents in this study had no access to a car. The remaining 75 per cent had shared or exclusive access to their own car.
9. Policies such as the introduction of the National Minimum Wage have benefited many low-paid female workers, including lone parents. In addition, wage supplementation has

become the responsibility of the tax, rather than the benefits system with the introduction of Working Families, Child and Childcare Tax Credits.

References

Atkinson, K., Oerton, S. and Burns, D. (1998), 'Happy Families: Single Mothers, the Press and the Politicians', *Capital and Class*, Vol. 64, pp. 1-11.

Bell, M. (1994), *Childerley*, University of Chicago Press, London.

Bullock, H., Wyche, K. and Williams, W. (2001), 'Media Images of the Poor', *Journal of Social Issues*, Vol. 57, pp. 229-246.

Bradshaw, N. Bradshaw, J. and Burrows, R. (1996), 'Area Variations in the Prevalence of Lone Parent Families in England and Wales: A Research Note', *Regional Studies*, Vol. 30, pp. 811-815.

Brush, L. (1997,) 'Worthy Widows, Welfare Cheats: Proper Womanhood in Expert Needs Talk about Single Mothers in the United States, 1900 to 1988', *Gender and Society*, Vol. 11, pp. 720-746.

Bryson, A. Ford R and White, M. (1997), Making Work Pay: Lone Mothers, Employment and Well-being, Joseph Rowntree, York.

Chambers, D. (2001), *Representing the Family*, Sage, London.

Cloke P (1997), 'Country Backwater to Virtual Village? Rural Studies and 'the Cultural Turn'', *Journal of Rural Studies*, Vol. 13, pp. 367-375.

Cloke, P. (1995), Rural Poverty and the Welfare State: A Discursive Transformation in Britain and the USA, *Environment and Planning A*, Vol. 7, pp. 433-49.

Cloke, P. and Milbourne, P. (1992), 'Deprivation and Lifestyles in Rural Wales', *Journal of Rural Studies* Vol. 8, pp. 359-371.

Cloke, P. and Little, J. (eds) (1997), *Contested Countryside Cultures: Otherness, Marginalisation and Rurality*, Routledge, London.

Cloke, P. Milbourne, P. and Widdowfield, R. (2000a), 'Homelessness and Rurality: "Out of Place" in a Purified Space?', *Environment and Planning D*, Vol. 18, pp. 715-736.

Cloke, P. Milbourne, P. and Widdowfield, R. (2000b),'The Hidden and Emerging Spaces of Rural Homelessness', *Environment and Planning* A, Vol. 32, pp. 77-90.

Cloke, P. Milbourne, P. and Widdowfield, R. (2001), *Rural Homelessness: Issues, Experiences and Policy Responses*, Policy Press, Bristol.

Denham, C. and White, I. (1998), 'Differences in Urban and Rural Britain', *Population Trends*, Vol. 91, pp. 23-34.

Duncan, S. and Edwards, R. (1999), *Lone Mothers, Paid Work and Gendered Moral Rationalities*, McMillan, Basingstoke.

Duncan, S. and Edwards, R. (1997), *Single Mothers in an International Context: Mothers or Workers?*, UCL Press, London.

Evason, E. and Robinson, G. (1998), 'Lone parents in Northern Ireland: the effectiveness of work incentives', *Social Policy and Administration*, Vol. 32, pp. 14-27.

Finlayson, L. and Marsh, A. (1998), *Lone Parents on the Margins of Work*, DSS Report No. 80.

Ford, R. Marsh, A. and Finlayson, L. (1998,) *What Happens to Lone Parents?*, DSS Report No. 77.

Ford, R. and Millar, J. (1998), *Private Lives and Public Responses: Lone Parenthood and Future Policy in the UK*, Policy Studies Institute, London.

Forrest, R. and Gordon, D. (1993), *People and Places: A 1991 Census of England*, School of Advanced Urban Studies, Bristol.

Gingerbread (2000), *The Future for Lone Parent Families*, Report of Gingerbread's Conference 28th March 2000.

Glenn, E. Chang, G. and Forcey, L. (1994), *Mothering: Ideology, Experience and Agency*, Routledge, New York.

Haskey J (1994), 'Estimated Numbers of One-Parent Families and their Prevalence in Great Britain in 1991', *Population Trends*, Vol. 78, pp. 5-19.

Hughes, A. (1997), 'Rurality and cultures of womanhood: domestic identities and moral order in village life', in P. Cloke and J. Little (eds) *Contested Countryside Cultures: Otherness, Marginalisation and Rurality*, Routledge, London, pp. 123-137.

Jencks, C. and Peterson, P. (1991), *The Urban Underclass*, Brookings, Washingston.

Little, J. (1986), 'Feminist Perspectives in Rural Geography: An Introduction', *Journal of Rural Studies*, Vol. 2, pp. 1-8.

Little, J. (1987), 'Gender Relations in Rural Areas: the Importance of Women's Domestic Role', *Journal of Rural Studies*, Vol. 3, pp. 335-42.

Little J and Austin (1996), 'Women and the rural idyll', *Journal of Rural Studies* Vol. 12, pp. 101-111.

Little J (1997), 'Employment marginality and women's self-identity' in P. Cloke and J. Little (eds), *Contested Countryside Cultures: Otherness, Marginalisation and Rurality*, Routledge, London, pp. 138-157.

Marsh, A. Ford, R. and Finlayson, L. (1997), *Lone Parents, Work and Benefits*, DSS Report No. 61.

Milbourne, P. (ed) (1997), *Revealing Rural Others: Representation, Power and Identity in the British Countryside*, Pinter, London.

Millar, J. (2000), 'Lone Parents and the New Deal', *Policy Studies*, Vol. 21, pp. 333-345.

Millar, J. (1992), 'Lone Mothers and Poverty' in Glendinning, C. and Millar, J. (eds.) *Women and Poverty in Britain: The 1990s*, Harvester Wheatsheaf, Hemel Hempstead, pp. 149-161.

Millar, J. and Rowlingson, K. (2001), *Lone Parents, Employment and Social Policy: Cross-National Comparisons*, Policy Press, Bristol.

Murdoch, J. and Day, G. (1998,) 'Middle Class Mobility, Rural Communities and the Politics of Exclusion', in K. Halfacree and P. Boyle (eds.) *Migration into Rural Areas*, Wiley, Chichester, pp. 186-199.

Murray, C. (1990), *The Emerging British Underclass*, IEA Health and Welfare Unit, London.

Murray, C. (1994), *Underclass: The Crisis Deepens*, London: IEA Health and Welfare Unit.

National Council for One Parent Families (2000) *One Parent Families Today: The Facts*.

Office of National Statistics (2002), *Social Trends* 32, Stationary Office, London.

Kiernan, K. Land, H. and Lewis, J. (1998), *Lone Motherhood in Twentieth Century Britain: From Footnote to Front Page*, Clarendon Press, Oxford.

Rake, K. (2001), 'Gender and New Labour's Social Policies', *Journal of Social Policy*, Vol. 30, pp. 209-231.

Reinhold, S. (1994), 'Through the Parliamentary Looking Glass: 'Real and 'Pretend' Families in Contemporary British Politics', *Feminist Review*, Vol. 48, pp. 61-79.

Rodriguez Sumaza, C. (2001), 'Lone Parent Families Within New Labour Welfare Reform', *Contemporary Politics*, Vol. 7, pp. 231-247.

Rowlington, K. and McKay, S. (2002), *Lone Parent Families: Gender, Class and State*, Prentice Hall, Harlow.

Short, J. (1991), *Imagined Country: Environment, Culture and Society*, Routledge, London.

Silva, E. (ed) (1996), *Feminist Perspectives on Lone Motherhood*, Routledge, London.

Smith, S. (1999), 'Arguing Against Cuts in Lone Parent Benefits: Reclaiming the Desert Ground in the UK', *Critical Social Policy*, Vol. 60, pp. 313-334.

Speak, S. (2000), 'Barriers to Lone Parents' Employment: Looking Beyond the Obvious', *Local Economy*, Vol. 15, pp. 32-44.

Song, M. (1996), 'Changing Conceptualisations of Lone Parenthood in Britain', *European Journal of Women's Studies*, Vol. 3, pp. 377-397.

Van Dereth A, Knijn, T. and Lewis, J. (1999), 'Sources of Income for Lone Mother Families: Policy Changes in Britain and the Netherlands and the Experiences of Divorced Women', *Journal of Social Policy*, Vol. 28, pp. 619-641.

Wallbank, J. (2001), *Challenging Motherhood(s)*, Pearson, Harlow.

Wilson, W. (ed) (1993), *The Ghetto Underclass: Social Science Perspectives*, Sage, London.

Winchester, H. (1990), 'Women and Children Last: The Poverty and Marginalisation of One-Parent Families', *Transactions, Institute of British Geographers*, Vol. 15, pp. 70-86.

Woodward, R. (1996), 'Deprivation and 'the Rural': An Investigation into Contradictory Discourses', *Journal of Rural Studies*, Vol. 12, pp. 55-68.

Chapter 8

Constable Countryside?
Police Perspectives on Rural Britain

Richard Yarwood and Caroline Cozens

Introduction

After a period of neglect, geographers and other social scientists have started to pay closer attention to crime in the countryside (Dingwall and Moody, 1999; Yarwood, 2001). Increasingly, evidence suggests that crime is causing concern in the countryside of developed world countries (Wiesheit and Wells, 1996). To counter these apparent threats, new methods of policing (Meyer and Baker, 1982; Friedmann, 1992) and surveillance (Phillips, 2000) are being introduced to these areas, with important social consequences. Following a period in which the study of rural crime was 'politically naive, methodologically simplistic and philosophically unengaged' (Moody 1999, p.15), its investigation is now emerging as an important area of rural studies (Yarwood and Gardner, 2000).

In the UK, crime rates and the fear of crime are lower in the country than the city (Mirless-Black et al, 1998; Aust and Simmons, 2002), yet recent work has demonstrated that the fear of crime can cause very real concerns for many rural residents (Anderson, 1997; Yarwood and Gardner, 2000). However, there has also been blurring between threats that may be labelled criminal and those that are more threatening to hegemonic rural ideas (Cloke, 1994; Yarwood and Gardner, 2000). In some cases the fear of crime reflects a fear of the 'other' rather than genuine criminal threat. Furthermore, what is deemed to be criminal is also socially constructed and has been contested over space and time (Sibley, 1994). The legislation, enforcement and policing of crime therefore reflect and enforce dominant values in rural society (Reiner, 1994; Bowling and Foster, 2002).

Although there has been some discussion of the *de jure* impact of legislation (Sibley, 1994, 1997), less consideration has been given to the *de facto* practice of rural policing and its implication for people in the countryside. Despite earlier interest in the police by geographers (Smith, 1986; Fyfe, 1991; Evans et al, 1992), more recent attention has tended to focus on particular aspects of policing, such as controlling New Age Travellers (Hester, 1999), partnership working (Yarwood, 2003) or the use of electronic surveillance (Williams et al, 2000). Relatively less attention has been given to the daily operation of the police force, which is a striking omission as policing is geographical in nature (Fyfe, 1991) and reflects much about power, social order and dominant ideologies in the countryside. However, very little is known about the ways in which the contemporary spaces of rural Britain are policed.

Recent events in Britain have demonstrated the need for a sharper focus on this topic. In 1999 the fatal shooting of an intruder by a farmer, Tony Martin, caused a widespread moral panic over a perceived lack of policing in the countryside (Lusher, 1999; Reid, 1999; Countryside Alliance, 2001). Although the rural public are generally more satisfied with the police than their urban counterparts (Dingwall, 1999; Kaufman, 1999; Aust and Simmons, 2002), their increasingly vocal concerns have prompted a series of policy measures to improve the visibility and effectiveness of rural policing (DETR, 2000). Little, though, is known about the practice of rural policing and its implications for rural society. Although some attention has been given to voluntary (Yarwood and Edwards, 1995), community (Shapland and Vagg, 1988) and private (Phillips, 2000) policing, there has been little discussion of policing from the perspectives of the police themselves.

Consequently, this chapter examines how the police view crime and policing issues in the British countryside. It is divided into two main parts. The first traces how competing ideologies have led to the police having multiple, contradictory and contested roles in late modern society. The second part uses empirical evidence to examine how the police are negotiating these tasks in a rural context. A key objective throughout the chapter is to trace the competing demands placed on the police and to understand how and why, and with what consequences, they respond to them.

What is Policing?

Reiner (1994) and Bowling and Foster (2002) have cautioned against the confusion of police with policing. Reiner (1994, p.1004) refers to *the police* as a particular institution, often assumed to be 'a state agency mainly patrolling public spaces in blue uniforms, with a broad mandate of crime control, order maintenance and service function' (Reiner 1994, p. 1003). *Policing*, on the other hand, is a much broader concept that refers to 'an intricate, almost unconscious network of voluntary controls and standards among people themselves and enforced by people themselves' (Bowling and Foster, 2002, p.981). Individuals have become coerced, often through surveillance, to conduct themselves according to prevailing norms in society (Foucault, 1973). If policing is viewed as a 'universal requirement of any social order' (Reiner, 1994, p.1003), then it is crucial to realize whose order is being policed (Bowling and Foster, 2002; Waddington, 1999), and by whom.

An understanding of policing should consider the way in which the police, public and other agencies regulate themselves and each other according to the dominant ideals of society. Such an analysis should consider the ways in which these ideas have been constructed, contested and transgressed over time and space. The following sections therefore consider the roles of the formal police, the public and the state in these contexts and how these lead to competing contradictions in the policing of rural spaces.

The Police and Policing

In 1829, Robert Peel created the state police in the United Kingdom by establishing the Metropolitan Police Force. Its original philosophy was to prevent crime (Crawford, 1997) through the visible presence of uniformed police officers who represented social order and respectability. However, two competing ideologies soon developed, initially in different national contexts (Bowling and Foster, 2002). On the one hand, the Metropolitan Force became the epitome of the 'liberal' approach to policing, valuing human rights, trust, partnership working and community safety. It aimed to work *with* people in order to keep peace and resolve conflict, deriving its legitimacy from public accountability. Conversely, the establishment of the Royal Irish Constabulary pioneered the 'military' approach through intrusive, often covert, methods and came to typify the 'colonial' approach to policing. The police worked *against*, rather than with, the public and derived its authority from the government and elite social groups.

Table 8.1 Multiple policing roles

Function	Mechanism
Public Reassurance	Visible police patrols; contact with individuals and community organizations; effective crime investigation and emergency service
Crime Reduction	Visible patrol; targeted policing; proactive policing; effective crime investigation and emergency service
Crime Investigation	Reactive detection work to arrest offenders and bring them to justice; proactive investigation
Emergency Services	Rapid response to disputes, disturbances, accidents and emergencies
Peace Keeping	Routine negotiation and problem solving in a range of neighbourhood disputes and disorders
Order Maintenance	Controlling crowds at sporting events, entertainment and demonstrations
State Security	Protection of public figures, state buildings, covert policing of dissident organizations

Source: Bowling and Foster (2002)

If the 'military' approach was to be avoided in Britain, a universal set of principles and regulations was needed to regulate the police (Crawford, 1997; Emsley, 2001). Consequently, emphasis was placed on developing professional rather than personal relationships with the public by means of an increasingly bureaucratic system of regulations. As the twentieth century progressed, complex divisions of labour emerged within and between forces to fulfil different operational tasks. As Emsley (2001, p.50) has noted, this led to a 'bewildering assortment of forces ... alongside a number of other individuals and groups who were performing police functions'. Similarly, Reiner (1994) has described the

British police as having an 'omnibus' role, which covers everything from community safety to state security (see Table 8.1).

'Hard' and 'soft' methods of policing continue to be employed to achieve these roles (Bowling and Foster, 2002). 'Harder' policing, or policing *against* people, includes activities such as riot control, the maintenance of public order and 'zero tolerance' policing. 'Softer' policing attempts to work *with* the public in a more pro-active manner to provide reassurance and reduce crime using pro-active methods. All forces are required to operate in both 'hard' and 'soft' terms, which is in itself something of a contradiction and, over time, the emphasis placed on these approaches has varied (Reiner, 1994; Crawford, 1997; Bowling and Foster, 2002). However, both approaches can contribute towards social inclusion or exclusion (Scarman, 1981; Sibley, 1994; Yarwood and Edwards, 1995; Young, 1999) and questions have remained about police accountability and power.

Policing, Power and Accountability

The relationship between the public and the police has varied widely over space and time. Widespread, often violent, resistance met the introduction of the police to rural areas because of fears that officers were being used to spy against people engaged in poaching, smuggling and other 'folk crimes' (Mingay, 1989; Jones, 1989a, 1989b). However, as the police grew in legitimacy and number (Emsley, 2001), they became widely accepted and an important component of established social order (Young, 1993).

Until the 1960s, the police followed 'softer', community-based strategies, which were aided by the organization of the police into small forces that were accountable to local populations (Smith, 1986; Fyfe, 1993; Baldwen and Kinsey, 1982). Although this should not be seen as a 'golden age' for the police, it has been viewed as a period of stability and legitimacy. As time progressed, it became widely accepted that the police alone were responsible for policing and, furthermore, were capable of fulfilling all the functions associated with it (Bowling and Foster, 2002; Reiner, 1994). This perpetuated a 'fetishism' (Reiner, 1994; Bowling and Foster, 2002) that policing was 'expert business' and that the police could and would control crime.

Since the 1960s, however, this belief has been increasingly questioned. Many commentators suggest that the 1964 Police Act was a watershed in police-public relations (Smith, 1986; Fyfe, 1991; Baldwin and Kinsey, 1982; Crawford, 1997). Under this legislation forces were rationalized and re-organized into larger units. Whereas there were over 130 local forces in 1964, there are now 42 forces in England and Wales (Fyfe, 1991). Local control and accountability was taken away from the public as the policing became increasingly bureaucratic. Furthermore, with larger territories to police, 'harder' methods of policing replaced policing *with* communities, with a policing *of* communities. Rising crime rates, a distanced police force and high profile examples of aggressive policing, such as in Brixton, London (Scarman, 1981), led to a decline in the public's confidence with the police. This was so much so that private policing has replaced state policing in many areas (South, 1988).

In response, a series of measures have been introduced in an effort to improve accountability and efficiency. First, a number of community-based policing initiatives were implemented in the 1980s and 1990s to improve police-public relations. Although some schemes, such as Neighbourhood Watch, have improved feelings of security for some residents (often those in low crime areas), community policing has been criticized as something of an uneven public relations exercise, often sidelined to a specific department, rather than being adopted as an underlying force philosophy (Crawford, 1997).

Second, there have been efforts to improve accountability by introducing performance targets. 'Citizen's Charters', introduced by John Major's Government in the early 1990s, aimed to improve the delivery of services by the state by switching emphasis from the provider to the consumer (Hill, 1994). Following recommendations by the Audit Commission, the National Audit Office and the Sheehy Enquiry (McLaughlin, 2001), the 1994 Police and Magistrate's Courts Act introduced 'free-standing, more business-like police authorities' (McLaughlin, 2001, p.93) that had a responsibility to devise an annual policing plan that detailed crime control targets, objectives and expenditure. Data used to monitor the performance of these plans were also used to assess national targets and compile public 'league tables' of police performance. Although chief constables and senior officers retained a degree of autonomy, the introduction of fixed term contracts aided the introduction of a 'performance culture'. Critics of the scheme argued that, amongst other things, it placed too much power in the hands of the Home Secretary, so that policing became even more centralized and less accountable at the local level. Its supporters argued that these measures made the police openly accountable to the public through a series of comparable measures of effectiveness and efficiency. This managerialist approach has continued under New Labour with the introduction of 'Best Value Practice' in 1999. This is a rolling programme of audits aimed at encouraging not only value and quality, but also 'joined up thinking' between different service providers in the public, voluntary and state sectors.

Third, this philosophy reflects a panacea that partnership working can deliver services and improve decision making at the local level. The 1998 Crime and Disorder Act placed legal responsibilities on the police and local authorities to co-ordinate the development of local crime and safety in partnership with local people and other local agencies in the public, private and voluntary sectors. The formation and operation of these partnerships has highlighted that the police are no longer viewed, if ever they were, as the sole agents of policing but one of many agencies who are responsible for maintaining law, order and public safety (Crawford, 1997; Hughes et al, 2002). As McLaughlin (2001, p.94) has stated 'local police work is being increasingly enmeshed in a complex network of overlapping relationship and interests'.

The role of the police in late modern society is therefore far from clear cut. The police have to provide a wide range of functions according to a number of competing ideologies. Furthermore, the position of police in policing networks has become increasingly contested and negotiated. As these positions have shifted, so too have their relationships with the public and the state. The following section considers how these tensions may impact on rural policing.

Rural Policing/Policing the Rural: a Question of Rural Space

An appreciation of how rurality is socially constructed and contested by different groups, including the police, is key to understanding the practice of policing in the countryside. The following section reviews existing literature to examine the relationship between rurality and policing.

Rurality as Distance

Although many geographers have abandoned attempts to measure rurality according to specific physical, environmental, demographic and social indicators (Halfacree, 1993), these kinds of definitions continue to have an important bearing in the policy and planning arena. In terms of emergency planning, rurality is frequently equated with distance. It is accepted that emergency response times will be longer in rural areas than in urban areas, as officers will need to travel further distances over wider areas. The need to meet performance targets (see above) has required forces to define urban and rural places and to position their personnel in areas where they can meet standard emergency response times most effectively.

Policing is viewed as a territorial activity, with forces organized on a spatial basis (with the exception of Transport and British Nuclear Police) to control particular counties or groups of counties. Within these constabularies, forces are in turn divided into hierarchies of divisions, sub-divisions and individual beats (Fyfe, 1993), each with their own spatial focus and responsibility. The territorial organization of space has been used to determine the numbers and roles of police based on area, population and crime rates. Many forces now operate a dual system of policing with some officers designated to responsive tasks and others (sometimes known as 'beat managers') dedicated to proactive, community policing (Yarwood, 2003).

However, a report published in 1999 (ORH, 1999) criticized the police funding formula used by the government to allocate resources to forces. It used simulation models to argue that full account had not been taken of the extra distances, longer travelling times and increased resources needed to respond to emergencies in rural areas. Subsequently the Crime Fighting Fund and Police Modernization Fund have been used to ring-fence money to recruit extra officers for rural forces and encourage innovation in rural policing (DETR, 2000).

Some studies have suggested that people in the countryside are more satisfied with the police than those in the city (Dingwall, 1999; Kaufman, 1999; Aust and Simmons, 2002). However, there has been criticism of the police by some rural pressure groups for failing to provide what they see as an efficient, effective and, above all, visible emergency service (Yarwood and Gardner, 2000). Young (1993) has also noted that there is a higher public expectation of the police in rural areas, placing even more demands on them. To understand the nature of these demands and the police's response to them, it is important to consider how rurality is perceived by these groups.

Rurality as a Social Construction

Although physical distances pose logistical problems for the police, especially in terms of performance, greater attention should be given to the way in which rural areas are imagined. It is widely held by geographers that rurality is socially constructed and that the way that the countryside is imagined helps to reproduce dominant social ideals (Halfacree, 1993). A key component of the rural myth is that the countryside is, or should be, a crime-free, safe place to live (Hanbury-Tenison, 1998; Cloke et al, 1995; Valentine, 1997). However, Jock Young (1999, 2002) has argued that insecurity and risk have increased due to structural changes in late modern society.

Since the 1960s, he argues, a 'Golden Age' of high employment, a strong welfare state and stable family structures has been replaced by a late modern society characterized by economic and social instability, together with a much weakened system of welfare. 'Where there was once a consensus of value, there is now a burgeoning pluralism and individualism' (Young, 2002, p.459). Society, now based on the sands of material and ontological insecurity, is pre-occupied with tackling the real and imagined risks facing it (Beck, 1992; Hughes, 2000). Its 'terrain has become a more risky place, both in terms of crime and disorder and in terms of demonization and scapegoating' (Young, 2002, p.485). Crime and fear of crime are, therefore, both products and processes of social exclusion.

Although Young's (2002) work is based in an urban context, the social and economic coherences of rural areas have also been shattered, producing a differentiated countryside (Cloke and Goodwin, 1993, Cloke and Little 1997). The insecurity wrought by these changes has led to an increased sense of risk and a desire to label and exclude particular groups, such as young people or Travellers, from sanitized rural spaces. There has been a blurring between criminality, which itself is open to contest, and cultural threat, leading to demands *to police the rural* rather than engage in *rural policing* (Yarwood and Gardner, 2000). Malcolm Young (1993) maintains that these ideas have increased expectations of the police in the countryside, adding to the pressures placed on the resources and operation of rural forces. Just as Young (2002, p.486) has suggested that there are 'numerous 'thems' and various 'us's', so too are there different *countrysides*, rather than one *countryside,* to police.

Summary

This section has briefly outlined the different demands placed on the police to enforce social order in an increasingly differentiated countryside. Not only is the role of the police contested, but so too is the nature of the countryside and its society that they are trying to police. Furthermore, the police are torn between accountability towards local people, who may be demanding that a particular and exclusive vision of rural life is maintained, and the state's demands for greater efficiency. Effective policing needs to encompass community roles, notwithstanding whose idea of community should be policed, and reactive roles. The rural police officer's lot is not, therefore, a happy one. The following section examines how different forces have attempted to police the real and metaphorical wilderness of the UK.

Research Foci

Some studies on the public's perception of the police in rural areas have concluded that levels of satisfaction are generally higher than in urban places (Dingwell, 1999; Kaufman, 1999; Aust and Simmons, 2002). Although there are indeed less officers per head of the population in rural areas (Table 8.2), these officers have to contend with significantly lower rates of crime (Aust and Simmons, 2002). Nevertheless, there is a widespread feeling amongst many members of the public that rural policing, especially its visibility, should be improved (Yarwood and Gardner, 2000; Yarwood 2003). Young (1993) suggests that idealized notions of rural life have increased public expectation of the police service. To rectify this situation, many police forces have undertaken initiatives aimed at improving community relations and reducing public concern with the apparent lack of police in the countryside (Yarwood, 2003). While this research has started to examine the public's relationship with the police little, if anything, is known about the police's relationship with the public in rural areas. If these initiatives are to succeed, then it is important to know more about how the police themselves view the everyday issues of rural policing.

To begin examining these issues, the remainder of the chapter focuses on rural policing from the perspective of the police themselves. It examines:

- how the police have constructed rurality;
- the organization of rural policing; and
- the policing of rurality.

Evidence is drawn from a survey of British police forces. In 2001, the Chief Constables of all thirty-nine non-metropolitan police forces were contacted with a request to identify a senior officer with responsibility for rural policing who would be willing to be interviewed about their constabulary's policy and practice. Twenty-nine forces agreed to take part in the survey and their officers were interviewed by telephone using a structured interview schedule. The quotations given below therefore reflect the viewpoints of individual officers, rather than the official position of the forces concerned. Consequently, forces and officers have been anonymized to preserve confidentiality.

Table 8.2 Policing in rural areas

Home Office Urban/Rural Classification	Forces	Police Officers per 100,000 Population, 2001	Ratio of Recorded Crime to Police Officers, 2001
Most Rural	Dyfed-Powys, Lincolnshire, North Yorkshire, North Wales	201	42.5
Less Rural	Cambridgeshire, Cumbria, Devon and Cornwall, Durham, Gloucestershire, Norfolk, Suffolk, West Mercia, Wiltshire	195	47.7
Middling	Avon and Somerset, Bedfordshire, Derbyshire, Dorset, Essex, Gwent, Hampshire, Humberside, Kent, Leicestershire, Northamptonshire, Sussex, Thames Valley, Warwickshire	196	57.5
More Urban	Cheshire, Cleveland, Hertfordshire, Lancashire, Northumbria, Nottinghamshire, South Wales, South Yorkshire, Staffordshire, Surrey, West Yorkshire	225	53.1
Most Urban	Greater Manchester, City of London, Metropolitan, Merseyside, West Midlands	298	55.5

Source: Aust and Simmons 2002. The definitions of rurality are based on Home Office 'Police Force Areas'

Rurality and Policing

Far from being an exercise in semantics, the ways in which different actors construct rurality have been shown to have important implications for rural society. In particular, the notion of an 'idealized' countryside can contribute towards the preservation of class interests in the countryside (Murdoch and Marsden, 1994) and the neglect of social problems (Cloke et al, 2002). Likewise, many academics, residents and policy makers have often supposed, with important implications, that the countryside is, or should be, a crime-free place to live (see Valentine, 1997; Moody, 1999; Yarwood, 2001). To date, however, no consideration has been given to the ways in which the police view rurality. It is important to consider this if an understanding is to be gained about their actions in the countryside.

The requirement to measure and audit policing performance has obliged forces to consider their service in urban and rural places. Consequently many forces have given some thought to distinguishing between these two environments. Although the Home Office has classified police forces according to a five-fold urban-rural continuum that ranges from 'most rural' to 'most urban' (see Table 8.2 and Aust and Simmons, 2002), there are no centralized definitions of 'urban' and 'rural' areas that are used within constabularies. Instead, many forces (16 out of 29) used their own set of indicators to determine differences between urban and rural areas. Population size was widely used, with settlement above or below a certain threshold, often 8,000 people, defined as urban or rural. Other forces classed entire local authority districts as 'urban' or 'rural' according to how built up they were deemed to be and, in some forces, this was calculated on a ward by ward basis. The range of these definitions reflects the subjective basis of these apparently objective attempts to define rurality (Halfacree, 1993) and also a requirement to publish performance data in particular forms for 'best value practice' audits.

It was clear that, in some cases, definitions were viewed as politically expedient rather than of practical use. Thus, some forces had based their urban-rural boundaries to reflect favourably their response to emergencies. One constabulary defined rural areas in 1993 yet modified them 'according to response times' and another had classed their whole constabulary as 'rural' 'for the purposes of response times'. Other respondents rejected efforts to define urban and rural places as unhelpful in the delivery of service provision:

> Whole force area is defined as predominantly rural and there is no differentiation between the two.

> Whole area of the constabulary is defined as urban and no areas are defined as rural.

> We considered carefully and looked at definitions applied by other agencies. Having done so we decided it would be unhelpful to draw, or attempt to draw a strict boundary. They blend one into the other and so we just promised both a quality service. .

So, where used, formal definitions were used mainly to classify space for Government audits. In everyday practice, some officers reported that their forces preferred to use 'common sense' definitions:

> We do not have a specific definition. However, we use the process of identifying a rural area as an area of low population that would clearly be judged by the general public as being rural and not urban.

> No scientific process has been used, we have relied on professional knowledge to define sectors as either urban or rural.

All respondents suggested that daily police operations in all forces were driven by 'professional' knowledge of rural areas. Social ideas, rather than statistical models, of rurality are therefore more important to the everyday practice of policing of the countryside and so it is important to consider what these 'professional' or 'common sense' ideals actually reflect. Two themes dominated: isolation and community.

Rurality as Isolation

Rurality was strongly associated with isolation by many officers. Although isolation is valued by some rural residents (Cloke et al, 1994), respondents felt it was problematic for three reasons.

First, isolation was thought to contribute to certain crimes. Vehicles were thought to be more prone to crime in isolated beauty spots, visitor car parks, and pub and hotel car parks. 'High class' residential dwellings were thought to be particularly vulnerable to burglary. Many respondents suggested that these 'hotspots' were targeted by criminals who travelled from urban areas in the hope of stealing antiques and other valuable goods. Isolated farms were also seen as vulnerable to thefts of plant machinery, vehicles and trailers. Furthermore, temporary 'hot spots' could temporarily emerge in tourist areas during the summer months.

Second, two-thirds of the officers suggested that certain places were harder to police or patrol due to their isolated nature. The borders of police areas were seen as particularly problematic in this respect:

> we have borders with seven other forces ... there are also major roads cutting through the county. We attract criminals from elsewhere because we are viewed as an 'easy touch'.

Another officer pointed out that some residents living near the border of his constabulary were actually closer to a police station in a neighbouring force. The physical geography of rural areas, especially extreme weather events, could also strain resources at certain times of the year:

> The flooding of the River Severn makes it a lot harder to reach certain parts of the area. The county also suffers from very spasmodic yet heavy numbers of fallen trees; again restricting access and burdening existing police resources.

Conversely, less isolated areas, such as towns or peri-urban rural areas, were generally thought of as easier to police, as resources were usually concentrated in or close to them. This said, other officers felt that some smaller villages were easier to police as it was possible to establish personal relations with their residents. One officer also noted:

> Urban policing is often more confrontational, this maybe because urban officers can afford to bring issues to a head, whereas their rural colleagues understand no back up will get to them within 20 minutes.

Finally, nearly all (80 per cent) of officers questioned felt that isolation could contribute towards the public's fear of crime:

> Properties in rural areas are often isolated (without neighbours) and often inaccessible: this could raise concerns.

> Isolation of small communities, the perception that the police are a long way away.

As these and other comments in this section have suggested, officers considered that people living in rural areas had particular needs and that policing had to take account of particular social relations in the countryside. The following section foregrounds these ideas.

Rurality as Community

The term community, so often seen as an ideal in the countryside (Bunce, 1994), was widely referred to by many officers, although the concept was often seen as problematic. One officer felt that the community ideal contributed towards a dangerous lack of awareness about crime:

> People in rural areas are too trusting and due to their 'personal' low fear of crime take fewer security measures... like leaving trailers and farm machinery insecure. Some rural communities still have an 'open-door' or 'latch-key' mentality.

Other officers felt that 'closer' communities were more likely to gossip about specific crime incidents, increasing feelings of insecurity.

Overwhelmingly, however, officers thought that rural communities were changing or that their residents felt that they were changing. Some noted that rural society was becoming fragmented and less personal in nature. For example, one described villages as 'dormitories or weekend playgrounds' whose inhabitants did not know each other. The loss of this (imagined) community contributed strongly towards fear of crime: one officer suggested that some 'locals' complained about strangers moving into their villages, reducing a sense of community and, therefore, security. Another noted that some rural communities felt 'under threat' by widespread public concerns over hunting, genetically modified crops and other rural issues. He thought that these could greatly contribute towards insecurity, especially if protests or actions were planned in their area.

Respondents felt that the fear of crime in rural areas was frequently exacerbated by sensationalist media coverage. This was not only in the national press but also in the local media, which was prone to reporting relatively minor incidents that the urban media would ignore. Relatively minor crimes were seen to be more shocking if committed in rural 'communities'. Indeed, the perceived loss of community was felt by one officer to make the public 'more reliant on TV rather than community activities' and therefore more prone to fear caused by the (over)reporting of crime by local and national media.

The insecurities of and threats to, what were often described as, 'traditional' populations were leading to increasing demands on the police. As Young (1993) suggested, many of the officers felt that rural communities had much higher expectations of the police and were more willing, and capable, of articulating them. These, as the next section illustrates, impacted on the way the rurality was policed.

Summary

The section above has highlighted some of the different ways in which the police view rurality. On the one hand, many forces have devised statistical definitions in response to government requirements to provide statistical data on performance and best value practice. There is a need for the police to use these definitions in order to achieve, and demonstrate to politician and citizen alike that they have achieved particular performance targets in rural areas. This might best be described as rural policing (Yarwood and Gardner, 2000). On the other hand, it is also clear that officers did not regard rurality as simply a measure of distance or isolation, although problems associated with the physical environment were acknowledged. More significantly, the perceptions of rural society, held by the police and residents, created particular challenges for the policing of the countryside. These extended beyond the prevention of crime and the reduction of fear of crime, and might best be described as policing the rural. The following section outlines the practices of policing these different ideas of rural space.

The Performance of Policing in the Countryside

All of the officers interviewed felt rural areas offered them particular challenges that were not encountered in urban places. Some of these were physical in nature, while others reflected what were perceived as differences in rural society (Table 8.3). The range of these issues reflects the multiple roles of the police in society (Bowling and Foster, 2002), the different ways in which rurality is perceived and also to whom the police are accountable. The following section examines how the police have attempted to meet the different demands associated with rural policing and policing the rural.

As this chapter has already noted, isolation was deemed to be a major obstacle to the policing of rural areas. Furthermore, two-thirds of the interviewees felt that, over time, policing had been withdrawn from remoter areas, exacerbating the problems found there. Despite these obstacles, the police are required to respond to emergency calls within specific time limits.

Table 8.3 Challenges to policing in rural areas

Issue	Number of Interviewees	Examples of Comments
Isolation	12	'Response targets can be challenging' 'There are high-value thefts at isolated locations' 'High visibility is hard to achieve' 'Unoccupied properties are targeted' 'Retail outlets and businesses (such as post offices) are particularly vulnerable' 'There is a lack of facilities for young people'
Lack of resources	7	'It is important to keep abstraction of beat officers in rural areas to a minimum' 'Owing to small numbers of officers, sickness and leave mean that urban stations need to supplement the rural stations. Less staff means less productivity as a presence at rural stations is required.' 'It is not possible to centralize fully some resources, such as custody' 'The demand is more irregular in rural areas'
High public expectation	4	'Rural communities have a higher expectation in terms of visibility, reassurance and service delivery' 'The communities perceive a disintegration of rural communities and are extremely influential'
Disproportionally high fear of crime	3	'There is a general public perception that crime is increasing in rural areas'
Special/Seasonal Events	3	'current rural issues such as hunting, shooting and GM crops require balanced policing' 'an influx of tourists and visitors can put pressure on police resources'
Travelling criminals	3	'many of the rural areas are bordered by areas of greater population density. As a result the county suffers from travelling criminals'

Rural Policing

Target response times were, on average, longer in rural areas than urban areas (Figures 8.1 and 8.2). Despite the concerns felt by the police and public, the respondents reported similar degrees of success in meeting these targets in both environments. Furthermore, there was a remarkable consistency in the physical resources deployed and methods used to police urban and rural areas (Table 8.4). However, there was some urban bias in the use of visible forms of policing, such as foot or bicycle patrols, CCTV and Neighbourhood Wardens. The lack of these schemes in rural areas may contribute towards feelings of isolation reported by some residents in other studies (Yarwood and Gardner, 2000).

schemes in rural areas may contribute towards feelings of isolation reported by some residents in other studies (Yarwood and Gardner, 2000).

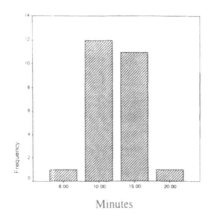

Figure 8.1 Target response times in urban areas[1]

Table 8.4 Policing resources and performance in urban and rural places

	Percentage of Target Times Achieved per constabulary	Average Number of Permanently staffed Police stations per constabulary	Average Number of Police stations staffed part time per constabulary	Average Number of Officers per constabulary	Average Number of Special Constables per constabulary
Urban	87%	7	11	692	119
Rural	87%	7	9	516	84

Source: Authors' interviews

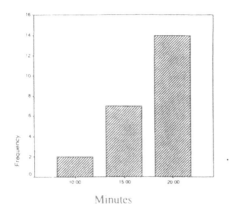

Figure 8.2 Target response times in rural areas

Although there were apparent similarities between urban and rural policing methods, officers highlighted a number of rural policing issues. First, 70% of the officers interviewed felt that more resources needed to be dedicated to the policing of rural areas. Various funding formulae and models have been used to allocate resources in rural places to take account of response times, terrain and distance (ORH, 1999). Despite improvements to these models and new monies from central government, many officers felt that the isolated nature of rural areas stretched their current resources:

> Overall, we have long argued that there are hidden costs and difficulties associated with sparse rural areas. This is now beginning to be recognized by government.

> Some rural areas do not get policed at all, owing to the demand of urban areas and the low staffing levels at two rural stations.

Furthermore, these limited resources were felt to be particularly vulnerable to redeployment or loss by emergencies or other 'out of the ordinary' activities:

> One event (such as a sporting tournament) can impact on the whole county. A recent example of this occurred when a football match took place and all officers in the constabulary had their rest days cancelled due to "suspected" or "likely" violence. This in turn impacted on overtime funds, which in turn reduces the funds available for other police initiatives, like rural policing, targeted patrols or organized police operations.

Second, special events, such as agricultural shows, concerts or festivals, sporting events, protests or large-scale evictions, were highlighted as particularly challenging for the police in rural places:

> Hunts are active across many of our divisions and are well attended by active protestors. Policing these is a labour intensive process.

Furthermore, large events, whether legal or illegal, were seen to cause more crime (such as pick pocketing and theft), anti-social behaviour (particularly drunkenness) and traffic congestion, all of which required well-organized and sensitive policing.

One officer noted that the requirement to police public disorder was increasing because he felt that the countryside had become the focus for civil protest activities against issues such as genetically modified crops, animal rights and hunting, which were unique to rural areas. Some officers were also concerned that efforts to ban hunting would lead to civil disobedience and policing difficulties. Other problems included illegal parties and encampments, complaints about Travellers and abandoned vehicles. The biggest challenge was to police disorder in an 'even handed' manner, even though it 'absorbed valuable resources' and accounted for a disproportionate amount of crime and police time in rural localities.

High numbers of visitors could also complicate the policing of some rural areas in the holiday season. One force had a:

> popular tourism area with huge influxes of tourists throughout summer and holiday periods. Policing levels are set on population levels-visitors do not count even though they reflect many of our victims and some of our offenders.

Furthermore, resource problems increased during the tourist season as:

> The resources go down (due to annual leave) as numbers of people to police go up.

Similarly, many officers noted that the numbers of police in rural places were particularly vulnerable to sickness and leave.

> Owing to small numbers of officers, sickness and leave mean that urban stations need to supplement the rural stations. Less staff means less productivity as a presence at rural stations is required.

Resource issues aside, it was also recognized by a number of interviewees that co-ordination within and between different forces could be improved in rural places. There were opportunities for forces to co-operate more effectively on the borders of their constabularies and, in some cases, there was a need to improve attitudes of rank and file officers to rural policing:

> there is a reluctance on the part of officers to cover more rural outposts of the force area.

All but two of the officers interviewed felt that the police were under pressure to improve the delivery of rural policing. There was evidence that many forces were adopting strategies to improve rural policing and performance. Five had adopted a rural policing strategy to co-ordinate and implement these changes, while another six were in the process of preparing one. Ten forces argued that rural issues were covered within existing strategies and so there was no need for a specific document that detailed these arrangements.

Table 8.5 reveals that the pressure to improve came mainly from central government and the rural public. In the former case, many officers had suggested that the 'performance culture' had driven change for more efficient and effective policing. However, community groups, parish (or community) councils and other local agencies were felt to be placing greater demands on the police, especially in light of the Tony Martin Incident. The media were seen to increase these pressures, and one officer noted that 'even The Archers had an anti-police story line'. To understand the impact of these issues, attention is now given to the policing of rurality.

Table 8.5 Sources of pressure to improve rural policing

Central Government	9
Media	6
Parish/Community Councils	2
Neighbourhood Watch Schemes	1
Local Authorities/Politicians	4
Rural Public	13
Other Community Groups	4
Landowners	1
The Police	1
Farmers	1

Source: Authors' Interviews

Policing the Rural

Although some previous studies have suggested that public satisfaction with the police is generally good (Kaufman, 1999; Aust and Simmons, 2002), a lack of police *visibility* in rural spaces causes concern (Yarwood and Gardner, 2000). Many rural residents want to see, and be seen by, the police. Surveillance is a powerful way of coercing groups of people to behave according to particular norms in society (Foucault, 1973). Although importance has been placed on the panoptic use of hidden surveillance in this process, Valentine (2001) has also pointed out the being surveyed can improve safety and provides a greater sense of security for the surveyed.

Public concern stems from a feeling that rural areas are not being watched. This is not strictly true as all beats within police constabularies are monitored through the collection and analysis of crime data and criminal intelligence information. In turn, these are used to control the deployment of police resources across the countryside (although not always effectively according to some officers, as discussed above). However, until the public know that they are being watched, feelings of insecurity may remain. A more visible police presence would provide assurance and allow surveillance to act as a two way process: not only would the watched come under surveillance, but so too would the watchers. If the police are seen to be patrolling, residents can monitor their actions (or lack of them) and improve local accountability. However, as Foucault (1973) has demonstrated, surveillance is a part of a power network that coerces people to behave in particular ways that are deemed to be acceptable. It is therefore important to consider some of the power relations associated with these demands for more surveillance.

Indeed, the interviews suggested that the 'public' demand for the police to improve their service (Table 8.5) represented the interests of particular groups. Although many officers felt that their relationship with the public was generally 'good' (Figure 8.3), this applied more to some groups than others. One interviewee suggested that the rural public, particularly farmers' organizations, 'Watch schemes' and parish councils, with 'traditional, pro-police viewpoints' were more willing to work with the police. Conversely, nearly half of the interviewees said that they had particularly poor relations with certain groups in rural areas, often 'travellers' and 'young people'. As other studies (Sibley 1994, 1997; Yarwood and Gardner 2000) have revealed, these 'others' are often the subject of demands to improve policing to counter their presence in sanitized rural spaces. There is a danger that demands to improve policing reflect cultural and class interests, rather than the need to counter criminal threats.

However, it would be over-simplistic to suggest that policing simply reflected the class interests of elites to the exclusion of 'others' or that the police viewed middle-aged, middle class rural residents 'easy' to work with and 'others' (Philo, 1993) as problematic. The respondents often felt that the expectations of middle-class residents were unreasonably high:

> the expectations of the public are totally different. Two burglaries in a village are a crime wave.

There is a general public perception that crime is increasing in rural areas and therefore there is a greater fear of crime. Rural communities have high expectations in terms of visibility, reassurance and service delivery.

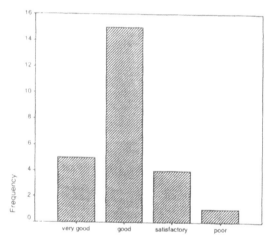

Figure 8.3 Police officers' perception of public relations

It was felt that these demands had been fuelled by media coverage that had distorted and over-exaggerated crime figures in rural areas (see Hoggart et al, 2002) to create a panic. Another officer thought that parish councils were 'ill-informed and unrepresentative'.

Likewise, it was realized that 'an element of prejudice on both sides' could develop if the policing of public disorder was not handled in an 'even handed way'. It was felt that poor relations with young people were caused by a lack of facilities or activities for young people, which could lead to boredom, anti-social behaviour and trouble with the police in some cases. Thus, most officers recognized some of the social issues behind their poorer relations with some sections of society and reported attempts to improve relations with these groups, particularly young people, through more pro-active policing methods.

However, the majority of rural policing initiatives were in response to the demands of middle class residents, which were re-enforced by government rhetoric and funding to encourage 'innovative' policing (Table 8.5 and DETR 2000). Many of these were aimed at improving surveillance. Most obviously, 'Watch' initiatives, such as Neighbourhood Watch, were employed in many rural areas (Table 8.6) and, in some forces, were tailored towards specific rural localities and issues, evident as 'farm' or 'horse' watch. These schemes operate by encouraging residents to survey particular territories and work with the police to enforce them against crime or perceived threats of crime (Yarwood and Edwards 1995).

Table 8.6 Policing methods in urban and rural areas

Policing Method	Percentage of Forces Using Particular Policing Methods in Urban and Rural Areas of their Constabulary		
	Urban and Rural Areas	Urban Only	Rural Only
Car Patrols	100%		
Foot Patrols	84%	16%	
Bicycle Patrols	52%	30%	
Neighbourhood Watch	88%	12%	
Farm Watch	12%	4%	76%
Horse Watch	17%	4%	56%
Business Watch	46%	30%	4%
Pub Watch	68%	24%	8%
Special Constables	96	4%	
'Crimestopper' Numbers	92%	8%	
Neighbourhood Wardens	22%	35%	4%
CCTV	48%	52%	
Traffic Cameras	59%	18%	4%
Traffic Calming	84%	16%	
Community Safety Partnerships	96%	4%	
Mobile Police Stations	28%		36%

Source: Authors' interviews

The interviewees thought that these kinds of 'Watch' schemes had important benefits for some elements of rural society as they provided a platform for dedicated officers to create closer links with local people and communities:

> Neighbourhood Watch has had some isolated good results but is an excellent way of rekindling community spirit.

> Farm Watch maintains close links with residents/farmers and local officers.

Neighbourhood Watch gives the population a sense of community, belonging and reassurance and also involves police correspondence and personal contact.

However, Neighbourhood Watch (NW) has been criticized for being socially selective, only benefiting certain groups and displacing crime to other localities (Bennett, 1999; Yarwood and Edwards, 1995; Williams, 1999). There is also a danger that the premise of enforcing space against threats of crime or suspicious activities, may be used to target activities that are culturally rather than criminally threatening and disturb hegemonic rural ideals (Yarwood and Gardner, 2000; Yarwood, 2002).

Despite this, many forces are increasing the use of surveillance in rural places, largely in response to the demands of (some) members of the rural public and also the release of monies available from the police modernization fund (DETR 2000). These have included a 'Rural Watch' to extend the principles of NW to wider communities, the use of mobile police stations to improve visibility and the use of rural safety initiatives. These are outlined below.

Rural Watch The Gloucestershire constabulary introduced the 'Rural Watch' scheme as one of ten initiatives from their Rural Policing Strategy. Its aim is to provide a framework for residents and participants from all areas of the county (and bordering counties) to work in partnership with the police, Neighbourhood Watch and other agencies to exchange information and actively assist in the prevention and detection of crime. The membership of the initiative includes residents but also landowners, employers, employees, those who use the countryside for recreational and leisure purposes and agencies such as the NFU, Forestry Commission and Ramblers Association. It works in co-operation with other Watch schemes that are located in rural areas. A similar 'Dale Watch' Scheme operates in North Yorkshire that encourages communities to patrol their local areas. It has been credited with the arrest of 'travelling criminals'.

Mobile police stations Mobile Police Stations (MPS) are an initiative introduced by many forces across the country. Gloucestershire first introduced a MPS in 2001 for the rural area of the Forest of Dean (West Gloucestershire) and 'has received very positive feedback from local people'. It aims to:

- reduce the fear of crime and provide reassurance;
- improve two-way communication and increase personal contact between members of the public and police officers;
- provide high visibility in rural areas on a regular basis and make it easier for people in rural areas to report crimes;
- support and promote crime prevention, Watch schemes, Crimestopper numbers and the constabulary's drug strategy;
- support divisional operations, increase intelligence and reduce actual crime; and

- assist in the implementation of the constabulary's Rural Policing Strategy.

Officers involved with their implementation have suggested that they 'give more access to rural communities, give officers a base to use for high visibility patrols and take staff and computer technology to rural locations'.

Rural Safety Initiative The Rural Safety Initiative (RSI) has been introduced by West Mercia Police to encourage residents to work in partnership with the police and other agencies to identify and tackle crime and fear of crime. The emphasis is on 'planning out' crime from the rural environment using principles of defensive space (Newman, 1972). Although based on urban initiatives such as 'safe schools' or 'safe businesses', the RSI is tailored to rural areas by emphasizing the importance of:

- maintaining the rural image;
- enabling the sustainability of rural communities;
- recognising the legitimate fears of those associated with the rural community and working to allay them;
- adopting a logical approach which allows the difference between perceived and actual problems to be identified;
- recognising the importance of cost effectiveness;
- understanding that any process must embrace the principle that every community's individual issues are site specific (Small, 2001).

These initiatives are enabled by local 'beat managers' who have responsibility for policing a collection of beats in a more holistic, reactive manner (Yarwood, 2003).

As well as improving surveillance, new policing initiatives frequently emphasize partnership working. This is in response to the requirements of the 1998 Crime and Disorder Act that obliges the police, local authorities and other relevant agencies to work in partnership with the public to identify and tackle crime and safety issues in particular places. As McLaughlin (2001) suggests, emphasis is being (re)placed on policing rather the police in the countryside. This offers both opportunity and danger.

Young (2002) has recognized that an 'exclusive society' can develop if particular groups are labelled and scapegoated by the elite. As the officers interviewed in this chapter have started to realize, much public insecurity is cultural, rather than criminal, in nature and has arisen from rapid and recent changes in rural society. Young (2002, p.487) has suggested that the way to an inclusive society is to recognize diversity within it: 'what we need is a reconceptualization of inclusion to embrace a politics that enhances social justice and welcomes diversity'. Partnership working may provide one way of doing this. The police, with their multiple roles and widespread contact with a wide cross-section of society are well placed to form, lead and manage partnerships to blend successfully resources from different sections of society (Kearns and Pattison, 2000). It is important that pro-active measures are taken to engage with groups currently excluded from partnership working, such as the young or nomadic. In turn, those lay, voluntary and state groups must be prepared to work with these groups in partnership, rather than using partnerships to exclude them further. The challenge for the police is to (re) identify their own roles within these changing structures and to balance the demands of competing groups. In this way better networks of *policing* can develop in rural areas to improve security for all.

Conclusions

The police, as much as the public, are being enrolled into networks of power that enforce elite discourses of rurality. Strategies aimed at improving visibility and surveillance are key to the successful development of these networks. The police are no longer simply the watchers of rural spaces, but are increasingly the watched as they are subjected to the twin searchlights of national audits and local accountability. The challenge for those researching rural society is to make visible those who are directing these spotlights and recognize the power relationships being played out in the discourses of policing rurality.

To begin doing this, two main research agendas can be suggested. First, more research is needed on the operation of new policing partnerships in rural areas. At an applied level, it would be valuable to assess the 'effectiveness' of these partnerships and whether they achieve their own goals, such as improving public safety or reducing crime. At a broader level, an understanding of the power relationships within and between these networks would contribute to knowledge about partnership working (Yarwood, 2002) and new forms of rural governance (Goodwin, 1998). As well as examining actors within these networks, special emphasis should be placed on examining the role of individual police officers. This chapter has examined policing from the standpoint of senior officers. It would be valuable to examine policing from the viewpoint of beat officers who are engaged in rural policing on a daily basis. In this way in would be possible to build up a fuller picture on the practice, as well as the policy, of policing and its impact on rural society.

Second, as this chapter has started to suggest, it would be valuable to explore the use of surveillance in a rural setting. This appears to be a key component in rural policing networks and, as the chapter has hinted, can reveal much about discourses of power in the countryside. Some current literature has started to examine the use of electronic surveillance in private spaces (Phillips, 2000), but greater attention should be given to the everyday surveillance of *public* space in the countryside. The active exclusion of groups from public, as well as private space, reveals just how exclusive (rural) society is becoming (Young 2002). As this chapter has started to show, a focus on who is watching whom will contribute to the understanding of some of these exclusionary power relationships.

Far from being a marginal topic, research on the policing of rural areas has the potential to reveal much about rural society and culture and, in doing so, contribute significantly to the theoretical and methodological development of rural geography.

Note

1. In this and subsequent figures and tables, the categories of 'urban' and 'rural' refer to the definitions used within particular forces. Thus in Figure 8.1, 12 forces respondents said that they had a response time of 10 minutes for urban parts of their constabulary. In Table 8.5, 16 per cent of interviewees said that foot patrols only operated in urban areas and 84 per cent said that they operated in urban and rural areas of their constabulary.

References

Anderson, S. (1997), *A Study of Rural Crime in Scotland*, Scottish Office, Edinburgh.
Aust, R. and Simmons, J. (2002), *Rural Crime: England and Wales*, Home Office, London.
Baldwin, R. and Kinsey, R. (1982), *Police, Power and Politics*, Quartet, London.
Beck, U. (1992), *The Risk Society*, Sage, London.
Bennett, T. (1990), *Evaluating Neighbourhood Watch*, Gower, Aldershot.
Bowling, B. and Foster, J. (2002), 'Policing and the Police', in M. Maguire, R. Morgan and R. Reiner (eds), *The Oxford Handbook of Criminology*, Third Edition, Oxford University Press, Oxford, pp. 980-1033.
Cloke, P. and Goodwin, M. (1992), 'Conceptualising Countryside Change: from Post-Fordism to Rural Structured Coherence', *Transactions, Institute of British Geographers*, Vol. 17, pp. 321-336.
Cloke, P. and Little, J. (eds) (1997), *Contested Countryside Cultures*, Routledge, London.
Cloke, P., Milbourne, P. and Widdowfield, R. (2002), *Rural Homelessness: Issues, Experiences and Policy Responses*, Policy Press, Bristol.
Cloke, P., Phillips, M. and Thrift, N. (1995), 'The New Middle Classes and the Social Construction of Rural Living', in T. Butler and M. Savage (eds), *Social Change and the New Middle Classes*, UCL Press, London, pp. 220-238.
Countryside Alliance (2001), *Policy Document, Summer 2000*, www.countryside-alliance.org/policy/handbook.pdf, accessed 19th December 2001.
Crawford, A. (1997), *The Local Governance of Crime: Appeals to Community and Partnership*, Clarendon Press, Oxford.
DETR [Department of Environment, Transport and the Regions] (2000), *Our Countryside: the Future – a Fair Deal for Rural England*, HMSO, London.
Dingwall, G. (1999), 'Justice by Geography' in G. Dingwall and S. Moody (eds), *Crime and Conflict in the Countryside*, University of Wales Press Cardiff, pp. 94-113.
Emsley, C. (2001), 'The Origins and Development of the Police' in E. McLaughlin and J. Munice (eds), *Controlling Crime*, Sage, London, pp. 11-53.
Evans, D., Fyfe, N. and Herbert, D. (eds) (1992), *Crime, Policing and Place: Essays in Environmental Criminology*, Routledge, London.
Foucault, M. (1973), *Discipline and Punish: the Birth of the Prison*, Pantheon, New York.
Friedmann, J. (1992) *Community Policing*, Harvester Wheatsheaf, Hemel Hempstead.

Fyfe, N. (1991), 'The Police, Space and Society: the Geography of Policing', *Progress in Human Geography*, Vol. 15, pp. 249-267.

Goodwin, M, (1998), 'The Governance of Rural Areas: some Emerging Research Agendas', *Journal of Rural Studies*, Vol. 14, pp. 5-12.

Halfacree, K. (1993), 'Locality and Social Representation: Space, Discourse and Alternative Definitions of Rural', *Journal of Rural Studies*, Vol. 9, pp. 23-28.

Hanbury-Tenison, R. (1998), 'Cultural divide', *Geographical*, Vol. LXX, p.67.

Hester, R (1999), 'Policing New Age Travellers: Conflict and Control in the Countryside', in G. Dingwall and S. Moody (eds), *Crime and Conflict in the Countryside*, University of Wales Press, Cardiff, pp. 130-145.

Hill, D. (1994), *Citizens and Cities*, Harvester Wheatsheaf, Hemel Hempstead.

Hoggart, K., Lees, L. and Davies, A (2002), *Researching Human Geography*, Arnold, London.

Hughes, G. (2000), 'Community Safety in the Age of the Risk Society', in S. Ballintyne, K. Pease, and V. McLaren (eds), *Secure Foundations: Key Issues in Crime Prevention, Crime Reduction and Community Safety*, Institute for Public Policy Research, Southampton, pp. 276-295.

Hughes, G. (2002), 'Crime and Disorder Reduction Partnerships', in G. Hughes, E. McLaughlin and J. Muncie (eds), *Crime Prevention and Community Safety: New Directions* Sage, London, pp. 123-141.

Hughes, G., McLaughlin, E. and Muncie, J. (eds) (2002), *Crime Prevention and Community Safety: New Directions*, Sage, London.

Jones, D. (1989a), *Rebecca's Children: a Study of Rural Society, Crime and Protest*, Clarendon Press, Oxford.

Jones, D. (1989b), 'Rural Crime and Protest in the Victorian Era', in G. Mingay (ed), *The Unquiet Countryside*, Routledge, London, pp. 111-124.

Kaufman, L. (1999), 'Crime in Rural Wales', in G. Dingwall and S. Moody (eds), *Crime and Conflict in the Countryside*, University of Wales Press, Cardiff, pp. 60-75.

Kearns, A and Paddison, R. (2000), 'New Challenges for Urban Governance: Introduction to the Review Issue', *Urban Studies*, Vol. 37, pp 845-851.

Lusher, A. (1999), 'Farmer Charged with Murder as Neighbours Pledge Help', *The Telegraph*, 24th August, p. 1.

Murdoch, J. and Marsden, T. (1994), *Reconstituting Rurality*, UCL Press, London.

McLaughlin, E. (2001), 'Key Issues in Policework', in E. McLaughlin and J. Munice (eds), *Controlling Crime*, Sage, London, pp. 53-100.

Meyer, F. and Baker, R. (1982), 'Problems of Developing Crime Policy for Rural Areas', in W. Browne and D. Hadwinger (eds), *Rural Policy Problems: Changing Dimensions*, Lexington Books, Lexington, pp. 171-179.

Mingay, G. (ed) (1989b), *The Unquiet Countryside*, Routledge, London.

Miirless-Black, C., Budd T., Partridge, S. and Mayhew, P. (1998), *The 1998 British Crime Survey*, HMSO, London.

Moody, S (1999), 'Rural Neglect: the Case against Criminology', in G. Dingwall and S. Moody (eds), *Crime and Conflict in the Countryside*, University of Wales Press, Cardiff, pp. 8-29.

Newman, O. (1972), *Defensible Space*, Macmillan, New York.

ORH (1999), *The Impact of Population Sparsity on the Cost of Provision of Police Services*, ORH Ltd, Reading.

Phillips, M. (2000), 'Landscapes of Defence, Exclusivity and Leisure: Rural Private Communities in North Carolina', in J. Gold and G. Revill (eds), *Landscapes of Defence*, Prentice Hall, Harlow, pp. 130-145.

Philo, C. (1992), 'Neglected Rural Geographies: a Review', *Journal of Rural Studies*, Vol. 8, pp. 193-207.

Reid, T. (1999), 'Soaring Crime and Drug Use 'Destroying Rural Life'', *The Times*, 24th August, p. 3.

Reiner, R. (1994), 'Police and the Policing', in M. Maguire, R. Morgan, and R. Reiner (eds), *Oxford Handbook of Criminology, Second Edition*, Oxford University Press, Oxford, pp. 997-1050.

Scarman, L. (1982), *The Brixton Disorders, 10-12 April 1981: Report of an Inquiry*, HMSO, London.

Shapland, J. and Vagg, J. (1988), *Policing by the Public*, Routledge, London.

Sibley, D. (1994), 'The Sin of Transgression', *Area*, Vol. 26, pp. 300-303.

Sibley, D. (1997), 'Endangering the Sacred. Nomads, Youth Cultures and the English countryside', in P. Cloke and J. Little (eds), *Contested Countryside Cultures: Otherness, Marginalisation and Rurality*, Routledge, London, pp. 218-231.

Small, I. (2001), *Rural Safety Initiative Seminar*, West Mercia Police, Worcester.

Smith, S. (1986), 'Police Accountability and Local Democracy', *Area*, Vol. 18, pp. 99-107.

South, N. (1988), *Policing for Profit: the Private Security Sector*, Sage, London.

Valentine, G. (1997), 'A Safe Place to Grow Up? Parenting, Perceptions of Children's Safety and the Rural Idyll', *Journal of Rural Studies*, Vol. 13, pp. 137-148.

Valentine, G. (2001), *Social Geographies: Space and Society*, Prentice Hall, Harlow.

Waddington, P. (1999), *Policing Citizens*, UCL Press, London.

Weisheit, R. and Wells, L. (1996), 'Rural crime and Justice: Implications for Theory and Research', *Crime and Delinquency*, Vol. 42, pp. 379-397.

Williams, K., Johnston, C. and Goodwin, M (2000), 'CCTV Surveillance in Urban Britain; Beyond the Rhetoric of Crime Prevention', in J. Gold and G. Revill (eds), *Landscapes of Defence*, Prentice Hall, Harlow, pp. 168-187.

Yarwood, R. (2001), 'Crime and Policing in the British Countryside: some Agendas for Contemporary Geographical Research', *Sociologia Ruralis*, Vol. 41, pp. 201-119.

Yarwood, R. (2002), 'Parish councils, partnership and governance: the development of 'exceptions' housing in the Malvern Hills District, England', *Journal of Rural Studies*, Vol. 18, pp. 275-291.

Yarwood, R. (2003), 'A (Rural), Policeman's Lot is Not a Happy One', in G. Higgs (ed), *Rural Services*, Pion Press, London, pp. 174-188.

Yarwood, R. and Edwards, W. (1995), 'Voluntary Action in Rural Areas: the Case of Neighbourhood Watch', *Journal Rural Studies*, Vol. 11, pp. 447-461.

Yarwood, R. and Gardner, G. (2000), 'Fear of Crime, Cultural Threat and the Countryside', *Area*, Vol. 32, pp. 403-411.

Young, J. (1999), *The Exclusive Society: Social Exclusion, Crime and Difference in Late Modernity*, Sage, London.

Young, J (2002), 'Crime and Social Exclusion' in M. Maguire, R. Morgan, and R. Reiner (eds), *The Oxford Handbook of Criminology, Third Edition*, Oxford University Press, Oxford, pp. 459-486.

Young, M. (1993), *In the Sticks: Cultural Identity in a Rural Police Force*, Clarendon Press, Oxford.

PART III
COMMUNITY AND GOVERNANCE

Chapter 9

Mobilizing the Local: Community, Participation and Governance

Bill Edwards and Michael Woods

Introduction

Social interaction and participation lie at the very core of any conceptualization or definition of 'community'. Such local mobilization in collaborative action and shared support stands as a defining construct in the interpretation of past rural life. In that 'gemeinschaft' world there has been a rather idealistic celebration of continuity in place, networks of family and kinship ties, a sense of mutuality, self-help and local, established leadership (Hillery, 1955; Bracey, 1959; Lewis, 1979; Harper, 1989). As both reality and stereotype, this engaged sense of the rural 'community' has been both reified and criticized in academic commentary: it has also been steadily modified and deconstructed by demographic, social and economic forces and the process of social change (Crowe and Allan, 1994; see Gardner, this volume). So much so, that it is now widely acknowledged that the contemporary rural in many parts of the developed world is largely a re-constituted, differentiated space in which people are far more diverse and share little of the past solidarity and experience that may once have bound them together.

Nonetheless within this contemporary, re-constituted rural space, whether accessible or remote, individuals and households still seek to interact locally. Whether this occurs out of a genuine commitment to support local activities or out of a desire to re-capture or re-create an imagined 'rural idyll' is an open question, but such interaction that does occur does not necessarily replicate past social relations. Instead the demographic re-composition that has occurred in such places means that new capacities-to-act are engaged, new agendas are followed and the nature and sense of 'community' that emerges is inevitably different from that which once existed (Gardner et al, 2004). Such changes have created a context for a re-consideration of the way 'community' is now constituted and enacted (Day and Murdoch, 1992; Day, 1998; Liepins, 2000a) and of how community participation is currently mobilized, practised and directed to particular ends in contemporary rural localities.

This is particularly pertinent as recent government policy and strategies for planned intervention in the British countryside have increasingly focused attention on encouraging local engagement and community participation in regeneration and development initiatives. Such an approach stands in stark contrast to the type of strategic intervention initially developed in the middle years of the twentieth century by planners and government agencies. Here is not the place to chart in detail such initiatives, but simply to record that State planning and regeneration intervention in rural areas was initially dominated by a top-down modernist form of spatial planning through local county planning policy or regional development agencies that made little attempt to engage local residents in the development process (Cloke, 1983; Robinson, 1990). While the Skeffington Report after 1969 did encourage local participation in the development of strategic policy options in County Structure plans, it did little more than mobilize elite participation. The continuing tension between these overarching strategies and local needs led from the late 1970s onwards to an experimental engagement at grassroots level through the involvement of local residents more directly in the regeneration process (Martin, 1976). This encouraged a shift to grassroots engagement and bottom-up involvement which was widely incorporated into policy initiatives from Europe such as LEADER and in governmental and agency strategy within the UK during the 1980s and 1990s that encouraged the mobilization of local residents in the solution of their own needs through community engagement. Actively supported by the Rural Community Councils, Action with Communities in Rural England and other public bodies, village appraisals and other strategies emerged which stressed the potential such initiatives offered to stimulate self-help, rural regeneration and development as an integral part of more centralized policies for rural intervention and change (ACRE, 1991a, 1991b; Sherwood and Lewis, 1994; Moseley et al, 1996; Asby and Midmore, 1996; Moseley, 1997; Edwards, 1998; Osborne et al, 2002).

Work on this theme, both national and European, has highlighted the success of such activities and examined the role of governmental policy in promoting these initiatives (Moseley, 2003; McDonagh, 2001). Such a strategy is not exclusively rural and has been effectively applied in a variety of geographical contexts in the developed and less-developed world in both urban and rural settings (Chambers, 1994). While the concern to improve local circumstances through self-help has a long pedigree, the recent promotion by policy makers of this mode of engagement, at a range of scales from the supra national to the local, indicates that further consideration now needs to be given to how this 'regime of practice'[1] should be interpreted. Particularly when such developments appear to reflect an explicit governmental strategy, applied to both urban and rural areas, which incorporates a rhetoric of participation, an emphasis on citizenship (rights and responsibilities) and places emphasis on the centrality of 'community'. An emphasis which is open to interpretation through either communitarian or neo-liberal lenses, or, governance and governmentality perspectives (Imrie and Raco, 2003; Herbert-Cheshire, 2000; Woods, 2004).

These new agendas require consideration to be given to a range of conceptual, methodological and policy questions, three of which are addressed in this chapter, namely:

•   How should 'community' be understood in this context?

- What issues arise from the proposed rural policies in England to engage the 'community' and measure its capacity-to-act as a vibrant community?
- What consequences arise as a result of the new strategies of governance that are emerging?

To address these questions we draw on research we have recently completed exploring participation in local community governance[2].

Re-positioning the Community:

In John Major's previous Tory Policy framed as 'active citizenship', and more recently under Tony Blair and New Labour, there has been a desire to re-engage local citizens in neighbourhood and community associations and action (Foley & Martin, 2000; Labour National Policy Forum, 2002). Whether the context is urban or rural there is now a conscious effort to encourage citizens to engage through active participation in civil society at a local level.

> [The] state should encourage individuals to play a full part in civil society. We disagree with Thatcher's famous statement that there is 'no such thing as society'. In contrast we believe that there is a role for the community and civil institutions within the framework of our political life (Labour National Policy Forum, 2002, p. 3).

In a rural context, this is expressed through a community involvement and a partnership rhetoric, which seeks to marshal and integrate civil participation through statutory parish institutions under the political mantle of the local and national State. Central to such participatory engagement is the presumption that it will be undertaken at a local scale, engaging people in that 'community'. However, the extent and population size of the geographical territories that are to frame these new mobilizations of the 'community' are rarely specified. That noted, the assumption underlying such strategy is that such local places may be characterized through sharing space as either a formal or functional 'community'[3]. The presumption is that through shared experience, recognized practice and local knowledge residents are best placed to determine an acceptable and appropriate response to need and future regeneration strategy for their area. An assumed corollary of this is that those who participate in public debate on these matters, constitute 'the community', and will be representative of all social fractions and interests in that place and hence all needs will be addressed. On such presumptions has practice been founded. However, it is now widely recognized that scale in geographical and social terms may influence effectiveness, and that those who do participate in such initiatives may not be representative or have the authority to speak on behalf of their 'community' (Warburton, 1997; Edwards et al, 2000, 2003; Osborne et al, 2002).

It is also increasingly self-evident that while such a territorially-based strategy addresses a place-based community of residence, such places may lack the social relations, institutions and practices associated with an archetypal, sociological 'community' or local civil society. In a rural context, the strategy is invariably described as community-based and this inevitably raises the shibboleth of traditional rural community studies and the baggage that is associated with that strand of rural enquiry (Bell and Newby, 1981; Harper, 1989; Crow and Allan, 1994).

These issues are particularly pertinent when the debate on these matters is all too often shrouded in unspecified presumptions that accord an 'iconic' status to 'community', when little is known of its contemporary constitution and cleavages. This may take one of two forms: either an unproblematic community stereotype is applied unrealistically to place-based 'communities of residence'; or 'communities of interest' are recognized, and within them, a comparably naïve presumption of a uniformity of attitude and behaviour amongst members appears to be assumed. In neither context is it safe to argue that this is the case and consequently there have been justifiable calls for a re-examination of the nature and form of 'community' now present in the contemporary countryside (Day and Murdoch, 1993). This clearly indicates that more care should be taken in the use of a 'community descriptor' in discussion of contemporary rural, participatory policy.

As Liepins (2000, p. 25) has astutely observed in her review of approaches to 'community', recent inquiries into local participatory action have taken a minimalist approach to defining 'community'. This she justifies by citing a recent special issue of the *Journal of Rural Studies* in which various articles addressed the themes of 'community governance', 'community development' and 'community action', but none defined 'community', the term was taken as self-explanatory and unproblematic (Day, 1998b; Edwards, 1998; however see disclaimer pp. 63-64, and Tewdwr-Jones, 1998, who defined community through Community Council areas). Consequently, as she noted, there is a real need to return to this issue and attempt to detail how in such contexts the term might be more appropriately employed.

That recognized, it is widely acknowledged that although local participation is primarily directed to engage residents 'in a place' (a definition utilized in the policy literature as 'the community'), those who participate may either be deeply embedded members 'of the community' with a developed sense of identity and belonging (see for example Cohen, 1985; Harper, 1989; Silk, 1999), or alternatively, participants are cast as representatives of 'communities of interest' in that place (see for example Rose, 1996; Hoggett, 1997; Day, 1998a; Edwards et al; 2001) reflecting the multiple membership of the more cosmopolitan, re-constituted rural space. Whether this leads to a position where a 'community of communities' is identified and accepted as the norm, or whether these insights generate a more multifaceted and nuanced reading of contemporary localities as 'communities', has yet to be resolved. What is now apparent is that these issues need to be addressed to establish how the rural 'community' is constituted and manifest in a contemporary context.

The approach adopted in this chapter towards such a definition is to see 'the mobilization of the local' as one manifestation of a 'nominalist' definition of how 'communities' are constituted at the present time. Liepins outlines an essential justification for such an approach. She proposes that consideration should now be given to a re-conceptualization of community as a social construct that 'embraces recognition of meanings, heterogeneity, spatial forms, dynamism and the relations associated with the uneven expression of power' (Liepins, 2000, p. 28). Of particular interest within this multifaceted approach is a concern with *the practices and activities* in which people participate and hence with 'the dynamic nature of "community" as a set of processes which are *"performed"* and *"contested"'* (Liepens, 2000, p. 31, emphasis added)[4].

This suggests that engagement with local participatory action provides one route through which to explore how 'community' is represented, en*acted* and currently constituted. Any examination of the characteristics and practices of those who participate in local action, both statutory and associational, reveals much of the political, social and civic relations in those places. The relations that emerge may be considered as a key indicator in the framing and definition of community through practice. In addressing such an agenda attention needs to be given, not only to which individuals and groups participate in the practice of 'performing community', but also to the degree to which such engagements are exclusive, how the leadership roles are acquired, on what basis and through what type of authority, and how power relations are played out through engagement with this or other local processes. For it is in the melting pot of such practice that the experiences generating solidarity and resolving contestation are cast and the shared (often mythical) memory and definition of 'community' created.

Interpretation of the characteristics of local participants and how they engage with local issues raises other challenges. It is often relatively easy to identify elite representation, sectional interest and community cleavages as part of this process, but far harder to reveal the routes and strategies through which a shared or collective 'community voice' emerges as a result of negotiation. This often leads to the foregrounding of a conflictual interpretation of contested rural spaces even though in many circumstances commonalities often remain and a more mature and reflective sense of 'community values' can emerge. Indeed the recognition of the presence and tolerance of 'different voices' might be viewed as a more welcome form of 'community', rather than the stereotyped, contested alternative.

If such issues can be addressed through a fresh examination of how local participation in community action is actually performed and practiced then an alternative and conceivably multifaceted definition of 'community' may emerge that is fine-tuned to the current social relations exhibited in the contemporary countryside. This is particularly pertinent when 'community' action is also seen to be an indicator of 'active citizenship' and communitarian engagement (Etzioni, 1995) and claimed to be a manifestation of new local expressions of participatory or deliberative democracy (Ray, 1999, 2001). This is further reinforced, when through the exercise of both rights and responsibilities, it is viewed by the current government as an appropriate realization of their expectations of ideal citizenship (Woods, 2004).

Consequently we would want to argue that engaging in the very act of participation defines the process through which 'communities' are constructed -

thus by focusing on the participatory or performative act of engagement, a community of local concern defines itself. This may not be the only rationale for, nor the only form through which, 'community' is manifest, as was recognized some time ago (Pahl, 1971), but it does seek to define 'community' through the process of coming together. Such civic engagement on behalf of the collective is central to the future outlined for rural areas in both political and policy terms and deserves much closer examination, particularly when so much is made of the need for active participation in local parish governance and of the role that may be played by local voluntary community action in the countryside. All of the issues outlined above cannot be addressed in this chapter, but it is appropriate to specify the approach to 'community' that informs our interpretation of participation and the role of leadership within it. Against this background, consideration can now be given to both the policy context and the practice of participation in statutory and voluntary associational activity that is occurring.

Mobilizing the Local: Policy, Local Participation and the Development of 'Vibrant Communities'

In late November 2000, Britain's Labour government unveiled its policy agenda for rural England with the publication of the white paper, '*Our Countryside: the future*'. The 176-page document marked a new milestone on the road of rural policy in Britain, consciously building on the format and conventions established five years earlier by the 'first rural white paper', '*Rural England*', published by the previous Conservative administration of John Major. Like its predecessor, '*Our Countryside*' represents an attempt to develop an integrated approach to rural policy, departing from the compartmentalization that had characterized British rural policy for most of the period since the Second World War. It is the product of an open consultative process, again contrasting with the closed policy-communities of the post-war era and, as in its predecessor, the policies and proposals contained within it seek to define the governing of rural England as not the responsibility solely of the State, but as a responsibility shared with rural people and rural communities.

> We will empower local communities, so that decisions are taken with their active participation and ownership. We will help communities map out how they would like their town or village to evolve and let them take on more responsibility for managing their own affairs (DETR/MAFF, 2000, p. 11).

> We want to enable rural communities to improve their quality of life and opportunity. We want to give them a bigger say in managing their own affairs and the chance to give everyone in the community a say in how it develops (Ibid, p. 146).

While these quotations clearly show recognition of the importance of local self-determination and participation, they also reveal that the impetus towards such engagement is clearly directed and facilitated by the State – '*we* will empower...*we* will let them take...*we* want to enable... etc.' Such statements may be read as either a genuine responsiveness by the State to the effectiveness of local community engagement, or as the specification of a path that it appears all must follow to achieve the desired outcomes (similar positions were advanced in DoE/MAFF, 1995; Scottish Office, 1995; Welsh Office, 1996).

One of the most notable elements of the second rural white paper for England (DETR/MAFF, 2000), was the introduction of an audit culture into British rural policy, through the specification of fifteen 'headline rural indicators' which would be used to 'monitor and report on our progress in achieving the objectives we have set [in this white paper]' (p. 165)[5]. Among the headline indicators included in this audit strategy is 'community involvement and activity', to be assessed through a measure of *'community vibrancy'*. This emphasis on community involvement and participation is not new, what is perhaps of greater significance is that evidence of community involvement and activity should be translated into a measure of 'community vibrancy' against which performance may be judged and by implication 'rewards' allocated: implicitly suggesting that the State expects its citizens to participate within and on behalf of their community, if rights and responsibilities are to be fully exercized.

It was proposed at this stage that an indicator would be constructed based on the numbers of meeting places, the number of locally based voluntary and cultural activities and participation in parish elections (turnout, seats contested etc) as a measure of 'community vibrancy'. The indicator would then be used to enumerate the percentage of parishes falling into four categories - vibrant, active, barely active, and sleeping - which in turn would be linked to policies for the reform of parish councils and the recognition of quality parish councils [QPC], to the promotion of community planning (through Village Design statements and Community Holistic Action Plans [CHAPs]) and to the provision of financial support for village services. Such activities are to be encouraged and developed by parish councils, who, it is anticipated, will bring together those active in local voluntary associations and community groups and facilitate a co-ordinated discussion of the needs of their constituents[6].

A preliminary assessment of community vibrancy conducted by the Countryside Agency in its *State of the Countryside 2001* report, used data available from the Rural Services Survey. It scored communities by a weighted calculation of the presence of a village hall or similar (weighting 3), contested parish council elections (3), local village traditions and annual events (2), the presence of a public house (2) and whether co-opted members served on the parish council (1) (Countryside Agency 2001). It perhaps unsurprisingly found a link between population size and 'vibrancy' with a majority of parishes above 5,000 population falling into the 'vibrant' category, but less than a quarter of parishes of less than 1,000 residents doing so. Over a fifth of parishes of a population of less than 1,000 were classified as 'barely active' (the Countryside Agency assessment did not use the 'sleeping' category). According to the government's own criteria stated in the white paper, its policies can be judged to have been successful if there is an upward movement of the balance of parishes from the sleeping category towards the vibrant category. Given the nature of these elements it is perhaps unsurprising that the assessment 'found' that larger communities were more likely to be 'vibrant'. The Countryside Agency also proposed that in the next updating of the community vibrancy indicator, proposed for 2003, 'formal voluntary activities, such as Village Appraisals and Local Agenda 21 activities, will be incorporated' (Countryside Agency, 2001, p. 27) - and employ data sources which are currently 'under investigation' (ibid, p. 78). Apart from the difficulties of measuring voluntary activity, a fundamental problem arises from the attempted aggregation of indicators relating to different types of participation and these are issues which we will return to later. The provisional nature of the methodology hence presents an opportunity for this measure and the assumptions which underlie it to be critiqued.

Three important strategic emphases were present in the 2001 community vibrancy measure. First, there was the recognition of a necessary link between the political and the civil - councillor and community group activity - to capture the range of active participation in place. Secondly, the proposal to create 'quality parish councils' offered more local responsibilities to those places that meet defined criteria providing, alongside other criteria, 'all parish councillors...have stood for election' (DETR/MAFF, 2000, p. 147) - and in so doing will seek to reward places characterized by competitive, democratic engagement. Third, parish councils are encouraged to draw other local voluntary groups into joint discussions, to act as local facilitators and to speak for and represent their community. While each of these emphases is laudable, they also appear to suggest a strategic governmental desire for strong and engaged community leadership and coordination to be mobilized through a statutory governance route, which will be rewarded by additional powers through successful mobilization of the community and through the successful incorporation of the civic.

Other actors within the rural policy sphere have made similar connections between community vibrancy and the appropriate structure for the expression of community leadership. The Rural Group of Labour MPs proposed in its *Manifesto for Rural Britain* 'a review of the current effectiveness of parish, town and community councils with a view to enhancing their roles and extending their discretion to lead, support or fund local initiatives' (RGLMPs, 2000, p. 12). Parishes which cannot meet these criteria will not only be denied access to additional powers and finance, but also may eventually find their form of governance coming under review, with the then Environment Minister, Michael Meacher, reportedly stating that 'we need to look at alternatives' to parish councils where an extension of powers might be inappropriate (Bevan, 1999). The implication of this kind of statement is that the 'poor leadership' of parish councils without an electoral mandate may hold back the development of particular places and that formal engagement through such a local political structure is the necessary route through which any local initiatives should be channelled.

Others might argue that in the last ten years, alternative community organizations have provided more effective opportunities for residents to engage with matters of local provision and indeed have been more appealing than parish council involvement. In a recent article, ACRE's[7] policy officer has linked issues relating to participation in the democratic process with the involvement of communities in creating their own solutions to problems of service provision, transport and housing. She proceeds to warn that,

> To date, although numerous communities have measured their needs, prioritized action and changed their circumstances, the Parish Council has not always been at the forefront of activities. As elected representatives of the community, Parish Councils should be instrumental in promoting approaches that aim to identify their community's needs, and should be responsive to their outcomes...[however] ... The perception of Parish Councils can be that they do not fulfil the role of representing and addressing the concerns of rural inhabitants (Sambrook, 2001, p. 15).

For Sambrook, the purpose of this observation is not to propose that neighbourhood fora or other innovative means of deliberative democracy might more usefully replace parish councils, but to call for an effort to inspire parishioners to participate in parish council elections (see Woods and Edwards, 2002; Woods et al, forthcoming). However, the sense of a problem of participation in rural community government compromising community revitalization is the same (see Fahmy et al, 2001). This quotation does however highlight that much of the recent engagement with rural initiatives has come directly from civil society/ community and voluntary organizations rather than existing political ones. So much so that in many rural districts it might be fair to argue that the most progressive leadership has come from the voluntary, civil sector.

The challenges the 2001 strategy faced were clearly identified in proposals made two years later in the *State of the Countryside Report 2003*.

> Because it is intangible, creating an indicator which monitors vibrancy directly is challenging. The range of activities across thousands of diverse rural settlements cannot be measured nationally or regionally. In the State of the Countryside 2001 the Countryside Agency reported on the number of parish level meeting places and contested parish council elections as a measure of people's opportunities to do things. We are now revising our approach, using a different, linked group of three indicators, each measuring something individually about the potential for, or reality of, participation in rural community life (Countryside Agency, 2003, p. 17).

The new indicators they propose are Community Space (local facilities and services), Community Engagement (social capital and participation) and Community Strength (community organizational capacity). The first of these causes little difficulty based as it is on availability and distance to facilities for different levels of provision across rural England drawn from the Countryside Agency's own survey data. Such provision is then to be scored and grouped into three categories based on access as limited, positive, or extensive. The second and third indicators seek to capture the extent to which such facilities do encourage people to participate and feel civically responsible and engaged, but are more experimental. The Community Engagement measure is to be based on material derived through the disaggregation of the national sample data in the General Household Survey Social Capital Module 2000. These data explore how people's feelings about their area, and their links with their neighbours, affect what they do in their community and create a sense of community or social capital which it is suggested contributes to active civic engagement. This measure will be derived from the four elements of local area satisfaction, civic participation, perception of local facilities and neighbourliness. These will later be used to explore how community or social capital within the locality relates to wider issues of social inclusion and cohesion. The final indicator, Community Strength, reflects a neighbourhood's ability to organize itself, its capacity to act to get things done, seek resources to do so and to deliver initiatives within the community through collective engagement. Drawing on work now underway in urban neighbourhoods in Bradford, the Countryside Agency is to develop a parallel strategy to identify how community vibrancy is created and sustained in a rural setting (Countryside Agency, 2003, p. 23). Although these second two indicators pose enormous challenges in terms of devising an adequate evidence base to categorize community involvement, they do seek to capture a more representative measure of the complex forces that generate community participation by stressing engagement within civil society in addition, it must be assumed, to monitoring involvement with statutory representative roles on parish councils. It will be interesting to see whether high scores on community engagement and community strength - measures essentially of how community is performed and enacted - correspond in any way with comparable levels of participation and competition for statutory representation on parish councils. Such an assessment may serve as an interesting measure of the relationship between civil and political participation that may challenge or confirm the efficacy of current governmental policy to promote the integration of both arenas of participation as the sites of action for a reinvigorated, participatory rural citizenry.

Problematizing Participation and Community Vibrancy

Some insight into the consequences of employing indicators of the type proposed in both 2001 and 2003 can be identified from our recent work exploring participation in community governance in England and Wales noted earlier. In terms of the criteria suggested to designate quality parish councils and to be indicative of community vibrancy, namely active interest and electoral competition for places on parish councils, our recent findings reveal considerable disinterest and apathy in engaging with this route to local leadership and participatory involvement. Contested elections for town, parish and community councils are the exception and not the norm, occurring in only 28 per cent of wards. In many areas there appears to be a significant problem with the recruitment of candidates. Over a third of wards had less candidates nominated than seats vacant in the last elections, whilst no candidates at all were nominated in around 3 per cent of wards. The proportion of wards requiring contested ballots has fallen sharply over the last decade, whilst the proportion with a shortage of candidates has doubled (Table 9.1).

Table 9.1   **Contestation of local elections in England and Wales 1998-2000 (Percentage of wards, 1998-2000: n=8573)**

	1964-67	1987-90	1998-2000
More candidates than seats	32	44	28
Same number of candidates as seats	46	38	32
Fewer candidates than seats	22	18	36
Uncontested (n.k.)	n/a	n/a	4

Notes and Sources:
Uncontested (n.k.) = Election uncontested where number of candidates or seats not known
1964-67 = Royal Commission, 1969; 1987-90 = DoE, 1992
1998-2000 Woods and Edwards ESRC Survey 2001

The sharpest decline in contested elections and the greatest increase in wards with insufficient candidates have both been for wards with less than 1000 electors. Only 18 percent of wards with less than 1000 voters had contested elections, whilst 40 per cent had insufficient candidates nominated in the 1998-2000 data. However, generally turnout is highest in small rural parishes and lowest in parished urban wards when elections occur. In two-fifths of elections the turnout is between 30 and 45 per cent, and the overall trend is downwards.

In spite of these concerns about the democratic mandate of parish, town and community councils, recent years have seen increasing pressure for the role of such councils in local government to be enhanced (Coulson, 1998). Following the 2000 Rural White Paper, the responsible government departments - the Department for Environment, Food and Rural Affairs (DEFRA) and the Department for Local Government, Transport and the Regions (DLTR) (later absorbed into the Office of the Deputy Prime Minister (OPDM)) – published proposals for the introduction of a new class of 'quality parish and town councils' in England that would be permitted to draw down additional functions to be discharged under delegated powers. To qualify for 'quality' status, councils will be required to pass an accreditation test with benchmarks including the completion of training by the council clerk, the publication of a regular newsletter, and, significantly, an electoral mandate. In the initial consultation document, this was proposed as all seats on a council to be filled by an election at the beginning of each four-year term (DEFRA/DLTR, 2001). Interpreted strictly, this requirement would have permitted only 28 per cent of parish and town councils in England to qualify for quality status, or around 2,300 councils in total (Table 9.2). However, as Table 9.2 shows, a relatively small reduction in the threshold of the proportion of councillors elected (as opposed to co-opted) to a council would substantially increase the number of councils eligible to qualify for the scheme.

Table 9.2   Number of parish and town councils in England that would pass an 'electoral mandate' threshold using different criteria

Criterion	% of councils in database in band	Projected total number of councils in band
All members elected in contested ballots	28	2,300
All members elected in full elections (either contested or uncontested)	63	5,200
80% of members elected in full elections (either contested or uncontested) (i.e. fewer than 20% co-opted)	78	6,400
50% of members elected in full elections (either contested or uncontested) (i.e. fewer than 50% co-opted)	97	8,000

Source: Woods and Edwards ESRC Survey 2001. Figures are based on a database of 6,400 councils projected to the full number of 8,285 parish and town councils in England.

The final proposals for the 'quality parish and town councils' scheme adopted an 80 per cent threshold for the proportion of elected members on a council, thus permitting over three-quarters of councils to potentially qualify (other 'tests' would also have to be met) (DEFRA/OPDM 2003). However, if the 'quality parish and town council' initiative is seen as a means of promoting or rewarding community vibrancy, two broader concerns can still be raised. Firstly, electoral contests do not necessarily indicate functioning, vibrant communities. There is a strong correlation between electoral contests and the population size of the community as the ratio of electors to councillors is greater in larger communities. Additionally, contested elections can reflect conflicts and divisions within communities that militate against the successful mobilization of community activity, whilst a council where all seats are filled without a contested election may indicate a stable community in which a functioning network of community leadership ensures the smooth transition of council members without incurring the cost of an actual ballot (Woods and Edwards, 2002; Woods et al, forthcoming). Secondly, examination of the motivations of serving councillors suggests that the increased responsibility that would result from 'quality' status might not be sufficient in itself to encourage more people to stand for election to parish and town councils and that additional measures to encourage participation might need to be run alongside the introduction of the scheme.

In contrast, while participation in the election of parish councillors is declining, voluntary associational activity in rural areas has grown, maintaining and developing existing provision and welfare support in rural areas. This has arisen partly as a result of individual enthusiasms and engagement and partly through encouragement from various funding programmes and community development initiatives. Rural areas have a larger number of voluntary groups relative to their population than urban areas, reflecting the community focus of many groups and the larger number of discrete communities, but clearly indicative of a willingness to participate and a high level of community 'vibrancy'. For example, in rural Wales' Unitary Authorities eleven voluntary organizations per 1000 population are registered on average compared with 5.7 per 1000 in each urban Unitary Authority with the Wales Council for Voluntary Associations. In the detailed case study localities that were part of our ESRC project drawn from Herefordshire, Gloucestershire, Dorset and Hampshire, between 7.5 and 24.4 voluntary organizations per 1000 population were recorded, with the greater range of provision occurring in the more isolated localities. Within each of these localities, small market towns are the site of most organizations, each with in excess of ten voluntary associations operating and in the most active with four times that number per 1000 population. In their more isolated hinterlands too there are between 12.8 and 6.5 organizations per 1000 population serving local voluntary associational needs for the more dispersed village, hamlet and farming populations. This level of activity suggests that while participation in the formal realm of community politics and parish representation still occurs largely through uncontested election and consensual selection rather than competitive election, many other residents are actively engaged in community participation in a host of alternative community activities. These may be seen as a *leit motif* of traditional rural self-help or alternatively as a measure of active, participatory citizenship and civic responsibility. As such they constitute a fuller record of vibrant rural communities. This suggests that the more recent 2003 Countryside Agency proposals to explore community engagement and community strength may have greater purchase in the identification of why, how and where community vibrancy occurs.

That noted, measuring community vibrancy through whatever measures of participation are adopted is not a straightforward process. Any attempt to do so is beset by conceptual questions about the type and nature of participation, by methodological questions about how measurement should be approached, what indicators should be used and at what territorial scale it can be effectively undertaken. The proposed indicators of community vibrancy raise questions in all these areas, not least because the issue of what they are supposed to be measuring and how remains unclear.

Put simply, the problem is this. Can the presence or absence of organizations or resources tell us much about the level of participation or the vibrancy of the community? Is participation in parish council elections a measure of other levels of participation in civil society? If not, which types of participatory act should be weighted as most critical in any measure of community vibrancy - participating as an elected parish councillor, or participating in some local form of voluntary action or participating in a village pub quiz league? If 'community vibrancy' relates to the total sum of communal activities taking place in a community, then the latter should certainly be included and the presence of a public house may act as an appropriate surrogate indictor. But so would the presence of a children's playground, where both children and parents might interact; whilst the role of churches and chapels as focal points for communal activity are equally important, but completely absent from the Countryside

Agency's indicator. The problem with seeking to measure total participation through surrogate indicators is that the variety of forms it may take makes producing an accurate, inclusive, assessment virtually impossible.

An alternative process-based way to understand vibrant communities is to ask *not* what is present or absent in a place, but *what* are local people doing, and *who* is doing *what* to improve circumstances in particular places, and *how* is 'community' being *performed*. In essence 'vibrant communities' are those involved in engaging with their own futures and shaping how this can be achieved. Consequently, it could be contended that community vibrancy relates to the capacity of a community to mobilize itself and to secure resources for its own benefit and manage them effectively. This has been appropriately recognized by the Countryside Agency in their most recent proposals, but they face an enormous methodological challenge in developing the surrogate indicators to capture this level of engagement. If this can be achieved, then the mobilization, participation and engagement by local people in local action or effectively in governing themselves, or in what might be termed community self-governance, becomes the focus of attention. This however raises further questions about what might be meant by community self-governance.

Positioning Participation in Community Self-Governance

As Rhodes (1997) has observed, the term 'governance' has been so widely and loosely employed in recent years that it risks becoming just another term which 'has too many meanings to be useful' (p. 15). Thus we start by positioning 'community governance' as a context of research rather than as an object of inquiry (Woods and Edwards, 2002). By employing the term 'governance' we seek to imply a focus on the processes of governing within rural communities - or, as activities which 'involve either the provision of public services within the community, or the representation of community interests to external agencies'. Furthermore, the preference for governance above government alludes to the 'complex set of institutions and actors drawn from but also beyond government' indicated by Stoker (1998; c.f. Goodwin, 1998), including those from local associational and community groups. For we would emphasize that the governing of rural communities involves not just the statutory parish, town or community council but also a disparate collection of voluntary organizations and committees, project-focused initiatives, ad hoc networks and individual action - and thus that the investigation of participation in the governing of rural communities requires an exploration of participation in all these activities, not just in the statutory institutions of government.

However, whilst at one - superficial - level 'governance' may work as a shorthand descriptor for the range of actors and organizations involved in governing, at other levels the conceptual development of the idea has imbued 'governance' with meanings that sit uncomfortably with a catch-all usage. Firstly, a common feature of concepts of governance is that the actors involved are not just drawn from across the spectrum of the public, private and voluntary sectors, but that governance results from the collective engagement of those actors in forms of networking of varying degrees of formality (see Rhodes, 1996; Stoker, 1997, 1998; Marsden and Murdoch, 1998; Goodwin, 1998; Murdoch and Abrams, 1998; Edwards et al, 1999; Jones and Little, 2000; Herbert-Cheshire, 2000). Whilst such networking in both formal and informal, deliberate and unintentional forms, is a feature of governing rural communities, we wish to remain open to the possibility that the governing of some rural communities may proceed through autonomous organizations which cannot be accommodated within Stoker's (1997, p. 3) statement that 'governance is about governmental and non-governmental organizations working together' (c.f. Jones and Little, 2000).

Secondly, much writing on governance stresses its - albeit contested - novelty: 'governance signifies a change in the meaning of government, referring to a new process of governing; or a changed condition of ordered rule; or the new method by which society is governed' (Rhodes, 1995, pp. 1-2; c.f. Stoker, 1997; Goodwin and Painter, 1996). Whilst the restructuring implied here can also be identified at the community scale, the 'governing' activities undertaken by many of the non-state actors that should be included are not the product of recent shifts in the process of governing, but are long-standing practices within rural settings (Bracey, 1959).

Thirdly, governance is a dynamic process, which operates through the 'tangled hierarchies' of public, private and voluntary sector actors, rather than an entity comprised of those actors. Whilst this observation is not inconsistent with our original usage of the term 'community governance', it does suggest that governance should be repositioned as the *consequence* of participation rather than as the *context* for participation - a distinction, which will become significant as we proceed to problematize participation.

Regardless of whether we wish to position community self-governance as a context for participation or as a consequence of participation, the operationalization of a research focus on this arena presents issues about what it means to participate in the governing of a community. This is a question of both breadth and depth. We would contend that such a question is more incisive than simply measuring 'community vibrancy'.

To take the former, the question relates to the breadth of activities which might be considered to constitute participation in the governing of a community. The term 'community self-governance' is employed to define an arena of participation, which extends beyond the parish or community council to embrace all activity that involves a determining role in providing either public services or resources within the community, as well as the representation of community interests to external agencies.

Thus, participants in community self-governing include not only parish councillors but also - for example - members of the village hall committee, school governors, officers of residents' associations, residents involved in Neighbourhood Watch and other similar schemes, organizers of applications to the National Lottery Charities Board for funding to provide community facilities, trustees of community-run shops or transport schemes, leaders of anti-development protest campaigns and so on. Whilst participants in many of these roles would describe their activity as apolitical, they could all however be construed as political acts in a broad sense in that they all seek to represent the community to some extent, and in that they all involve mediation between the community and the State or its agencies.

By this token, there are numerous forms of participation in community life which do not qualify as participation in community self-governing (but they may instead be critical in the performance of community) - for example running a scout troop, organising a babysitting circle, playing on the cricket team, or contributing to the church flower rota. Other activities are more ambiguous. For example, being a member of a Women's Institute would not constitute participation in community self-governing, but being the president of the WI may present the opportunity to participate as a representative in governing through the incorporation of the holder of such a role into wider local networks of governance. Similarly, running a youth football team is not participation in community self-governing, but the football team coach becoming part of a bid for funding to enhance the village's sports ground may be, in that such action shapes the future for others. The focus of such activity is also significant here. Here the focus is essentially with community self-governing - participatory activities, which are performed by residents of a community within that community for the collective benefit of that community.

In parallel with such activity, there are many agencies which contribute to the governing of a community that are external to that community and which have responsibilities beyond the geographical territory of that community across a wider area. Participation by community residents in these agencies - as district and county councillors, members of health authorities, magistrates, members of partnerships etc. - is not included in our current definition of participation in community self-governance, unless they hold local positions, independent of their exogenous roles, and engage as resident participants in local governance.

In addition, there are residents who are involved in political and voluntary participation in organizations, which are targeted at a specific group of people - either within the community or across a wider territory. As such they belong to particular 'communities of interest' and are not necessarily included as participants in wider community self-governing, although they may play such a role within their 'communities of interest'. These include, for example, volunteers in meals-on-wheels services and youth group leaders, for their main interests are focused on the needs of special interest groups. Such individuals, if serving on a representative community forum, would, however, be part of a group, network or partnership that might collectively be concerned with community self-governing.

Turning to questions of depth, the issue becomes one of how much engagement is required for an individual to qualify as participating in the governing of the community.

- Does completing a community appraisal questionnaire or voting in a parish council election constitute participation, or is a more sustained and involved engagement required associated with an official role in local organizations?

- Are candidates for parish councils participants, or just those successfully elected?
- Are all members of a residents' association participating in community governing, or just the officers?

Such questions lead us to draw distinctions between different forms of participation in community governing.

Participation in the governing of rural communities may be distinguished by a number of, often-related, criteria. The most notable of these distinctions is between mass participation by community residents and elite participation in positions of community leadership. Within the former category participation may be further plotted along an axis from fairly passive participation as, for example, a member of a residents' association who attends meetings but rarely speaks, through voting in parish elections, to active participation in public meetings or planning for real exercises. Each of these activities, however, is clearly differentiated by time commitment, involvement, information required for effective participation, and the potential individual influence arising from participation in positions of community leadership (see Edwards et al, 2002; Gardner et al, 2004). Clearly those who lead are critical to community vibrancy because they create the possibility for others to participate. Consequently, a distinction should be drawn between mass participation as a background indicator against which more detailed investigation of participation in community leadership positions can be contextualized.

That noted, the nature and demands of participation also vary between different community leadership roles. For example, one might draw a distinction between the statutory responsibilities, public accountability and breadth of scope associated with participation as a parish councillor and the more focused and self-defining role played by the unelected co-ordinator of a community development project or leader of a campaign against housing development or the organizer of a local fete or festival. Given these differences, one might anticipate that the motivations for participation and the experiences of participation will differ and inevitably those individuals involved in these roles will have different interpretations of why and to what end they participate. Yet, as observers of rural life know, frequently it is the same individuals who occupy several of these positions - so how do they conceive of their multi-faceted participation?

These types of questions seem to us to be critical in understanding the nature of community vibrancy and the level of participation that is actually occurring. They go much further than simply auditing the presence or absence of certain opportunities for local participation. They also widen attention away from a concern with the formal channels of local participation such as the quality parish council and provide an opportunity for the consideration of a far wider range of engagement in the emerging organizations developed by civil society, and focus on issues that require attention to be given to local context, practice, power and consequence. They constitute, therefore, an important research arena for future work that interrogates more closely the nature of participation and its consequences. Alongside these enquiries, and implicated in them, is the role that policy initiatives play in shaping the authority and legitimacy placed on such action and this requires some further comment in our final section.

Concluding Reflections

The 2000 Rural White Paper resonates strongly with the tone of the previous white papers published in 1995/96, where Murdoch (1997) identified the emphasis on community and individual action as representing a shift in the mode of governmentality of rural England. Following Rose (1993, 1996a, 1996b) – who in turn adopts the concept of governmentality (or the means by which society is rendered governable) from Foucault (1991) - Murdoch positions the first rural white paper as part of a transition from a welfarist-governmentality characterized by state intervention in a nationally-configured social and economic space of government, to a new rationality of 'governing through communities' which,

> does not seek to govern through 'society', but through the regulated choices of individual citizens, now construed as subjects of choices and aspirations to self-actualization and self-fulfilment. Individuals are to be governed through their freedom, but neither as isolated atoms of classical political economy, nor as citizens of society, but as members of heterogeneous communities of allegiance, as 'community' emerges as a new way of conceptualising and administering moral relations amongst persons (Rose, 1996b, p. 41, quoted by Murdoch, 1997, p. 112).

As we have discussed elsewhere (Edwards et al, 2001), the second rural white paper continues to reinforce the strategy of 'governing through communities', but does so by subtly modifying the means through which such a strategy is envisaged as operating. In Rose's original coinage, 'governing through communities' did not necessarily refer to geographical communities, but related to the acting out by individuals of their responsibilities within settings which may be communities of residence but could equally be occupational communities, interest communities or other communities of identification. Thus in the 1995 Rural White Paper emphasis was placed on detailing actions and duties to be performed by individuals not just within rural communities, but potentially as farmers or business owners or ramblers or as members of a range of other interest communities. In the 2000 Rural White Paper, in contrast, the ideological influences of post-Thatcherism have been superseded by the principles of the Blairite Third Way such that individual action has been discursively marginalized and a renewed emphasis placed on collective action by rural communities, through established rural community institutions.

A participatory strategy is now clearly expected to proceed through statutory political representation in and co-ordination between the established hierarchy of government institutions, and with a joined-up partnership incorporating a range of agencies and actors, rather than simply through the previously implied, independent, public engagement of individual 'active citizens'. This might be interpreted as a deliberate attempt to channel or steer the enthusiastic engagement and vibrancy of participation in the voluntary and associational sector in rural areas through the established, but representative and elected route of local governance, the parish council and also as a deliberate attempt to encourage citizens to engage more directly in both civil and political activities.

The English Rural White Paper in prioritising an audit culture, in stressing community vibrancy as a key indicator and in emphasising a future key role for quality parish councils begs the question of what encourages a community to be vibrant. Most local action springs from need, reaction to threat or in response to locally promoted opportunities. However, for these factors to mobilize local residents into action there must be leadership. This is not easily achieved – a fact widely recognized by those involved in rural development.

Reflecting on these policy proposals two normative assumptions emerge that merit closer inspection. The first is that both 'community vibrancy' and 'participation' are viewed as good things, and that communities that score highly on these measures are therefore 'better' communities than those which do not. The danger with this assumption is that the normative expectation of a vibrant community may reflect more an 'idyllized' stereotype of an old-style English village rather than the reality of the modern countryside criss-crossed by highly fluid patterns of work, travel, family and social relations, consumption and leisure that generate multiple communities of interest. The second normative assumption is that it would be a beneficial process to distil 'community activity' and 'participation' into indicators that it is presumed, could give an unambiguous and neutral measure of community vibrancy. This raises a number of questions. Namely:

- Are the criteria adopted appropriate?
- Can 'participation' be adequately enumerated and assessed through parish council elections, the community space, community engagement and community strength surrogate indicators?
- At what scale and for which units can appropriate statements about 'community vibrancy' be made, particularly when actions and facilities often are developed to serve a wider territorial area than one parish?
- If the parish alone is the base, how are such wider initiatives accommodated?
- What support or consequence will follow from a political agenda that frames the reading of such participatory activities in social capital terms and chooses to audit how communities perform?

Given these questions there is much that might be asked about the proposed new practices that are emerging that can form the basis for future research enquiry.

We are tempted to suggest in conclusion that in reviewing these strategies to enumerate and codify 'community vibrancy' we may be parties to seeing new forms of community self-governance enrolled through new strategies of governmentality. That recognized however we would want to argue further that understanding community vibrancy therefore is not simply achieved through an audit or enumeration of what is going on. Such an intention begs the key question, which is to ask how has this activity been achieved? Which circumstantial factors (of need, of reaction or opportunity) may prompt engagement? To mobilize the community to action, to create local organizations requires individuals genuinely committed to taking responsibility towards shaping the future for others – i.e. a form of citizenship or civil responsibility – through leadership. Far too often such players have been seen to operate on the local stage for self-interest - in a vibrant community the question may be, how is such a stage or set assembled so that the world can be changed for others? At its essence community vibrancy is about 'capacity to act or power to', 'not power over' or 'power through'. Exploring these issues will constitute an interesting and challenging future research agenda.

Notes

1. The term 'regime of practice' is used here simply as a descriptor of the established instruments and practices used in planning and policy engagement over a given period. While community action has been a feature of rural and urban areas for a considerable period of time, it has certainly been the case that since the late 1980s the engagement of community in policy consultation and delivery has increasingly become an established practice and requirement at sub-national, national and European levels. This has been accompanied by the requirement for 'partnership working' co-ordinating action between different public, private and voluntary or community groups across different scales of engagement. Hence we choose to describe this period as dominated by 'a regime of practice' in policy delivery terms. There is an obvious parallel here with Stone's (1989) notion of 'regimes', with the way Dean (1999) employs Foucauldian ideas of 'regimes of practices' in conceptualising governmentality, but we are not using it with such associations, though we recognize that these resonances need further exploration in future research.
2. Woods, M. and Edwards, W.J. *Participation, Power and Rural Community Governance in England and Wales*, a project funded by the Economic and Social Research Council as part of the Democracy and Participation programme, running from 2000-2005, Award No. L215 25 2052.
3. The exact definition of 'community' in the context of the White Papers is not clear. The Concise Oxford English Dictionary (1990) defines community as *'all people living in a specific locality'*. The definition of 'community' used by the UK Government is *'"Community" is taken to mean any group of people with a common bond above the family unit and below the first stage of municipal administration'* (DETR, 1998).
4. This is one of the three approaches advocated by Liepens (2000) alongside a concern with 'exploring the meanings' and 'mapping of the spaces and structures' of community to generate fresh understandings of the nature of the contemporary community.
5. The fifteen rural indicators that were proposed were equitable access to services; tackling poverty and social exclusion; better education for all; an affordable home; better rural transport; safer communities; high and stable levels of employment; prosperous market towns; thriving rural economies; a new future for farming; protecting and enhancing the countryside; restoring and maintaining wildlife diversity; protecting natural resources; increased enjoyment of the countryside and the promotion of community involvement and activity. The intention is that the performance of various policy instruments can be formally measured across this range of indicators to evaluate progress and territorial variations in success. This would allow the effectiveness of policy to be judged and, in the longer-term, alternative strategies to be explored and developed.

6. 'Quality parish council' the white paper suggests should be, representative of all parts of its community; meet a quality test – show that it is effectively managed; with audited accounts and a trained clerk; be committed to work in partnership with principal authorities; in proportion to size and its skills, deliver local services for principal authorities; work closely with voluntary groups in the town or village; lead work by the community on the Town or Village plan; working with its partners, act as an information point for local services. In order to qualify for 'quality parish council' status a council would have to pass a 'quality test' which is likely to include holding at least three meetings a year, having a trained clerk, publishing an annual report, presenting properly audited accounts and being composed of councillors who have all stood for election (as opposed to being co-opted). Having passed this test it is envisaged that a quality parish council could assume a number of roles including working with partners to undertake services funded from its own resources, including looking after the village environment (litter, bus shelters, village green, cemeteries etc) and providing facilities such as playgrounds and village halls; helping to draw up a town or village plan and supporting local biodiversity action plans; supporting community transport schemes and childcare provision; seeking suitable sites for affordable housing and developing youth activities and services for the elderly: and taking on the delivery of some services (e.g. facilities management, litter collection, street lighting) on behalf of principal local authorities (district and county councils) and other partners (DETR/MAFF, 2000, p. 147).

7. ACRE is the acronym for the charity Action with Communities in Rural England established in 1987 to serve as a co-ordinating, facilitating and lobbying body for rural communities in England.

References

ACRE (1991a), *Taking Stock of Your Parish*, Action with Communities in Rural England, Cirencester.

ACRE (1991b), *Village Appraisals*, Action with Communities in Rural England, Cirencester.

Asby, J. and Midmore, P. (1996), 'Human Capacity Building in Rural Areas: The Importance of Community Development', in P. Midmore and G. Hughes (eds), *Rural Wales: An Economic and Social Perspective*, Y Lolfa, University of Wales, Aberystwyth, pp. 105-124.

Bell, C. and Newby, H. (1971), *Community Studies: An Introduction to the Sociology of the Local Community*, Allen & Unwin, London.

Bevan, S. (1999), 'Labour Threatens to Abolish Parish Councils', *The Sunday Times*, 3rd October 1999.

Bracey, H.E. (1959), *English Rural Life: Village Activities, Organisations and Institutions*, Routledge and Kegan Paul, London.

Chambers, R. (1994), 'The Origins and Practice of Participatory Rural Appraisal', *World Development*, Vol. 22, pp. 953-969.

Cloke, P.J. (1983), *An Introduction to Rural Settlement Planning*, Methuen, London.

Cohen, A (1985), *The Symbolic Construction of Community*, Tavistock, London.

Coulson, A. (1998), 'Town, Parish and Community Councils: The Potential for Democracy and Decentralisation', *Local Governance*, Vol. 24, pp. 245-248.

Countryside Agency (2001), *The State of the Countryside*, Countryside Agency, Cheltenham.

Countryside Agency (2003), *The State of the Countryside*, Countryside Agency Publications, Wetherby.

Crow, G. and Allan, G. (1994), *Community Life*, Harvester Wheatsheaf, London.

Day, G. (1998b), 'Working With the Grain? Towards Sustainable Rural and Community Development', *Journal of Rural Studies*, Vol. 14, pp. 89-106.

Day. G. (1998a), 'A Community of Communities? Similarity and Difference in Welsh Rural Community Studies', *The Economic and Social Review*, Vol. 29, pp. 233-257.

Day, G. & Murdoch, J. (1993), 'Coming to Terms with Place', *Sociological Review*, Vol.41, pp. 82-111.

Dean, M. (1999), *Governmentality: Power and Rule in Modern Society*, Sage, London.

DEFRA/DLTR (2001), *Quality Parish and Town Councils: A Consultation Paper*, DEFRA/DLTR, London.

DEFRA/OPDM (2003), *Quality Parish and Town Councils*, The Stationary Office, London.

DETR, (1998), *Community-based Regeneration: A Working Paper*, DETR, London.

DETR/MAFF (2000), *Our Countryside: The Future - a Fair Deal for Rural England*, Cm. 4909, The Stationery Office: London.

DoE/MAFF (1995), *Rural England: A Nation Committed to a Living Countryside*, Cm. 3016, HMSO: London.

Edwards, B. (1998), 'Charting the Discourse of Community Action: Perspectives from Practice in Rural Wales', *Journal of Rural Studies*, Vol. 14, pp. 63-78.

Edwards, B., Goodwin, M., Pemberton, S. and Woods, M. (1999), *New Forms of Rural Governance? Partnerships, Power and Regeneration in Rural Wales*, Paper presented to the RGS/IBG Annual Conference, Leicester.

Edwards, B., Goodwin, M., Pemberton, S., and Woods, M. (2000), *Partnership Working in RuralRegeneration*, Policy Press & Joseph Rowntree Foundation, Bristol.

Edwards, B., Woods, M., Anderson, J. and Fahmy, E (2001), *Governing Through Communities: Emerging Geographies of Rural Citizenship*, Paper presented to 'Spaces of Voluntarism' session of the RGS/IBG Annual Conference, Plymouth.

Edwards, B., Woods, M., Anderson, J. and Gardner, G. (2002), *Democracy and Participation: Deconstructing the Multiple Authorities of 'Local' Leadership*, Paper presented to RESSG Conference, Cardiff.

Edwards, B., Goodwin, M. and Woods, M. (2003), 'Citizenship, community and participation in small towns: a case study of regeneration partnerships', in R. Imrie and M. Raco (eds), *Urban Renaissance? New Labour, Community and Urban Policy*, Policy Press, Bristol, pp. 181-204.

Etzioni, A. (1995), *The Spirit of Community: Rights, Responsibilities and the Communitarian Agenda*, Fontana, London.

Fahmy, E., Edwards, B., Woods, M. and Anderson, J. (2001), *Patterns and Permutations of Participation: Community Leadership in Rural Britain*, Paper presented to the Association of American Geographers Annual Meeting, New York.

Foley, M. and Martin, S. (2000), 'A New Deal for the Community? Public Participation in Regeneration and Local Service Delivery', *Policy and Politics*, Vol. 28, pp. 479-491.

Foucault, M. (1991), 'Governmentality', in G. Burchell, C. Gordon and P. Miller (eds), *The Foucault Effect*, Harvester Wheatsheaf, London, pp. 87-104.

Gardner, G., Edwards, B., Woods, M. and Anderson, J. (forthcoming), 'Endogenous Leadership: Process and Practice in Rural Community Governance', *Sociologia Ruralis*.

Goodwin, M. (1998), 'The Governance of Rural Areas: Some Emerging Research Issues and Agendas', *Journal of Rural Studies*, Vol. 14, pp. 5-12.

Goodwin, M. and Painter, J. (1996), 'Local Government, the Crises of Fordism and the Changing Geographies of Regulation', *Transactions, Institute of British Geographers*, Vol. 21, pp. 635-648.

Harper, S. (1989), 'The British Rural "Community": An Overview of Perspectives', *Journal of Rural Studies*, Vol. 5, pp. 161-184.

Herbert-Cheshire, L. (2000), 'Contemporary Strategies for Rural Community Development in Australia: A Governmentality Perspective', *Journal of Rural Studies*, Vol.16, pp. 203-217.

Hillery, G. (1955), 'Definition of Community: Areas of Agreement', *Rural Sociology*, Vol. 20, pp.111-123.

Hoggett, P. (ed.) (1997), *Contested Communities: Experiences, Struggles, Policies*, Policy Press: Bristol.

Imrie, R. and Raco M. (eds) (2003), *Urban Renaissance? New Labour, Community and Urban Policy*, Policy Press, Bristol.

Jessop, B. (2000), 'Governance Failure', in G. Stoker (ed), *The New Politics of British Local Governance*, ESRC Local Governance Programme, Macmillan Press, Basingstoke, pp. 11-32.

Jones, O. and Little, J. (2000), 'Rural Challenge(s): Partnership and New Rural Governance', *Journal of Rural Studies*, Vol. 16, pp.171-83.

Labour National Policy Forum (2002), *Democracy, Citizenship and Political Engagement*, Labour Party, London.

Lewis, G. J. (1979), *Rural Communities: A Social Geography*, David and Charles, Newton Abbot.

Liepins, R. (2000a), 'New Energies for an Old Idea: Reworking Approaches to "Community" in Contemporary Rural Studies', *Journal of Rural Studies*, Vol.16, pp. 23-36.

Liepins, R. (2000b), 'Exploring Rurality Through "Community": Discourses, Practices and Spaces Shaping Australian and New Zealand Rural "Communities"', *Journal of Rural Studies*, Vol. 16, pp. 325-342.

Marsden, T. and Murdoch J. (Guest Editors) (1998), 'Rural Governance and Community Participation', *Journal of Rural Studies*, Vol. 14, pp. 1-119.

Martin, I. (1976), 'Rural Communities', in G. E. Cherry (ed), *Rural Planning Problems*, Leonard Hill, London, pp. 49-84.

McDonagh, J. (2001) *Renegotiating Rural Development in Ireland*, Ashgate, Aldershot.

Moseley, M., Derounian, J. and Allies, P. (1996), 'Parish Appraisals – a Spur to Local Action?', *Town Planning Review*, Vol. 67, pp 309-29.

Moseley, M. (1997), 'Parish Appraisals as a Tool of Rural Community Development: An Assessment of the British experience', *Planning Practice and Research*, Vol. 12, pp.197-212.

Moseley, M. J. (2003), *Rural Development: Principles and Practice*, Sage Publications, London.

Murdoch, J. (1997), 'The Shifting Territory of Government: Some Insights from the Rural White Paper', *Area*, Vol. 29, 109-118.

Murdoch, J. and Abram, S. (1998), 'Defining the Limits of Community Governance', *Journal of Rural Studies*, Vol. 14, pp. 41-50.

Osborne, S.P., Beattie, R.A. and Williamson, A.P. (2002), *Community Involvement in Rural Regeneration Partnerships in the UK*, Joseph Rowntree Foundation, Policy Press, Bristol.

Pahl, R.E. (1970), 'Community and Locality', in R. E. Pahl (ed), *Patterns of Urban Life*, Longmans, London, pp. 100-114.

Ray, C. (1999),'Towards a Meta-framework of Endogenous Development: Repertoires, Paths, Democracy and Rights', *Sociologia Ruralis*, Vol. 39, pp. 521-537.

Ray, C. (2001), *Culture Economies*, Centre for Rural Economies, Newcastle.

Rhodes, R. (1996), 'The New Governance: Governing Without Government', *Political Studies*, Vol. 44, pp. 652-667.

Rhodes, R. (1997), *Understanding Governance: Policy Networks, Reflexivity and Accountability*, Open University Press, Buckingham.

Robinson, G.M. (1990), *Conflict and Change in the Countryside*, Belhaven, London.

Rose, N. (1993), 'Government, Authority and Expertise in Advanced Liberalism', *Economy and Society*, Vol. 22, pp. 283-299.

Rose, N. (1996a), 'Governing "Advanced Liberal Democracies"', in A. Barry, T. Osbourne and N. Rose (eds), *Foucault and Political Reason*, UCL Press, London, pp. 37-64.

Rose, N. (1996b), 'The Death of the Social? Re-figuring the Territory of Government'. *Economy and Society*, Vol. 25, pp. 327-356.

Rural Group of Labour MPs (2000), *A Manifesto for Rural Britain*, RGLMP, London.

Sambrook, L. (2001), 'Democracy, Participation and Choice', *The Rural Digest*, Vol. 1, pp.14-15. A.C.R.E: Cirencester.

Scottish Office (1995), *Rural Scotland: People, Prosperity and Partnership*, Cm. 3014, HMSO: Edinburgh.

Sherwood, K.B. and Lewis, G.J. (1994), 'Local Appraisals and Rural Planning in England: an Evaluation of Current Experience,' in A. Gilg (ed), *Progress in Rural Policy and Planning*, Wiley, London, Vol. 4, pp. 69-88.

Silk, J. (1999), 'Guest Editorial: The Dynamics of Community, Place and Identity', *Environment and Planning A*, Vol. 31, pp. 3-17.

Skeffington, A.M. (1969), *People and Planning*, HMSO, London.

Stoker, G. (1997), 'Local Political Participation', in R. Hambleton, H. Davis, C. Skelcher, M. Taylor, K. Young, N. Rao and G. Stoker (eds), *New perspectives on Local Governance*, Joseph Rowntree Foundation, York, pp. 157-196.

Stoker, G. (1998), 'Governance as Theory - Five Propositions', *International Social Science Journal*, Vol. 155, pp. 17-28.

Stoker, G. (2000), *The New Politics of British Local Governance*, ESRC Local Governance Programme, Macmillan Press, Basingstoke.

Stone, C. (1989), *Regime Politics*, Kansas University Press, Lawrence, K.A.

Tewdwr-Jones, M. (1998), 'Rural Government and Community Participation: The Planning Roles of Community Councils', *Journal of Rural Studies*, Vol. 14, pp. 51-63.

Warburton, D. (1997), *Participatory Action in the Countryside: A Literature Review*, CCWP 07, Countryside Commission, Cheltenham.

Welsh Office (1996), *A Working Countryside for Wales*, Cm. 3180, HMSO: London.

Woods, M. (1997), 'Discourses of Power and Rurality: Local Politics in Somerset in the Twentieth Century', *Political Geography*, Vol. 16, pp. 453-478.

Woods, M. and Edwards, W.J. (2002), *Participation, Power and Rural Community Governance in England and Wales*, Final report to ESRC, Award No. L215 25 2052.

Woods, M., Edwards, W.J., Anderson, J. and Fahmy, E. (forthcoming), 'Electoral participation in town, parish and community councils: a new critical analysis', Paper submitted to *Local Government Studies*.

Woods, M. (forthcoming), 'Political Articulation: The Modalities of New Critical Politics of Rural Citizenship', in P. Cloke, T. Marsden and P. Mooney (eds) *The Handbook of Rural Studies*. Sage, London.

Chapter 10

A Sense of Place: Rural Development, Tourism and Place Promotion in the Republic of Ireland

David Storey

Introduction

The sizeable social and economic transformations which have occurred in much of rural Europe in recent decades have precipitated significant changes in rural development policy and practice. One obvious consequence is the evolution of more integrated rural policies which emphasize the importance of non-agricultural sources of employment in rural areas and the need for a multi-sectoral approach to rural development. Within this, tourism and related place promotional activities assume prominence. Alongside this there has been a shift towards the promotion of a partnership approach in which interested groupings – statutory, voluntary, community-based – are seen to come together and participate in the development process (Curry, 1993; Goodwin; 1998; Storey, 1999; Edwards et al, 2000; Yarwood 2002a). Community-based tourism and heritage projects have been major beneficiaries of rural development funding in Ireland in recent years. This chapter deals with the realm of rural place promotion in Ireland; more specifically it focuses on the utilization of elements of local heritage in promoting or regenerating rural localities. The arguments surrounding recent trends in rural development are examined before outlining key issues which emerge in the debate surrounding the utilization and promotion of heritage. The chapter briefly focuses on two different types of heritage-related projects in Ireland which have been recipients of European Union LEADER funding and concludes by suggesting potentially useful avenues for future research.

Trends in Rural Development

A partnership approach has been utilized in a wide variety of rural development programmes in the Republic of Ireland (Moseley et al, 2001). The development rhetoric emphasizes the empowerment of local people through active involvement in local-decision making. Local people are not just seen as passive participants but as active agents in implementing change. More broadly these changes have been interpreted as reflecting a shift away from government and towards governance (Rhodes 1996; Goodwin 1998; Kearns and Paddison 2000) and have been characterized as a 'new localism' (Moseley, 1999). Such partnerships are said to *'empower themselves* by blending resources, skills and purposes with others. The capacity to get things done no longer lies (if it ever did) with government power or authority in one place' (Kearns and Paddison, 2000, p.847).

Various developmental programmes, such as the EU LEADER initiative, place great emphasis on partnership working arrangements involving statutory, voluntary and community-based organizations. The LEADER programme is aimed at fostering economic development in rural areas utilising a partnership approach and emphasising innovation in the types of projects to be encouraged. It also encourages animation and capacity building. LEADER is now in its third phase. LEADER I operated from 1991 through to 1994, LEADER II from 1995 to 1999 and the current phase, LEADER +, will run through to 2006. LEADER operates through a series of geographically based Local Action Groups consisting of representatives of the appropriate local authorities, other development agencies and community groups. These groups can provide part-funding (normally 50 per cent) to local group or partnership projects which can be seen as innovative.

Developments such as this have met with a range of criticisms including suggestions that they represent a recognition of the failure of 'professionals' to deal with the problems associated with rural restructuring. In addition community-based responses provoke serious questions about the ways in which we conceptualize community. It is certainly both unwise and inaccurate to assume that all members of a geographically defined community have shared interests and aspirations. These consensual models of community run the risk of eliding very real fractures amongst the rural population. More significantly, perhaps, there are suggestions that, far from being a new paradigm, these apparent changes in developmental practice remain more illusory than real with power still residing with centralized agencies and little effective empowerment occurring (Storey, 1999, 2003; McDonagh, 2001).

With the decline in importance of agriculture both in terms of its economic output and employment levels, there has been a need to re-package the countryside in different ways. In some respects this change is seen as a move from an arena of production to one of consumption. One obvious way in which the countryside is 'consumed' is through visitors coming to 'gaze' on particular landscapes. Here it is suggested that particular places may become 'centres of spectacle and tourist consumption rather than places of material production' (Mordue, 1999, p.631). It is perhaps unsurprising that a sizeable portion of LEADER-funded projects in Ireland and elsewhere have been tourist-oriented (Kearney et al, 1995, Kearney and Associates, 2000, Moseley et al, 2001). In this way, rural development funding is being fed into projects designed to attract visitors. Indeed the authors of an evaluation of LEADER II in Ireland argue that 'tourism is and will remain the main opportunity for agricultural diversification in many disadvantaged or remote rural areas, and should remain central to any comprehensive rural development agenda' (Kearney and Associates, 2000, p.118).

EU and government funding for rural development projects thereby plays a key role in the continuing commodification of the countryside.

This process of commodification, it is suggested, is a logical outcome of a capitalist mode of production. Everything - landscapes, people and places – is deemed to have an exchange value (Mordue, 1999). This heavy emphasis on tourism as a pillar of rural development means that local development groups are increasingly drawn towards the arena of place promotion. Within rural Ireland, rural locations have proved popular attractions to both domestic and international visitors, although the phenomenon of the rural idyll is perhaps less marked in Ireland than in Britain due to a long and well documented history of rural poverty, deprivation and sustained and heavy rural depopulation (Curtin et al, 1996; Pringle et al, 1999). However, tourism patterns display considerable geographic variation and this has led to efforts in many rural areas, especially those not automatically seen as tourist destinations, to identify and market local distinctiveness.

Cloke (1992) identified five features used in rural place promotion (or what he termed the production of imagined rural spaces) in Devon. These included what might be seen as the traditional rural emphasis on nature and landscape as well as the promotion of rural places as ideal family holiday venues. One of Cloke's categories was 'history'. While 'heritage' (as in a celebration or representation of the past) rather than history (the study of the past) might be a more appropriate categorization, this tendency appears to reflect, as Cloke notes, the growing importance of the cultural arena in consideration of rural change.

Increasingly rural places are engaged in forms of place promotion and place marketing which focus on elements seen as unique to that locality. One way in which rural places can create a niche for themselves is to turn to elements of their past and put these in the service of contemporary place promotion. In this way there is an emphasis on place distinctiveness through which a particular a sense of place is portrayed (Holloway and Hubbard, 2001). This essentially means that heritage becomes a commodity like any other, one which is sold to visitors. According to an evaluation of LEADER II in Ireland the promotion of heritage and culture (along with village enhancement) was the key type of tourism-related activity funded under the programme (Kearney and Associates, 2000) reflecting the broader global trend of a burgeoning 'heritage industry'.

Rural Development, Place Promotion and the Utilization of Heritage

This section provides a distillation of key arguments surrounding the utilization of heritage as a place promotional strategy. It is notable that the bulk of the literature on place marketing and place promotion has dealt with urban areas (Kearns and Philo, 1993; Gold and Ward, 1994; Ward, 1998). Heritage has now become a key element in regeneration strategies in both urban and rural areas and some researchers have recently begun to explore these ideas within a rural context (Cloke, 1992; Hopkins, 1998; Mitchell, 1998; Mordue, 1999).

The idea of preserving elements of the past, be they landscapes or buildings, has a long history, becoming a relatively popular practice in the nineteenth century. It is also reflected in the idea of countryside conservation, and the creation of the National Parks and AONBs in the UK reflects the 'national' value placed on certain landscapes (Aitchison et al, 2000). This interest has expanded enormously in recent decades as evidenced by the vast increase in the number of museums, heritage centres and interpretative centres in Britain and Ireland leading one author to comment that 'heritage is everywhere' (Lowenthal, 1998, p. xiii).

It has been suggested that a concentration on elements of the past reflects a fear of the loss of particular traditions and elements of distinctiveness (Lowenthal, 1998; Graham et al, 2000). An interest in rural traditions might be seen to reflect a fear of the disappearance of the rural. Certainly, in Britain this appears to have an influence. In Ireland, which remains a much more rural place, the 'disappearance of the rural' seems less relevant but, within the context of a changing society, economically, socially and culturally, the preservation of local traditions may exert a considerable influence. In the Irish context the economic boom of the past few years has been accompanied by a continued growth in heritage attractions. It could be argued that this represents an attempt to hang on to a rapidly disappearing past.

While this upsurge of interest in rural heritage might be seen to offer economic, social and cultural benefits to rural localities it is useful to reflect on some key debates surrounding the use and abuse of heritage. For some, heritage is seen as synonymous with history. However, as David Lowenthal forcefully argues history and heritage are not one and the same thing:

> Heritage is not history at all; while it borrows from and enlivens historical study, heritage is not an inquiry into the past but a celebration of it, not an effort to know what actually happened but a profession of faith in a past tailored to present-day purposes (1998, p. x).

While heritage can be seen as a means of representation (through which particular sets of meanings are transmitted) it is also a commodity which, like all commodities, is both produced and consumed (Graham et al, 2000). It is both an economic and cultural product, a duality which can cause considerable tension. As Johnson (1999) suggests, 'heritage tourism is not just a set of commercial transactions, but the ideological framing of history and identity' (p.187). As such it is a strand in the broader process of the commodification of place and is, in turn, a mechanism whereby a sense of place may be produced or reproduced. To some this can be seen as the ultimate logic of a capitalist mode of production in which everything becomes commodified (Brett, 1996). It becomes a resource to be exploited like any other and put into the service of urban or rural regeneration, thereby serving economic, social, cultural and political purposes.

It should be reasonably obvious that one motivation behind the promotion of local heritage is the generation of income and employment in a changing rural economy. Visitors are drawn to locations, they spend money locally and employment is generated either directly in the particular site or indirectly in other local service outlets (restaurants, bars, hotels, etc.). On the negative side, questions may well be asked about the ability of the heritage industry (like tourism more generally) to generate sustainable, full-time and well paid employment (Shaw and Williams, 2002). It is also clear that, through income generation, the promotion of place through heritage can generate revenue which may then be available for further rounds of preservation and reconstruction. Particular sites can be further enhanced through the revenue gained from initial visitors. Many developments of this nature proceed in a piecemeal fashion with further enhancements occurring when additional revenue has been generated.

Although, as noted above, heritage is not the same as history, it does reflect an interest in the past and, importantly, it can serve to promote a still wider interest in that past through encouraging those who had not previously given much thought to such issues. Heritage projects afford local people, as well as visitors, an opportunity to learn more about the locality and to make that knowledge available to a wider audience both local and external. They can, therefore, clearly fulfil an educational role. However, a series of issues arise here as to the educative value of heritage.

In the process of identification and development of a particular project some elements are included, others, (consciously or unconsciously) are excluded. Decisions must be made as to what elements of the local are promoted; what is included, what is excluded, what is to be displayed, what is not to be displayed. The significance of individual items, events or specific locations as well as their potential as exhibits must be assessed. However, this is more than a straightforward 'technical' difficulty related to resource availability, time, financial or other constraints. There are also wider conceptual considerations involved in the selection and presentation of particular items pertaining to local heritage. In critiquing such developments there is a need to explore such questions. Why focus on some individual(s), event(s) or location(s) and not on others? The consequences of these decisions mean that some people's heritage is considered while that of others is not (Walsh, 1992; Lowenthal, 1998).

There is also the issue of timelessness. The past should not be viewed as a constant. Rather, it should be seen as a situation that was continually changing (over short time spans as well as long ones). Linked to this is the problem of viewing the past of a locality as though it were entirely separate from the present. For example, displays of 'historical' farm implements may mask the fact that the equipment on display was until recently a very real part of local farmers' lives (and may still be in some cases).

The above considerations also mean that there is a very real risk that heritage presentations may tend to sanitize the past. There are risks of presenting a view that does not offend the sensibilities of local people or particular interest groups. People struggled against land, machines, and other people, but quite often a rose-tinted view of the past is presented. In part this might reflect a feeling of nostalgia, a harking back to a supposed golden age when there was more sense of community (Wright, 1985). This ignores the very real hardships experienced by rural dwellers in the past. In an Irish context poor rural living conditions remain contemporary problems (Curtin et al, 1996, Pringle et al, 1999). In part of course it is necessary to present the past in a way that is palatable to a present-day audience. Indeed it is here that, perhaps, commercial considerations come to the fore. Many rural initiatives, even if they are recipients of some form of grant assistance, need to be seen to be self-sustaining. In order for this to happen visitors are necessary. As a result there is a need to be entertaining. The entertainment objective can impede more purely educational objectives though this does not have to be the case. People, it might be supposed, do not wish to visit locations where they will be confronted with uncomfortable or disturbing reminders of the past. In recreating the past, it may be seen as more marketable to present enjoyable and uplifting experiences rather than ones which make people depressed, angry or upset. This means there is a risk (or a deliberate strategy in some cases) of promoting a romanticized version, one which people are comfortable with. Recreating battle scenes may contribute to a renewed interest in local or even national history but it is unlikely to recreate a sense of terror, pain or those other negative features of violent conflict. Of course it must be recognized that there are difficulties in presenting a 'warts and all' account. As Yarwood observes, while 'the public might like to try their hand at milking a cow or making cheese, it is unlikely that slaughtering a pig would be quite so appealing' (2002b, p. 45). Here the duality of heritage as both conduit of cultural meaning and an economic resource emerges most clearly.

These ideas mean that a recurrent feature in many debates surrounding heritage is the issue of authenticity. In using selected senses of place questions arise over the historical accuracy of the version presented. Of course, as Johnson (1996) points out, it does not always have to be this way. The past can be critically explored while still proving to be a popular tourist attraction. Johnson, utilising the example of Strokestown House in County Roscommon, Ireland concludes that 'not all heritage landscapes need be bogus, sanitized, and hypnotic renderings of an invented past' (1996, p. 564). However, it must be recognized that the (almost inevitably) selective nature of heritage promotion means that only partial views can be presented and there is no guarantee that the version of events depicted is accurate. As Lowenthal points out 'the sheer pastness of the past precludes its total reconstruction' (1985, p. 214).

Underpinning these arguments is the idea of dissonance (Tunbridge and Ashworth, 1996; Graham et al, 2000). Heritage may well be interpreted in different ways by different people and by different groups. There are multiple meanings attaching to place with no one single unified perspective that can be agreed by all. The 'product' may well be 'consumed' in many different ways. In short it is highly unlikely that there will be universal agreement over what constitutes an authentic representation of the past.

In general terms, it could be argued that this commodification of place constitutes an element in a broader process of socialization. There is a naturalization of certain selected elements of the past. These attempts can be seen as a way of sending messages to local residents telling them their place matters, that everything is OK, that the future is bright. They may provide an uplifting image, one which perhaps hides harsher realities for some local residents. A celebration of the past (or selected elements of it) can serve to divert attention away from deep-seated social and economic problems (Hewison, 1987; Harvey, 1989). In this sense heritage serves a present-day political function while ostensibly being concerned with the past. The promotion of local events and the creation of heritage centres may serve many positive functions but it may also, through a focus on the past, mask present-day problems. None of this is to impugn the motives of those involved in local initiatives. Rather it is to broaden the ways in which we view these.

Ultimately, the key issue here is one of packaging of place as a commodity rather than emphasising its depth and complexity. In part this is understandable. It is easier to portray a small number of selected elements; complexity is a little more difficult! Complexity can also be confusing thus rendering the visitor experience in a more negative light. It is also likely to be cheaper to focus on selected elements. Ultimately there is usually a commercial imperative to make the project a success. To a large extent this will depend on visitor numbers, thereby events/sites/ideas which are enjoyable and entertaining are more likely to be developed rather than ones which are complex and searching.

As indicated earlier the EU LEADER programme, together with other rural initiatives, has part-funded many heritage-based activities in rural Ireland. Given the emphasis on community and partnership it might be suggested that this reflects an appropriate attempt to democratize heritage, a heritage 'owned' by local residents, those who might be best placed to make judgements as to what constitutes important components of local heritage. In the view of Timothy and Boyd (2003), 'heritage is a community resource, and thus all sectors of a community should be involved in its planning and development' (p. 279). Place promotion, it can be argued, may take on a community-building role. Local people may become mobilized around the creation of a local heritage project. Such projects may provide a focal point around which groups and individuals coalesce, thereby engendering a sense of dynamism in the locality. Much has been made of the problematic idea of rural places exhibiting close communal bonds, shared values and a general sense of togetherness. As suggested earlier these idealized representations of rural society may be somewhat wide of the mark (Crow and Allan, 1994; Edwards, 1998; Storey, 1999; McDonagh, 2001). Ironically, the idea of rural community may in itself be perceived as a significant element in a place's distinctiveness and, by extension, be an important component in encouraging urban dwellers to visit it. Community based heritage projects pre-suppose that the residents of the locality have a shared sense of that past and that they can agree on what are legitimate elements to portray.

Examples of Rural Heritage Projects

In promoting place a number of key elements may be utilized. Information derived from field research, brochures, publicity material and websites produced by local groups and material provided from EU LEADER sources indicates a number of apparent foci in rural heritage projects in the Republic of Ireland. These include place, landscape, specific events, key individuals, traditions (e.g. music) and architecture. The range of heritage projects which obtained LEADER II support include the re-enactment of a 16th century siege, the opening of an Irish music centre celebrating the work of a local fiddler and the renovation of an old mill.

Two examples of the utilization of local heritage are outlined below. These very different projects serve to illustrate the diverse nature of what is considered local heritage and illustrate some of the debates surrounding its use as a mechanism for rural development.

Political Heritage: The 1798 Rebellion

In Ireland, 1998 marked the 200th anniversary of the 1798 United Irishmen rebellion. This had aimed to create an Irish Republic following the apparent success of the French revolution (Keogh and Furlong, 1996; Whelan, 1998). While a series of commemorative events occurred across the country in 1998 (coupled with an outbreak of academic publications) there were particular events in the south-east and, more specifically, in County Wexford, a major centre of rebellion in 1798. In this way a national event of historic significance is also remembered as a local event, and particular places which assumed an importance in the original event utilized the bicentenary as an opportunity to remember the past and, hence, to obtain project funding.

Throughout Wexford, historical events such as past battles provided fertile material for the promotion of place. Many local organizations and community groups participated in a variety of commemorative events. Battle re-enactments and related activities occurred in locations throughout the region. Of more lasting significance perhaps was the opening of a 1798 visitor centre at Enniscorthy and projects such as the erection of a statue commemorating 'the pikemen' (pikes being the main weapon used by the rebels) outside Wexford town (Figure 10.1). Funding from LEADER II was successfully obtained for many of these activities.

As part of the rebellion commemorations Wexford Organization for Rural Development (the LEADER group for the county) sponsored the institution of the Father Murphy Centre which opened in 1998 in the tiny village of Boolavogue in north Wexford (Figure 10.2). The centre commemorates Father John Murphy, a local priest who played a leading role in the insurrection in the area. After leading insurgents at a number of locations, he was subsequently captured by militia and hanged (Furlong, 1991). As a consequence of both his activities when alive and the nature of his death he became something of a local folk legend and has become the titular hero of a famous rebel song. The development includes the refurbishment of the priest's house (a thatched stone-built dwelling) together with a number of 'authentic' outhouses featuring displays of Irish farm life and old farm implements. The complex also contains an archive of historical material. It is clear that the centre serves the function of attracting visitors to the area but is also designed with educational purposes in mind. It is also a very clear example of the utilization, within contemporary rural development, of the historic role of a key individual associated with a particular place.

Linked to this development local women (part of another community initiative which has gained funding from a variety of sources, including LEADER), under the tutelage of a local tapestry artist, created a tapestry depicting scenes from Irish history. This now hangs on a wall in the local church. (Figure 10.3). The

Figure 10.1   'Pikemen' memorial near Wexford town

Figure 10.2   Entrance to Fr. Murphy Centre, Boolavogue, County Wexford

Figure 10.3   Boolavogue community tapestry, County Wexford
(Image courtesy of Boolavogue Tapestry Group)

Figure 10.4   Belturbet railway station, County Cavan

Boolavogue Textile Studio continues to function as a voluntary group engaged in tapestry making. These types of development can be seen as serving both economic functions (through the generation of income) but they also can be seen in a broader cultural sense. Local history is (re)presented and mediated through these activities. In becoming focal points for local activity, they serve, in some sense at least, to promote feelings of community identity. More significantly they also encourage the development and rediscovery of particular skills (in this case, tapestry weaving), a key objective of the LEADER programme.

The celebration of 1798 raises a whole set of questions concerning the political context in which heritage is located. It might be suggested that the cessation of the political conflict in the north of Ireland helped to make it easier or somehow more acceptable to celebrate events from Irish history. At this moment in time, in the light of an on-going peace process and the ceasefires by paramilitary groupings, commemorations of past rebellions may be seen as relatively safe whereas they would previously have been seen in many circles as giving succour to present-day rebels. Thus, contemporary political circumstances have a profound effect on the permissibility of the nature and form of celebrations of the past. Particular forms of heritage are invoked under specific circumstances. They not just place-specific but also time-specific.

The 1798 commemorations are a clear case of an element of national history being re-worked to serve both contemporary economic and community-building purposes at a local level. The manner in which these events are interpreted is a moot point. The 1798 rebellion, like many others in Ireland, was ultimately unsuccessful. Indeed, shortly after the island of Ireland became fully incorporated into the United Kingdom under the 1800 Act of Union. To some extent the celebrations of these events allowed particular places to 'cash in' on the past, but in a broader sense they can be seen as renewing or encouraging interest in that past. They also allow opportunities for local people to involve themselves in what can be seen as broad community-based activities and to acquire skills. Employment generation and income raising, the more overt criteria that might be applied to the measurement of the success of development projects, may be very minor components in all of this.

Transport Heritage

A somewhat different project in County Cavan involves the restoration of an old train station as a visitor centre in the border town of Belturbet. This is a classic example of the utilization of an interest in old steam trains and in narrow-gauge railways (Figure 10.4). The centre opened in 2001 and is a multi-functional project comprising a small heritage centre within the old station building, containing railway memorabilia, accommodation and conference facilities. The latter is located in the old railway goods store. Here, a form of transport, now obsolete, returns to serve another function. Belturbet station closed in 1959 having been in operation since 1885. The station was a junction between the standard gauge Great Northern Railway and a local narrow gauge track. A place which served as a transport node to travel elsewhere now becomes an end-point in itself. Such developments can be seen as symbolising both technological and economic change in rural localities.

LEADER funding, together with other sources, was utilized in the original restoration work by a local community development association. The town of Belturbet is located close to the border with Northern Ireland and for much of last few decades the main road linking it to the north was closed. This contributed to its economic stagnation, serving to cut it off, to some extent, from part of its natural hinterland. This particular development will link to the increased visitor number as a consequence of the re-opening of the Shannon-Erne waterway in 1994 as part of joint north-south tourist venture, a navigable canal linking the Rivers Erne (on which Belturbet is located) and Shannon. Two seemingly redundant modes of transport are re-invented in the cause of rural development.

In many senses such a development may seem ironic. Most of Ireland's narrow-gauge railways were closed in the 1950s and 1960s when the proliferation in car usage was interpreted as a sign that the age of railway transport was over. Now these items, once seen as redundant in a modernising Ireland, become centre-points in the burgeoning heritage industry. The reconstruction of railway stations may sit a little uncomfortably in places with limited public transport availability. In the early 1920s Ireland had some 3,000 miles of track, a high per capita figure relative to many other European countries at that time. Extensive line closures occurred in the 1950s and 1960s due to the increases in car ownership. Rail travel, which had accounted for almost a quarter of passenger movement in Ireland in the late 1930s dropped dramatically (Lee, 1969; Barrett, 1982). This fits into a broader pattern of service rationalization in rural Ireland (Storey, 1994; Cawley, 1999).

Discussion and Conclusions

The brief overview here suggests a need to explore more fully the various ways in which local heritage plays a role in rural place promotion and, hence, acts as a mechanism for the regeneration of rural localities. More specifically it suggests a need to explore these developments within the broader context of heritage, place promotion, place marketing and the re-imagining of place. The upsurge in place promotional activities utilising elements of local heritage is a multi-faceted phenomenon. While Cloke (1992) referred to the 'pay as you enter' countryside experience, the reality is somewhat more complicated. The dual nature of heritage as both economic and cultural product means that its outcomes are highly complex. Although many developments can be seen primarily in terms of economic regeneration such initiatives also have cultural outcomes. The examples touched on in this chapter indicate that, through the various devices used, a number of end products are being sought. Chief amongst these are the economic regeneration of the locality, the education of both local people and visitors in aspects of local culture, heritage or tradition, the promotion of a sense of local community, and the provision of place promotional skills for local residents.

The chapter raises issues which clearly require much more in-depth and systematic research. Firstly, there is a need to examine how successful or effective these projects actually are. If they are designed to attract visitors have they succeeded? If they were developed with a view to educating people, there is a need to evaluate their effectiveness. If the intention is to enhance a sense of local community then there is a need to explore the extent to which this has been achieved. In brief, it would be useful to evaluate projects in the light of their own stated objectives.

In addition, it might be useful to explore the ways in which these various place promotional efforts are interpreted. How do visitors make sense of these projects? What sense of place is conveyed? This is a seriously neglected aspect of research (see Rose, 2001). By the same token it is also important to understand how local people interpret these promotional efforts. Questions which arise include the extent to which local people identify with, support or feel comfortable with these projects. Some of the heritage literature falls into the trap of appearing to assume that people will always and everywhere interpret presentations in the way intended by the promoters. A key point here is not to assume that people interpret things in a simplistic manner. As Urry (1995) and others have argued, we are not 'cultural dupes'. It is vital to remember that visitors do not naively interpret what they see in front of them.

This knowingness on the part of the visitor has led some writers to postulate about the postmodern tourist - someone who is aware of the ambivalence of what is presented to them and who does not assume there is something authentic about what is seen or even that something authentic exists or can be 'captured' (Mordue, 1999). This has led some to suggest that we live in an era where 'post-tourists' partake of the 'post-rural' (Hopkins, 1998). In summary the myriad ways in which people interpret these developments needs to be explored rather than simply assumed. Together such studies should uncover some of the inherent conflicts and tensions inevitable in utilising the past as a mechanism for facilitating rural development.

References

Aitchison, C., Macleod, N. and Shaw, S.J. (2000), *Leisure and Tourism Landscapes. Social and Cultural Geographies*, Routledge, London.

Barrett, S. D. (1982), *Transport Policy in Ireland*, Irish Management Institute, Dublin.

Brett, D. (1996), *The Construction of Heritage*, Cork University Press, Cork.

Cawley, M. (1999), 'Poverty and Accessibility to Services in the Rural West of Ireland', in D. Pringle, J. Walsh and M. Hennessy (eds), *Poor People, Poor Places. A Geography of Poverty and Deprivation in Ireland*, Oak Tree Press, Dublin, pp. 141-156.

Cloke, P. (1992), 'The Countryside as Commodity', in S. Glyptis (ed), *Leisure and the Environment. Essays in Honour of Professor J.A. Patmore*, Belhaven, London, pp. 53-67.

Crow, G. and Allan, G, (1994), *Community Life. An Introduction to Local Social Relations*, Harvester Wheatsheaf, Hemel Hempstead.

Curry, N. (1993), 'Rural Development in the 1990s - Does Prospect Lie in Retrospect?', in M. Murray and J. Greer (eds), *Rural Development in Ireland: a Challenge for the 1990s*, Avebury, Aldershot, pp. 21-39.

Curtin, C., Haase, T. and Tovey, H. (eds) (1996), *Poverty in Rural Ireland. A Political Economy Perspective*, Oak Tree Press, Dublin.

Edwards, B. (1998), 'Charting the Discourse of Community Action: Perspectives from Practice in Rural Wales', *Journal of Rural Studies*, Vol. 14, pp. 63-77.

Edwards, W., Goodwin, M., Pemberton, S. and Woods, M. (2000), *Partnership Working in Rural Regeneration: Governance and Empowerment*, Policy Press, Bristol.

Furlong, N. (1991), *Father John Murphy of Boolavogue 1753-1798*, Geography Publications, Dublin.

Gold, J.R. and Ward, S.V. (eds) (1994), *Place Promotion: The Use of Publicity and Public Relations to Sell Towns and Regions*, John Wiley, Chichester.

Goodwin, M. (1998), 'The Governance of Rural Areas: Emerging Research Issues and Agendas', *Journal of Rural Studies*, Vol. 14, pp. 5-12.

Graham, B, Ashworth, G.J. and Tunbridge, J.E. (2000), *A Geography of Heritage. Power, Culture and Economy*, Arnold, London.

Harvey, D. (1989), *The Condition of Postmodernity. An Enquiry into the Origins of Cultural Change*, Blackwell, Oxford.

Hewison, R. (1987), *The Heritage Industry. Britain in a Climate of Decline*, Methuen, London.

Holloway, L. and Hubbard, P. (2001), *People and Place. The Extraordinary Geographies of Everyday Life*, Prentice Hall, Harlow.

Hopkins, J. (1998), 'Signs of the Post-rural: Marketing Myths of a Symbolic Countryside', *Geografiska Annaler*, Vol. 80 B, pp. 65-81.

Johnson, N. C. (1996), 'Where Geography and History meet: Heritage Tourism and the Big House in Ireland', *Annals, Association of American Geographers*, Vol. 86, pp. 551-566.

Johnson, N.C. (1999), 'Framing the Past: Time, Space and the Politics of Heritage Tourism in Ireland', *Political Geography*, Vol. 18, pp. 187-207.

Kearney, B., Boyle, G.E. and Walsh, J.A. (1995), *EU LEADER I Initiative in Ireland. Evaluation and Recommendations*, Stationery Office, Dublin.

Kearney, B. and Associates (2000), *Operational Programme for LEADER II Community Initiative 1994-1999. Ex Post Evaluation*, Final Report prepared for the Department of Agriculture, Food and Rural Development, Dublin.

Kearns, A. and Paddison, R. (2000), 'New Challenges for Urban Governance', *Urban Studies*, Vol. 37, pp. 845-850.

Kearns, G. and Philo, C. (eds) (1993), *Selling Places. The City as Cultural Capital, Past and Present*, Pergamon Press, Oxford.

Keogh, D. and Furlong, N. (eds) (1996), *The Mighty Wave: The 1798 Rebellion in Wexford*, Four Courts Press, Dublin.

Lee, J.J. (1969), 'The Railways in the Irish Economy', in L. M. Cullen (ed), *The Formation of the Irish Economy*, Mercier Press, Cork, pp. 77-87.

Lowenthal, D. (1985), *The Past is a Foreign Country*, Cambridge University Press, Cambridge.

Lowenthal, D. (1998), *The Heritage Crusade and the Spoils of History*, Cambridge University Press, Cambridge.

McDonagh, J. (2001), *Renegotiating Rural Development in Ireland*, Ashgate, Aldershot.

Mitchell, C. J. A. (1998), 'Entrepreneurialism, Commodification and Creative Destruction: A Model of Post-modern Community Development', *Journal of Rural Studies*, Vol. 14, pp. 273-286.

Mordue, T. (1999), 'Heartbeat Country: Conflicting Values, Coinciding Visions', *Environment and Planning A*, Vol. 31, pp. 926-946.

Moseley, M. J. (1999), 'The Republic of Ireland. The New Localism as a Response to Rural Decline', in E. Westholm, M. Moseley, and N. Stenlås (eds), *Local Partnerships and*

Rural Development in Europe. A Literature Review of Practice and Theory, Dalarnas Forskningsråd, Falun, pp. 25-43.

Moseley, M.J., Cherrett, T. and Cawley, M (2001), 'Local Partnerships for Rural Development: Ireland's Experience in Context', *Irish Geography*, Vol. 34, pp. 176-193.

Philo, C. and Kearns, G. (1993), 'Culture, History, Capital: A Critical Introduction to the Selling of Places', in G. Kearns and C. Philo (eds), *Selling Places. The City as Cultural Capital, Past and Present*, Pergamon Press, Oxford, pp. 1-32.

Pringle, D., Walsh, J. and Hennessy, M. (eds) (1999), *Poor People, Poor Places. A Geography of Poverty and Deprivation in Ireland*, Oak Tree Press, Dublin.

Rhodes, R. (1996), 'The New Governance: Governing Without Government', *Political Studies*, Vol. 44, pp. 652-667.

Rose, G. (2001), *Visual Methodologies. An Introduction to the Interpretation of Visual Materials*, Sage, London.

Shaw, G. and Williams, A.M. (2002), *Critical Issues in Tourism. A Geographical Perspective* (2nd edition), Blackwell, Oxford.

Storey, D. (1994), 'The Spatial Distribution of Education and Health and Welfare Facilities in Rural Ireland', *Administration*, Vol. 42, pp. 246-268.

Storey, D. (1999), 'Issues of Integration, Participation and Empowerment in Rural Development: The Case of LEADER in the Republic of Ireland', *Journal of Rural Studies*, Vol. 15, pp. 307-315.

Storey, D. (2003), 'Changing Approaches to Rural Development', *Geography Review*, Vol. 16, pp. 31-33.

Timothy, D.J. and Boyd, S.W. (2003), *Heritage Tourism*, Prentice Hall, Harlow.

Tunbridge, J.E. and Ashworth, G.J. (1996), *Dissonant Heritage: The Management of the Past as a Resource in Conflict*, John Wiley, Chichester.

Urry, J. (1995), *Consuming Places*, Routledge, London.

Walsh, K.T. (1992), *The Representation of the Past: Museums and Heritage in the Post-modern World*, Routledge, London.

Ward, S.V. (1998), *Selling Places. The Marketing and Promotion of Towns and Cities 1850-2000*, E&FN Spon, London.

Whelan, K. (1998), *Fellowship of Freedom. The United Irishmen and 1798*, Cork University Press, Cork.

Wright, P. (1985), *On Living in an Old Country. The National Past in Contemporary Britain*, Verso, London.

Yarwood, R (2002a), Parish Councils, Partnership and Governance: The Development of Exceptions Housing in Malvern Hills District, England, *Journal of Rural Studies*, Vol. 18, pp. 275-291.

Yarwood, R. (2002b), *Countryside Conflicts*, Geographical Association, Sheffield.

Chapter 11

'Community'-Based Strategies for Environmental Protection in Rural Areas: Towards a New Form of Participatory Rural Governance?

Susanne Seymour

Introduction

The Rio Earth Summit of 1992 lent new support to ideas of participatory governance in the promotion of global 'sustainable development'. Informed as the conference was by neo-liberal as well as democratic impulses, post-Rio participatory governance has been characterized by a withdrawal of state services as well as an increased role for non-state actors, the so called 'major groups' (Thomas, 1994), market-based mechanisms and voluntary environmental agreements (Karamanos, 2001). In many Western democracies there has been a subsequent shift in the role of the state from provider towards enabler (Murdoch, 1997) and from regulator towards moderator, together with a focus on '"the community" as the new terrain of governing' (Higgins and Lockie, 2002, p. 421). Government agencies from a range of Western countries are beginning to enrol non-state actors as participants rather than consultees in environmental management, whilst a greater variety of knowledges are beginning to be recognized as legitimate in environmental decision-making (O'Riordan, 2000). This, however, is still a process in its infancy.

The emphasis placed at Rio on the importance of locally-based actions, in addition to the role of major groups, has strengthened the renewed focus on 'community'- based actions. Warburton (1998, p. 7) traces extensive references to 'community' in Agenda 21, noting that Section 3.7 states, 'Governments, in cooperation with appropriate international and non-governmental organizations, should support a community-driven approach to sustainability'. Such principles have been followed through by commitments at regional, national and local levels, most notably through Local Agenda 21.

Yet while understandings of 'community' and how such communities may operate are fundamental to these strategies, it has been recognized that many recent policy initiatives and academic studies have failed to interrogate these aspects sufficiently (Agrawal and Gibson, 1999; Warburton, 1998; Liepens, 2000).

This omission is particularly striking in relation to rural studies as ideas of 'community' have a longstanding influence on both the representation and constitution of 'rurality' (Harper, 1989; Short, 1992). How rural communities are constituted and represented and who or what may be excluded from them has not been sufficiently considered. How strategies are agreed within such communities is often neglected and conflicts glossed over. Furthermore, the practices adopted by governments in relation to community involvement and evidence of movement towards enabling or moderating roles needs further examination.

This chapter will examine the interactions between ideas of community and rurality and assess the implications these have for a range of government or agency-led initiatives which ascribe environmental protection functions to rural communities. In particular, the focus will be on measures to target diffuse pollution problems. These will include Community-Based Environmental Protection (CBEP) initiatives in the USA, the Australian *Landcare* programme; and a pilot *Landcare* scheme in England.

Community and Rurality: Identifying Rural Communities?

Community

The term 'community', whose use dates back to at least the 14th century, has a history of multiple interpretations (Williams, 1983) and discussion of the term in academic circles has generated considerable dispute (Cohen, 1989). Much academic effort was expended in the early post-war period in attempts to define 'community' precisely, usually by reference to structural dimensions of social relations. These attempts have since been superseded by approaches that seek to interpret community formation through consideration of the construction of cultural meaning alongside social processes (Cohen, 1989; Harper, 1987). Others have gone further, with Day (1998, p. 251) asserting, 'There is broad endorsement by now of the view that at both local and national levels, community is something which is essentially "imagined" (Anderson, 1983) or "constructed" (Cohen, 1985)'. Focusing on practice, Cohen (1989, p. 12) has argued that in broad terms 'community' suggests a group of people with 'something in common with one another' which 'distinguishes them in a significant way from the members of other putative groups'. Moseley (2003, p. 74) takes this further, adding 'people in communities must share something in common and, through sharing that something and *being conscious that they do so*, interact' (emphasis added).

Examining the construction of community in detail, Cohen (1989) has suggested that communities delineate membership by the drawing of social and/or spatial boundaries which serve to exclude a variety of 'others' and to add exclusivity to the term (see Young, 1990). Yet contrary to assumptions of commonality, Cohen (1989) also highlights that communities are typically internally heterogeneous, with multiple notions of 'community' in operation within a particular community group. Communities, he argues, are held together by commonality of symbol rather than a uniformity of meaning and the 'triumph of community is to so contain this variety that its inherent discordance does not subvert the apparent coherence which is expressed by its boundaries' (Cohen, 1989, p.20). For Day (1998) this recognition of internal variety suggests the potential for community change which is not totally reliant on exogenous factors, while Warburton (1998) welcomes forms of community which recognize difference whilst retaining a sense of attachment.

The boundaries of community suggested by Cohen (1989) may nonetheless create pressure for homogeneity within them, resulting in complex managements and performances of identity in the context of 'community' expectations. Attempts to manage internal variation have led to criticisms of community as a reactionary concept, founded on conservative views of social relations, in terms of class, gender, and sexuality (see Agrawal and Gibson, 1999; Little and Austin, 1996). Thus while 'community' in popular discourse has usually carried positive undertones (Williams, 1983) scholars from a number of different persuasions have questioned the value of working with ideas of community due to the complex and deflective qualities of the term. Community thus emerges as a strategically deployed and thus highly politicized concept (Day, 1998). However, it can be argued that it is precisely because of these flexible qualities of 'community' that it is necessary to examine the ways in which it is symbolized and practised and the influences which these enactments have. In Moseley's (2003, p. 75) terms 'despite being an elusive, exploited and somewhat fuzzy concept, community *matters*'.

Approaches to community based on premises of either 'structure' or 'construction' fall into at least six major styles. First, arguably the most popular approach is to define communities by reference to geographical area. The most common scale is that of the locality because of a connection between scale and a particular style of social relations focused on everyday encounters. Indeed, in many cases small settlements or areas are equated with 'communities', as in the case of the French commune (Williams, 1983). Warburton (1998, p. 16) celebrates a strongly locality-based notion of community in her examination of sustainable development strategies, arguing that locality is 'where environmental issues matter most to most people'. Crucial here are senses of the local, how far a locality is seen as distinct from other places, how far it embraces what Massey (1991) has termed 'a global sense of place'. Narrow interpretations of the local result in exclusions of non-residents or those who have recently moved to the area. Place-defined communities may thus have strongly insular agendas.

A second approach is to define community by reference to shared social or economic characteristics, for example by religious adherence, ethnicity, sexuality or occupation (Agrawal and Gibson, 1999), distinguishing the 'Muslim community', 'the Black community', the 'gay community', the 'mining community', or the 'farming community'. However, while such labels may give power to minority groups fighting against dominant identities in society, they imply an unrealistic homogeneity as a shared set of values and norms cannot necessarily be read from shared social or economic characteristics (Warburton 1998).

Communities defined by reference to 'interest' highlight explicitly the political nature of the term. In academic circles, the conceptualization of 'policy communities' within policy networks highlights such an approach (Marsh and Rhodes, 1992). It is argued that the networks forming around certain policy sectors come to resemble a 'community' which displays 'stability of relationships, continuity of a highly restrictive membership, vertical independence based on shared service delivery responsibilities and insulation from other networks and invariably from the general public' (Rhodes, 1988, p. 78, cited in Winter, 1996, p. 26). Agricultural policy making in both the UK and Australia has frequently been conceptualized as a 'policy community' (Winter, 1996; Smith, 1992; Martin and Ritchie, 1999). Communities of interest are not, however, limited to economic groupings and may also form around issues, such as environmental concerns.

Allied to communities of interest are 'communities of commitment'. This conceptualization of community is founded on the notion that those who commit to the ideals of 'community' or who work for community goals can win community membership. This is an idea akin to idea of the 'justice of earned deserts' (see Jennings and Moore, 2000, p. 187).

A fifth style of community conceptualization is that highlighted by the Brundtland definition of sustainable development as 'development that meets the needs of the present without compromising the ability of future generations to meet their own needs' (WCED, 1987, p. 8). Such a sense of generational communities, in which present and future generations in particular are tied together, is highlighted in calls for sustainability. However, the scope of future generations is left unclear, with interpretations ranging from future humanity as a whole to specific family descendants, and from the next generation to all future generations. Certainly sensitivity to generational aspects can broaden ideas of community to those unborn or more usually to consideration of children, but it can also act in an exclusionary manner as in the rural discourse of 'locals' and 'incomers'.

A less commonly articulated notion of community is one which embraces the non-human, inspired by deep ecology and Actor Network Theory. Latour's 'Parliament of Things' (1993, pp. 142-5), for example, calls for the recognition and representation of human and non-human, machines as well as living organisms. Such a sense of 'community' would recognize not only the importance of human-ecology linkages, as Duane argues (1997, p. 772) 'communalities in how they [people] relate to a particular ecosystem, or resource as beneficiaries of that place or contributors to its condition', but also the roles of inanimate objects in networks of connection.

Rurality and Community

With emerging industrialization in 19th century Europe, community began increasingly to be associated with rural contexts, particularly villages where it was argued that distinctive community relations persisted in the face of urbanization and industrialization (Williams, 1983). The most obvious expression of this was in the work of the nineteenth century German sociologist, Tönnies and his concepts of *Gemeinschaft* ('community' characterized by deep, customary, long-term associations) and *Gesellschaft* ('society' characterized by impersonal, contractual relations). Newby (1977) argues that although originally devised as styles of social relationships, Tönnies, influenced by idealized views of country living prevalent during his lifetime, reinterpreted them as 'a taxonomy of settlement patterns' with the rural village seen to epitomize *gemeinschaft* (p. 95). In contrast to the disconnected, individualized life of the city, rural society was characterized as 'personalistic, traditionalistic, stable, religious, familial, ... the classical repository of community' (Cohen, 1989, p. 27). In reciprocal fashion, community has been linked to idealizations of rural living and has formed one of the cornerstones of rural idylls and understandings of the rural over the centuries (Short, 1992). Imaginings of 'rurality' and 'community' have thus become mutually constitutive. Idealized assumptions of community in rural areas characterized British rural community studies of the 1930s-early 1950s, with a legacy reaching to the present-day. Such studies have been criticized by numerous authors as ahistorical, treating communities as self-contained, coherent, isolated units (Bell and Newby 1971; Day, 1998; Harper, 1989; Wright, 1992).

Emerging from this critique have been two major approaches to rural community studies (Cloke, 2000), both of which recognize the importance of 'performance'. One emphasizes definitions of rural communities based on the notion of 'centredness' (see Harper, 1987) in which the researcher defines both rural people and rural places by the focus of physical, social and symbolic relationships on a single place. The second approach focuses on the deployments of community in rural social and institutional relations. For example, in his work on middle class formations in rural areas Phillips (1998) stresses how middle class migrants may both anticipate and perpetuate notions of 'community'. Some migrants actively engage in the creation of community to fulfil their ideals about country living and to differentiate themselves from other new residents who do not engage in the same way. Likewise in their investigation of women's identity and the rural idyll, Little and Austin (1996) highlight the importance of senses of community to the identity of rural women and the maintenance of traditional gender relations. Other investigations of rurality, however, reveal different styles of rurality, with more emphasis on qualities of peace, beauty and greenness than community in rural imaginings (Halfacree, 1995).

Recognizing the importance of the 'performance' of community suggests the potential for community to work differently in different places and for the conservative aspects of the notion to be undermined (Day, 1998). For example, Day and Murdoch's (1993) institutionally-led study of constructions of community in rural Wales shows how boundaries may be differentiated and negotiated. They found different attitudes towards 'outsiders' in relation to formal local politics than with respect to social activities. As Day (1998) reports 'people can be enrolled into networks or pushed away from them, for example, according to whether they show behaviours and attitudes thought to be appropriate or inappropriate to local styles of interaction. "Pushy" attempts at premature integration may lead to rejection, while a readiness to play a "helpful" role may bring acceptance' (p. 252). Such research suggests a complex and locally-specific social interplay exists in the formation of community in which adapting to the rules of the community game are vitally important.

Shifting Powers? Rural Governance, Public Participation and the Environment

A number of recent initiatives have encouraged a greater role for rural publics in the governance of rural areas. These often draw upon the language of 'community'. The most prominent of these in England have been the two Rural White Papers issued in 1995 and 2000. The 1995 White Paper was the first major review of rural policy in England since 1942 and, as Murdoch (1997) drawing on governmentality perspectives has pointed out, embodied the retreat of the state from provider to partner or enabler through indirect 'government through community' (a term derived from Rose, 1996). However, both White Papers deploy idealized, organic notions of rural community (all who live or work in rural areas are regarded as part of this community) which appear to ignore recent academic scholarship and which serve to legitimate the withdrawal of government services. For example the 2000 White Paper states that 'The community strength of rural England is an important part of the character of the countryside' (DETR/MAFF, 2000, p. 9), while the 1995 version emphasizes a style of community which is specifically rural and local in emphasis, with both the desire and capacity for such communities to help themselves:

> Self-help and independence are traditional strengths of rural communities. People in the countryside ... do not expect the Government to solve all their problems for them ... In any case, local decision-making is likely to be more responsive to local circumstances than uniform plans. Improving the quality of life in the countryside starts with local people and local initiative (DoE/MAFF, 1995, p.16).

Whilst such statements project unproblematized visions of rural communities, they are legitimated by reference to consultation with a range of rural and urban people and organizations, and to the ideals of sustainable development through the promotion of 'active communities'. The latter, it is suggested, have been facilitated by a range of initiatives spanning social, economic and environmental concerns, from village appraisals and Local Exchange and Trading Systems (LETS), to national and EU sponsored programmes such as Rural Action and LEADER, action only bolstered by a long established involvement in volunteering in rural areas (DoE/MAFF, 1995, pp. 10, 17, 19; see also Marsden and Murdoch, 1998; Lowe et al, 1998).

It has been argued that community governance also extends to 'regulation' of rural areas, through involvement in the planning system (Murdoch and Marsden, 1994; Walker, 1995) or through more informal networks. Such work argues that different patterns of regulation are found according to varying local social and economic conditions and different constructions of the rural. In England, there is evidence of local environmental regulation of agricultural practices occurring through local public pressure in relation to agricultural pollution (Ward et al, 1995), pesticide and manure application and footpath management (Tsouvalis et al, 2000).

Alongside such explicitly rural initiatives, environmental and sustainability discourses have also called for greater community participation in the hope of creating more sustainable lifestyles. The environmental movement tended in its early years to attract either those with scientific or technical expertise (Warburton, 1998, p. 10) or those committed to adversarial politics, and participatory and community approaches were downplayed until the 1980s. However, since the early 1990s a gradual convergence of environmental and community policy has been reported in Britain (Warburton, 1998, p. 14).

A local 'community' basis for action to protect the environment has been claimed to have a number of advantages. Firstly, it is seen to be more equitable, allowing local people to influence solutions rather than the state imposing them and avoiding conventional adversarial approaches (see Environmental Protection Agency (EPA), 1999). This is a particularly heightened issue in contexts in which indigenous or aboriginal groups are involved and the state embodies colonizing values. Secondly, it has the potential to draw on local knowledges of the environment. These may take the form of traditional, non-Western knowledges, or site-specific knowledges (Lane and McDonald, 2002). Thirdly, such an approach allows environmental problems to be addressed at a broader scale than that of the individual property. Place-based approaches are also seen to have the potential for consideration and coordination of a variety of problems and issues (as in Local Agenda 21). Finally, there are thought to be benefits in terms of efficacy.

However, it is generally agreed that many attempts at 'community'-based management of natural resources have been based upon naïve, even "mythic" (Lane and McDonald, 2002, p. 11), models of community and that many of these potential advantages have not been attained (Agrawal and Gibson, 1999). Indeed, Lane and McDonald (2002) argue that such projects 'appear to take little heed of significant theoretical and empirical literature that unpacks the community concept, questions its viability as a focus for policy and planning, and critiques the outcomes that emerge from community-based action' (p. 9).

Firstly, contrary to widespread opinion, there is evidence of continued inequity at local level following the instigation of community-based strategies due to unequal access to relevant 'resources' for community involvement (including property, skills and influence). Indeed, the powers of existing elites may become further entrenched and 'outsiders' further excluded (Lane and McDonald, 2002). For example, both Selman (1998) (in relation to Local Agenda 21 in Gloucestershire, England) and Martin (1999) (in relation to Rural Action in England) found that participation was dominated by white, middle-class participants and that the young, poor and ethnic minorities tended to be under-represented. Lane and McDonald (2002) also argue that threats to democratic representation increase as the unit of political organization becomes smaller.

There is also evidence of disputes over legitimate knowledges. Local knowledges are often not taken seriously by government officials (Agrawal and Gibson, 1999). On the other hand, reliance on local knowledges alone has been recognized as limiting. It cannot be assumed that local communities are in harmony with their environments or have the relevant knowledge or capacity (in terms of personal and social attributes, organizational skills and financial resources) to effectively manage the environment (see Lane and McDonald, 2002).

A reliance on local decision-making is also subject to naïve localism and the exclusion of national level stakeholders (Lane and McDonald, 2002; Singleton, 2002). An obvious expression of this is represented by NIMBYism (Selman, 1996). This has led to some gross cases of environmental injustice whereby 'undesirable' developments are deflected by more powerful local communities to places where there is less local resistance. Thus decisions which are good for one community and its local environment may have adverse impacts on people and the environment more widely (Lane and McDonald, 2002). Win-win solutions cannot be assumed (Singleton, 2002).

In cases where co-management strategies are adopted they may be susceptible to 'capture and corruption' (Singleton, 2000, p. 8). Evaluative work on community-based or collaborative approaches in Western countries is in its infancy, with the implication that policy shifts have preceded balanced evaluation (Bellamy et al, 1999; Singleton, 2002). Singleton (2002) has highlighted that such approaches may lead to further withdrawal of government input in areas where local and producer interests already have substantial influence, leading to inaction.

These problems with the internal functioning of rural communities are exacerbated by problems of government-public interaction. In many cases there are problems with the capacity of state agencies, particularly those with environmental concerns, in the facilitation of community involvement. There is a tendency to employ people with technical, scientific expertise rather than with community development skills. Alongside weaknesses in institutional capacity, there may also be institutional reluctance to devolve powers to communities (Lane and McDonald, 2002).

Initiatives for Rural Community Protection of the Environment

In this section I examine a range of initiatives designed to enrol rural communities into the protection of the environment, focusing particularly on those dealing with diffuse pollution problems. Such diffuse problems have attracted 'community'-based voluntary approaches for a number of reasons. Diffuse pollution problems pose difficulties for conventional legal regulation which is based on identification of individual polluters and polluting substances. Usually it is impossible to identify an individual land or enterprise manager responsible for diffuse pollution due to its widespread nature and multiple sources. Neither is it always possible to identify a pollutant and take precautionary action or use market mechanisms, such as taxes, to address the problem substances (as for example one might do in relation to specific pesticides). For example, problems of nutrient enrichment, soil erosion and salinity cannot easily be addressed by the banning or taxing of the offending substances since they are widespread in the environment and derive from a range of origins, both manufactured and 'natural' in the case of nutrients. Instead, there has been a focus on 'regulating' land management practices to address such diffuse problems, often with the aim of achieving measurable environmental targets. Changing such practices may be approached by legislative means, such as the European Union's Nitrate Directive and the designation of Nitrate Vulnerable Zones. However, several countries, to a greater or lesser extent, have encouraged the use of voluntary approaches based around a rhetoric of 'community' to address diffuse problems. The following sections examine examples of such voluntary, 'community' schemes in the USA, Australia and England.

Diffuse Pollution and the US Environmental Protection Agency's Approach to Community Based Environmental Protection

Evidence of harm from a range of diffuse environmental impacts, including soil erosion and more recently habitat destruction and water pollution, is well-established in the US (Potter, 1998; Singleton, 2002). Taking diffuse water pollution as an example, under legislation instigated by the Clean Water Act of 1972 a framework has been established for its control. States are required to establish water quality standards, set as Total Maximum Daily Loads (TMDLs) under the 1977 Clean Water Act, and to develop programmes to ensure such standards are complied with. However, while identification and programme development are legal requirements, current approaches to controlling diffuse pollution in the US seek to achieve voluntary compliance with water quality goals through collaborative, usually place-based initiatives which draw on a language of 'community' (EPA, 2003a; Sansom, 1999).

   Early initiatives addressing diffuse water quality issues focused on watershed approaches and date mainly from the 1980s (Leach and Pelkey, 2001). By placing TMDL assessment in a specific time-space context such measures enhanced the ability to set standards for diffuse pollution control (Haith, 2003). However, the major impetus for the formation of most watershed councils, particularly in the Western and Pacific North-West where these approaches have the longest history, has been concerns over water use rights and the implications of water use for habitat destruction (Singleton, 2002). Water quality issues have subsequently been accommodated within the watershed thinking of most such initiatives, with schemes specifically focused on water quality emerging in the 1990s. They have been supported by a range of federal government institutions, most notably the Environmental Protection Agency (EPA) and the Bureau of Land Management (Kenney, 2000, cited in Singleton, 2002). Cline and Collins (2003) cite estimates of around 3600 grassroots river or watershed based groups and 1000 watershed partnerships. However, it is generally agreed that evaluation of such partnerships, and collaborative natural resource management more broadly, is lacking (Singleton, 2002; Cline and Collins, 2003). Recent research, nonetheless, has linked the existence of such watershed groups with positive impacts on the most serious problems (Leach et al, 2002) and with greater levels of protection activities and funding (Cline and Collins, 2003). Thus, despite some criticisms, community-oriented watershed approaches in the US have been viewed in generally positive terms.

In 1994 the EPA launched a Community-Based Environmental Protection (CBEP) initiative which covers both rural and urban areas. CBEP complements a whole suite of EPA initiatives, including watershed approaches, and wider framework policies and pressures emanating from discourses of sustainable development and neo-liberalism. For example, in 1990 the Pollution Prevention Act encouraged a shift in the EPA ethos from pollution control towards pollution prevention (Burnett, 1998), a move sympathetic to the increased use of voluntary measures. In addition, from 1991-96 the EPA initiative on 'reinventing government' set out to encourage greater involvement of communities or regulated groups (Singleton, 2002) and there have been further moves to devolve EPA powers to private companies and local authorities in return for better than average environmental performance (EPA, 2003a).

The Agency has promoted CBEP as a new approach to environmental protection, regarding it as a set of principles to be integrated into all its programmes rather than seeing it as a separate initiative (EPA, 1999). It recommends CBEP as 'a holistic and collaborative approach to identifying environmental concerns, setting priorities, and implementing comprehensive solutions' (EPA, 1999, p. 3). This provides the potential to co-ordinate air, land and water pollution issues, to avoid adversarial stalemate over such matters (Colvin, 2002, pp.447-8) and to address diffuse environmental problems, over some of which the Agency still has no legislative authority (EPA, 1999). The EPA's commitment to CBEP is illustrated by its investment in 1995 of 14 per cent of its budget and 15 per cent of its staff in such schemes (Sansom, 1999, p. 13). However, it recognizes that it does not have the resources to be directly involved in all such programmes and positions itself mainly in a supportive role, with targeted direct involvement in priority areas (EPA, 2003b). Thus the EPA has been instrumental in establishing community-based initiatives, particularly in situations where federal statutory standards were not being met, and has supported and publicized grass roots initiatives (for example the Applegate Partnership) (EPA, 1999, pp. 13-14).

The core principles of the EPA's CBEP are set out as follows:

- Focus on a definable geographic area.
- Work collaboratively with a full range of stakeholders through effective partnerships.
- Assess the quality of the air, water, land, and living resources in a place as a whole.
- Integrate environmental, economic, and social objectives and foster local stewardship of all community resources.
- Use appropriate public and private, regulatory and non-regulatory tools.
- Monitor and redirect efforts through adaptive management (EPA, 1999, p. 6).

These core principles of CBEP raise key issues about its operation in practice. Due to a lack of explicit evaluations of schemes specifically established under the 1994 CBEP initiative there is not a well-developed body of literature which helps clarify these matters. The reports by the EPA (2003b and c) and Sansom (1999) tend to celebrate the successes of the programme and its adopted schemes, emphasizing a 'win-win' philosophy, with little reflection on problems. For example, in the EPA assessment of the San Miguel Watershed initiative, included as one of five projects evaluated in its CBEP review, the Agency highlights how CBEP has helped with the integration and co-ordination of efforts, securing funding, capacity building, developing public education and support in relation to environmental issues and the early recognition of emerging environmental problems (EPA, 2003b, 2-6, 2-7). More broadly the review concludes that its investigation suggests that CBEP 'can be very effective, provided that the process is carefully designed to organize input of participants and delineate clear roles and responsibilities' (EPA, 2003c, p. 2).

There are thus several issues which remain to be clarified in relation to the EPA's CBEP initiative. Firstly, there are questions raised by its engagement with the term 'community'. In its framing policy, the EPA indicates an awareness that 'community' is a contested and constructed term, declaring 'Intrinsic to CBEP is an understanding of "community"', and sets out a working definition of 'community' in CBEP which is flexible and inclusive (EPA, 1999, p. 5). This definition 'includes places and people that are associated with an environmental issue(s) ... defined by either natural geographic or political boundaries' (EPA, 1999, p. 5). The key element, according to the EPA, 'is that the people involved have a common interest in protecting an identifiable, shared environment and quality of life' (EPA, 1999, p. 5). However, there remains a certain tension between this stated inclusive concept of community, attuned to context, and the operationalization of 'community' involvement in which there is an emphasis on organization and clarification of community input by 'CBEP practitioners' who are 'encouraged to define and understand the appropriate scope of "community" for each particular place' (EPA, 1999, p. 5; EPA 2003b). A further implication of this wide interpretation of 'community' in CBEP is that it is a style of collaborative governance in which government, publics and interest groups work together on a partnership basis. However, the 2003 EPA evaluation of CBEP outlines a rather limited role for the Agency in these collaborative arrangements. While the review is clear that the Agency is important as a funding body and key targeter of funding, it is less certain about the nature of its involvement in CBEP, advocating several different styles of engagement. The review seems most supportive of the EPA as a 'niche' player, usually through the provision of information and specific technical expertise, or as a facilitator. However, the review also acknowledges that it is sometimes appropriate for the EPA to take a lead role as the key mobilizer of a project and relevant groups, although even here it is recommended that such central involvement should soon be handed over to 'local stakeholders as procedures and roles are established' (EPA, 2003, pp. 7-9). There is also a problem with the scope of 'community' involvement caused by different, unexplained deployments of the term, including confusion between uses of 'stakeholder' and 'community'. The latter terms at times are used interchangeably, while at others 'community' is used as a subset of a wider group of stakeholders (EPA, 1999, p. 6).

It is also apparent that there can be tensions between the use of legal and voluntary mechanisms in EPA-sponsored CBEP. In 1997, the EPA declared that: 'Full protection of our nation's ecosystems requires communities and individuals to conserve or restore habitats and solve other environmental problems not specifically addressed by traditional regulatory approaches' (cited in Sansom, 1999, p. 22). However, Freeman (1997) has argued that the established conventions under which the EPA operates, of legalistic approaches and discretion constraint, undermine Agency attempts at collaborative governance (p. 91). She argues that the EPA does not have a sophisticated approach to dealing with stakeholders in regulatory negotiation and that 'smaller, community-based environmental groups' are most likely to be excluded (pp. 77-8). This is further supported by Colvin's (2002) assessment of a forest-based initiative, the Albany Pine Bush Preserve Commission (Washington State), set up under legislation in 1988, but promoted subsequently by the EPA as a good example of CBEP. Even here where the EPA was not directly involved, Colvin found that a major local environmental interest group, Save the Pine Bush, reported feeling so excluded from the decision-making processes of the Commission that it had become reliant on threatening lawsuits against other stakeholders in order to have any influence over the management of the preserve (p.451). In addition, despite the involvement of a range of stakeholders, as promoted by EPA CBEP models, Colvin (2002) found only limited support for the Preserve Commission as 'valuable as a forum for exchanging information and knowledge' (p. 451). Stakeholders had not heard of CBEP and did not feel the Commission had instigated any 'adaptive management or new participation techniques' (p. 451).

Finally, there are tensions apparent within CBEP in relation to the environmental protection aims of the EPA and the sustainable development scope of CBEP. CBEP aims at being place-based rather than media or issue based. This helps to address and co-ordinate a whole range of issues locally and tends to attract a greater number and diversity of stakeholders (EPA, 2003a). The Agency also sees merits in going beyond coordination between environmental issues towards the integration of wider quality of life issues into initiatives. CBEP is thus informed by the Local Agenda 21 process of Sustainable Communities in the US (EPA 2003a) and 'considers environmental protection along with human social needs, works toward achieving long-term ecosystem health, and fosters linkages between economic prosperity and environmental well-being' (EPA, 1999, p. 5). Whilst the EPA is approaching quality of life issues from an environmental perspective, there is the potential that its desire to 'develop solutions that help to sustain social, economic, and environmental well-being' (EPA, 1999, p. 3) may dilute its environmental protection functions and pull it in different directions as Bell and Gray (2002) argue is the case with the Environment Agency in England.

Landcare in Australia

The Australian National Landcare Programme (NLP) was launched in 1989, primarily to address growing problems of soil and water degradation, with the 1990s designated by the Australian Prime Minister, as the 'Decade of Landcare'.

Australian Landcare has been 'hailed internationally as one of the most significant participatory environmental programs ever developed' (Martin and Ritchie, 1999, p.118). It has exerted considerable influence in Australia, spawning a whole series of 'care' initiatives perpetuated under the Natural Heritage Trust (NHT), established in 1996 (Crowley, 2001), and internationally, with the Landcare concept being fundamental to the UN Convention to Combat Desertification (Ewing, 1996). The successes of the programme are usually related to the large scale of landholder involvement, over 30 per cent of farm businesses by 1998 (Higgins and Lockie, 2002), and the generation of large-scale tree planting. However, critics highlight problems with the nature of community involvement, the environmental impacts of the scheme and a lack of effective monitoring and evaluation (Bellamy, et al, 1999, Martin and Ritchie, 1999).

As Ewing (1996) has pointed out, the 'idea of "community" is one of the fundamental ideological constructs of Landcare' (p. 264). Indeed, it forms the basis for action with the Programme aiming to give '"communities ownership of their problems and control over their solutions"' (Alexander, 1992, p. 14, cited in Martin and Ritchie, 1999, p. 119). A positive aspect of this focus on 'community' is that it has helped shift perspectives beyond individual landholding/owning boundaries to landscape and catchment scales more relevant to the recognition and management of natural resource processes and problems (Higgins and Lockie, 2002).

However, the deployment of the rhetoric of 'community' under Landcare has been problematic in a variety of ways. Firstly, it has been selective in its scope at a number of scales. Nationally, the Landcare vision promoted awareness of environmental degradation amongst the whole Australian 'community' including urban citizens and corporations, embodied in Landcare Australia Limited, the non-profit making public company set up by the Commonwealth in 1989 to market and raise funds for Landcare (Ewing, 1996, p. 271). Despite this, Landcare and the subsequent NHT programme as a whole have retained a strong rural focus, with rural or regional areas receiving 90 per cent of NHT funding between 1996 and 2001 (Crowley, 2001, p. 256). Indeed, Ewing (1996) found that the wide, initial remit of Landcare had generated tensions amongst farmers, fearful of urban, industrial interests, and it is likely that such attitudes shaped the subsequent development of the programme. The rural and regional focus has prompted allegations of 'green barrelling' from the Labour opposition (Crowley, 2001, p. 264) and concerns from the Australian Conservation Foundation that the NHT was becoming more of a regional development fund than an environmental programme (Crowley, 2001). The latter critique is the more compelling coming as it does from the environmental organization that entered an unprecedented alliance with the National Farmers Federation to lobby for the establishment of a national landcare programme (Martin and Ritchie, 1999). Nonetheless, there have been notable successes from industry involvement in Landcare, such as the Alcoa World Alumina Australia sponsorship with Greening Australia of the Living Landscapes project in the Wheatbelt of Western Australia. This has facilitated longer term, in-depth strategies for integrated ecological and agricultural management at the catchment scale resulting in continued agricultural viability, extended habitats and considerable social learning (Dilworth et al, 2000).

Moving to the scale of rural and regional areas, engagement in Landcare by rural communities has been characterized by Martin and Ritchie (1999) as dominated by farming or major landholding interests such that an agri-centric view of the rural is embodied in Landcare. They report that 'representation on local (Landcare) and regional (TCM) groups is dominated by land-holders' and that '"Stakeholders" in practice refer mainly to local farmers' (p. 129). Indeed, there is a sense that Landcare has become a programme geared mainly towards farmers, akin to European agri-environmental programmes, but that this is obscured by a language of rural community involvement (see Higgins and Lockie, 2002). The general omission of non-agricultural elements of rural community in Landcare initiatives is highlighted by the establishment of a separate Rural Towns programme in the Wheatbelt of Western Australia to address salinization problems (Beresford et al, 2001). Furthermore, Landcare's focus on 'farming communities' has also been selective, with lower rates of involvement from indigenous and non-English speaking groups (Higgins and Lockie, 2002). Landcare's focus on agricultural communities, Crawley (2001) suggests, has fed back into the design of the whole NHT programme, influencing the nature of expertise on assessment panels, with that relating to biodiversity matters notably lacking.

The community focus of Landcare has also presented problems in terms of the balance between government and citizen responsibilities. At the local level, Ewing (1996), drawing on her case study from Glenelg, Victoria, argues that official projections of both 'community' and 'participation' were not sustained. On the one hand participating landholders questioned the currency of appeals to community asking, '"What is this community we hear of night and day?"' (p. 262). On the other hand they were critical that government officials retained ultimate control over decisions, privileging expert over lay knowledges. Thus Ewing concludes that the 'experiences of Landcare suggest the danger of government seeing "participation" as an opportunity to transfer the responsibility for complex, long-term and costly problems to local people, without the necessary resources to make a significant difference' (p. 274).

This locally-based argument links to wider critiques of the community construct in Landcare. A number of authors argue that government sponsored discourses of community participation in Landcare have helped legitimate neo-liberal regimes which perpetuate voluntaristic styles of environmental protection and strengthen individual entrepreneurial activity (Martin and Ritchie, 1999; Higgins and Lockie, 2002). Indeed, Martin and Ritchie's (1999) account of the development of Landcare in Australia has much in common with accounts of the development of the voluntary agri-environmental programme approach in Britain (Winter, 1996). They conclude that such 'Local forms of participation cannot solely be the basis for sustainable rural environments. The state has the responsibility to develop adequate co-ordinating and regulatory structures to enable local action and to reflect the wide range of legitimate interests in the quality of rural environments' (Martin and Ritchie, 1999, p. 132). Higgins and Lockie (2002, p. 420), however, hold that programmes such as Landcare 'reveal a number of subtle and novel ways in which governments are attempting to influence the practices of land managers'. They reject simplistic interpretations of neo-liberalism, in which 'non-interventionist market-based' approaches replace 'interventionist political rationality', and argue that an '*advanced liberal*' way of governing is being established which seeks to 'govern land managers in ways that emphasize the social causes and consequences of their practices' (p. 420). Yet, central to this argument is the role of knowledges embedded in what Higgins and Lockie (2002) term 'technologies of agency'. Many authors have highlighted the scarcity of robust empirical data on which to evaluate programmes such as Landcare, particularly in terms of social criteria (Martin and Ritchie, 1999; Bellamy et al, 1999). With the agri-centric orientation of Landcare at both local and national levels it is highly likely that quantifiable agriculturally-oriented knowledges of natural resource management are being privileged over environmental and social understandings. Thus, while it can be argued that advanced liberal governance may be operating through Landcare, it is style of governance that remains strongly marked by agricultural communities of interest.

The Environment Agency and English Landcare

The Environment Agency (covering England and Wales) was created in 1996. It is the state agency with responsibility for the protection and regulation of the soil, air and water environments, with the principal aim to protect and enhance the environment in ways that contribute to sustainable development (Bell and Gray, 2002). In rural England it works alongside the Department for Environment, Food and Rural Affairs (DEFRA), its parent Department, and a range of state, private and NGO groups. The Agency and its predecessors, however, have had little direct influence over agricultural policy and practices (Lowe et al, 1997). Until 2001 these matters were the remit of the Ministry of Agriculture, Fisheries and Food (MAFF) whilst Agency operations were overseen by a succession of environmental ministries. MAFF had a tradition of working closely with farmers' representatives and other agricultural interests in a closed style of 'policy community' in which voluntarism and agricultural exceptionalism were pursued in relation to agri-environmental issues, and environmental bodies were generally excluded (Smith, 1992; Winter, 1996). The Agency and its main predecessor, the National Rivers Authority (NRA), both contributed to and benefited from the gradual opening of this policy community in response to a series of environmental, social and policy challenges. With the most recent of these, the Foot and Mouth Disease crisis, culminating in the formation of DEFRA and MAFF's incorporation within it, the Agency now has the potential to exert greater influence than ever before over agri-environmental policy and practice. However, as Winter (2003, p. 51) points out, it will be striving to do so in a Department strongly influenced by the staff and culture of the former MAFF and in which the wider context of 'a highly diverse rural economy and society' is not strongly to the fore. This institutional context is likely to have a significant bearing on the Agency's approach to community-based environmental protection strategies in rural areas.

The Environment Agency has shown interest in community-based initiatives used in the US and Australia to combat diffuse pollution and soil erosion in rural areas. Such interest is partly linked to wider agendas emerging from the

Rio Earth Summit processes, including a commitment to the ideals of stakeholder involvement, dialogue and consensus building (Environment Agency, 1998a; Burgess, 2000). On the other hand such interest is also linked to the marginalized institutional position in which it has found itself in relation to agricultural matters. Both the Agency and the NRA thus made early efforts to enlist public support for, and even involvement in their regulatory activities, notably through public reporting of pollution incidents (Ward et al, 1995).

Despite early public statements of toughness, particularly towards polluting farmers, NRA pollution control strategies tended towards persuasive, conciliatory approaches (Seymour et al, 1999). The Environment Agency continued with these approaches in its early years, occupied with institutional reform related to its cross-medium remit (Bell and Gray, 2002). Such informal approaches can be seen as legitimate, embodying as they do an ethos of prevention rather than punishment after the event. They can also be interpreted as a response to decreasing levels of point source water pollution from agriculture from the late 1980s to 1998. However, in the late 1990s as levels of point source agricultural pollution began to rise again and evidence grew of problems with diffuse pollution from agriculture, the Agency began to express renewed concerns over agricultural pollution (Environment Agency 1999, 2001a).

An early move to encourage self-regulation, in line with the Agency's preference for informal approaches towards agricultural pollution, was the establishment of a pilot *Landcare* scheme in 1996. This single pilot was promoted in the South West, a leading region in shaping the Agency's agricultural and rural policy. The project was located in the upper Avon catchment in south Wiltshire, an area of international nature conservation importance with valuable wild and commercial fisheries. The land use is dominated by mixed farming and the area is subject to pressures from commuter housing, rural industrialization and tourism. The Agency suspected that soil erosion problems were affecting water quality and fisheries in particular and that agricultural practices were largely to blame. Due to the landscape-based, diffuse nature of the problem and lacking funds or authority to implement a scheme which would pay farmers for practice change, a catchment-based approach was adopted.

Landcare in Wiltshire, Stage 1 (1996-1998): Farmers and the Environment Agency
The initial *Landcare* group focused specifically on the issue of soil erosion in the upper Hampshire Avon and the promotion of Best Farm Management Practices (BMPs) which would help prevent this. Despite the tacit recognition, embodied in the earlier casting of rural publics as 'pollution watchdogs' (Lowe et al, 1997, p. 197), that rural areas contain more than farming populations, the target membership for this group was farmers. Following Australian rather than US models, other local stakeholders were not actively courted and the focus was on the single issue of soil erosion from agricultural land. The major strategies used were firstly to encourage farmers to join the Landcare Group, serviced by a Project Officer (from the Environment Agency) and circulated with a group Newsletter. A second strategy was to establish a demonstration site for BMPs on the farm of one of the leading farmer members of the Landcare group (Paul Bryson, personal

communication, 1998). In early 1998 the Environment Agency funded a short study by the University of Nottingham on farmers' attitudes to environmental issues in the area and the use of Best Management Practices (Seymour et al, 1998)[1].

From this work with the farmers and analysis of the policy, a number of problems can be highlighted with the initial Environment Agency strategy. These can be identified with farmers' responses; the design of the project; and the context of the project. Firstly, farmers questioned whether soil erosion was an issue in their area, highlighting instead the over-abstraction of water as a key environmental problem. Such contestation was informed, not only by their judgements of priority environmental issues in their area, but also by a more general questioning of the legitimacy of public and 'expert' views of farming impacts and a privileging of knowledge coming from within farming and local agricultural communities. In addition to a feeling of unfair targeting in the midst of more prominent problems, such as low river flows, farmers felt their expertise over land management was being ignored. There is also evidence that negative responses by farmers were exacerbated by the *Landcare* project design which limited involvement to farmers and the Environment Agency, arguably leading to less effective regulation. For example, if the local water company had been involved in the group, a more balanced approach to causes of and solutions to environmental problems might have been pursued. In addition, the project focused on a single issue which exacerbated the feeling amongst farmers that they were being unfairly targeted. Thirdly, the initiative was not sufficiently supported by wider policy and legislative frameworks. The initial project suffered from a lack of money and staff support within Agency. Little practical support for the initiative was forthcoming from outside the Agency, most notably from other state agencies or government departments. At a more strategic level, the initiative was hampered by the context of a continuing productivist agricultural policy, despite the 1992 CAP reforms (Winter, 1996).

Landcare in Wiltshire, Stage 2 (1999-2002): stakeholder representatives and farmers In December 1998 the Environment Agency began to change the emphasis of the *Landcare* Project by seeking to draw representatives from a wider range of interested 'stakeholder' groups into a Landcare Partnership (Environment Agency, 1998b). This sat well with Agency policy on stakeholder involvement and consensus building (Environment Agency, 1998a). Paul Bryson, the Environment Agency's *Landcare* project officer made an open call for participation: "'We hope that everyone will pitch in and share their ideas on how we can all pull together and improve the River Avon'" (Environment Agency, 1998b). Stakeholders included representatives from farming organizations, agricultural advisors, water companies, conservation groups, landowners, statutory bodies and local authorities[2]. This move helped broaden the scope of the Wiltshire *Landcare* initiative and facilitated the provision of new funds for its operation (Environment Agency, 2000). By 2002, DEFRA had become a Landcare partner and was providing funds for complementary initiatives such as the Wessex Outdoor Pig Partnership (Environment Agency, 2001b). New initiatives were also established which ascribed value to and encouraged the use of farmers' expertise. A prominent example was the establishment of a workshop on a trial farm at which farmers were invited to design and monitor trials in relation to maize production (Environment Agency, 2001c).

The Environment Agency's *Landcare* initiative to 2002 had some positive achievements in relation to farmer and wider stakeholder group involvement. Firstly, it established a grouping of farmers who identified with the *Landcare* initiative and its practices. It also established pilot approaches to encourage use of farmers' expertise through the establishment of demonstration practices on members' farms and through the practice workshop. Secondly, at a more strategic level, it achieved the involvement of stakeholder group representatives leading to a wider range of viewpoints being involved in the strategy and the generation of more funds for the initiative, including funding from DEFRA. In addition, the wider policy and legislative frameworks governing agricultural production, environmental protection and community participation provided a more sympathetic context for the initiative. The Agenda 2000 reforms, the creation of DEFRA and the EU Water Framework Directive (2000) give implicit or explicit support for such approaches.

Yet despite these steps forward, the initiative remained hampered in a number of ways. Although there was a broadening of support from organizations beyond the Environment Agency, the *Landcare* initiative remained a pilot scheme. It also remained a single issue initiative in which integrated thinking was underplayed, a feature which may link to ongoing problems reconciling the regulatory cultures of predecessor organizations (Bell and Gray, 2002). There was only a weak sense of a self-perpetuating local group which helped set agendas and locally-sensitive knowledges of farming and soil were still not firmly incorporated into the scheme (Julie Ingram, personal communication, 2003). In addition, local, non-agricultural publics were not encouraged to join in the *Landcare* initiative, despite evidence that farmers do respond to informal regulatory pressures from their neighbours (Sansom, 1999). However, perhaps the most fundamental problem with this *Landcare* initiative related to the lack of integrated monitoring of the initiative and its impact on the area. The Agency itself recognizes this, highlighting the impossibility of evaluating the project effectively without such monitoring, and ascribes the problem to a lack of money and resources (Environment Agency South West Region, 2002). Since 2002, however, *Landcare* has developed positively in terms of greater farmer involvement, attempts to draw in other local stakeholders and the rolling out of the initiative to other areas (Chris Westcott, Environment Agency, personal communication, 2003).

Participatory Rural Environmental Governance?

This chapter has reviewed a range of government or agency-led initiatives which make explicit or implicit appeals to ideas of 'community' to protect the rural environment. The focus has been on official deployments of different styles of 'community' in initiatives addressing diffuse pollution problems. Local, place-based interpretations of 'community' dominate the case study examples but they do not always operate in isolation. In many cases social, interest, commitment, generational and non-human 'communities' co-exist alongside 'place communities' although there is considerable variation in the ways in which they interact. In the USA example, place-based approaches draw in a range of rural social groupings whereas both Australian and English Landcare schemes show less awareness of spatial or social differences in 'rural communities' at the local level. In both attention has focused implicitly on 'the farming community' which is treated as representative of rural society and generally seen as a coherent socio-economic grouping. The dangers of such complete reliance on local, lay governance have been highlighted by Singleton (2002) and Martin and Ritchie (1999) who suggest this may lead to a further damaging withdrawal of government influence in situations where local and producer interests already hold substantial sway.

Place-based approaches, however, may draw in wider 'communities of interest' and 'communities of commitment'. In CBEP in the USA, place-focused initiatives aim to engage a range of stakeholder interests located beyond the locality. This process also occurs within Australian and English Landcare but in a more limited manner. For example, in England the Environment Agency has pursued a stakeholder partnership approach at a strategic but not local level to draw on a range of expertise and resources. Thus government agencies themselves may fall under the scope of 'community'. When they do so there remains a tendency for 'communities of interest' forged around producer or environmental interests to persist, with strong allegiances still apparent between farmers and agricultural or rural ministries. Nonetheless, the model of collaboration between government and non-governmental groups and publics holds promise. It has scope for the backing of voluntary measures by legislative frameworks, as occurs in the USA, it allows the input of significant technical expertise and information and government involvement adds legitimacy and support to local or lay group decisions (Lane and McDonald, 2002). However, collaboration poses a number of key challenges. The first is the need to develop the capacity for collaboration, particularly within *government* institutions. To be effective, as Warburton (1998, p. 26) points out, 'capacity-building must be a two-way process' between government and other groups. The second challenge is to move away from adversarial approaches and the privileging of a narrow range of technical expert knowledges. Burgess (2000) suggests one way of doing this is through discussion-negotiation and multi-criteria analysis in which stakeholders work to understand one another's perspectives and knowledges and use this understanding to agree priorities for action. A third challenge is to retain the capacity for effective government in collaborative situations. To achieve this Singleton (2002, p. 9) advocates strong, well-informed and properly resourced government agencies to allow them to 'simultaneously cultivate social trust and demonstrate the capacity and willingness to be tough'.

Engaging with ideas of sustainable development has the potential to draw generational and non-human aspects into place-based communities. Indeed, notions of community built purely on ideas of human associations appear increasing untenable in light of sustainability concerns. While there is a danger that consideration of environmental protection alongside economic and social issues may lead to the non-human environment losing out, focusing on place-based qualities and networks of human/non-human association helps extend the notion of community. The designs of CBEP in the USA and the Living Landscapes initiative in Australia (Dilworth et al, 2000) pose challenges to the idea of community as a social and cultural space (Liepins, 2000), and promote a wider notion of community constructed through socio-ecological networks.

Despite the lack of integrated evaluations, approaches which focus on 'community' involvement are only likely to increase in importance in future modes of rural and environmental governance. Governments and other state agencies advocating such approaches need to develop more nuanced ideas of 'rurality' and 'community', ideas which encompass an awareness of the contested nature of the terms and their enactment through specific material and imagined spaces.

Notes

1. The Discussion Group involved 8 farmers. The Postal Questionnaire elicited 118 replies, a response rate of about 34 per cent.
2. The members of the Landcare Partnership in 2000 were listed as: National Farmers Union, Country Landowners Association, MAFF's Farming and Rural Conservation Agency, Royal Society for the Protection of Birds, Wiltshire County Council, Salisbury District Council, Wessex Water, Bournemouth and West Hants Water Co, British Independent Agricultural Consultants, Game Conservancy Trust, Ministry of Defence (Defence Estates Organisation – a major landholder in the area), English Nature, Wiltshire Fisheries Association, Wessex Salmon and Rivers Trust and the Environment Agency (Environment Agency 2000).

References

Agrawal, A. and Gibson, C.C. (1999), 'Enchantment and Disenchantment: the Role of Community in Natural Resource Conservation', *World Development*, Vol. 27, pp. 629-649.

Bell, C. and Newby, H. (1971), *Community Studies*, Allen and Unwin, London.

Bell, D. and Gray, T. (2002), 'The Ambiguous Role of the Environment Agency in England and Wales', *Environmental Politics*, Vol. 11, pp. 76-98.

Bellamy, J.A., McDonald, G.T., Syme, G.J. and Butterworth, J.E. (1999), 'Evaluating Integrated Resource Management', *Society and Natural Resources*, Vol. 12, pp. 337-353.

Beresford, Q., Bekle, H., Phillips, H. and Mulcock, J. (2001), *The Salinity Crisis: Landscapes, Communities and Politics*, University of Western Australia Press, Crawley, Western Australia.

Burgess, J. (2000), 'Situating Knowledges, Sharing Values and Reaching Collective Decisions', in I. Cook, D. Crouch, S. Naylor and J. Ryan (eds), *Cultural Turns/Geographical Turns*, Prentice Hall, London, pp. 273-287.

Burnett, M. L. (1998), 'The Pollution Prevention Act of 1990', *Environmental Management*, Vol. 22, pp. 213-224.

Cline, S.A. and Collins, A. (2003), 'Watershed Associations in West Virginia: Their Impact on Environmental Protection', *Journal of Environmental Management*, Vol. 67, pp. 373-383.

Cloke, P. (2000), 'Rural Community', in R.J. Johnston et al (eds), *The Dictionary of Human Geography*, 4th edn, Blackwell, Oxford, pp. 718-719.

Cohen, A. (1989), *The Symbolic Construction of Community*, Routledge, London.

Colvin, R.A. (2002), 'Community-Based Environmental Protection, Citizen Participation and the Albany Pine Bush Preserve', *Society and Natural Resources*, Vol. 15, pp. 447-454.

Crowley, K. (2001), 'Effective Environmental Federalism? Australia's Natural Heritage Trust', *Journal of Environmental Policy and Planning*, Vol. 3, pp. 255-272.

Day, G. (1998), 'A Community of Communities? Similarity and Difference in Welsh Rural Community Studies', *The Economic and Social Review*, Vol. 29, pp. 233-257.

Day, G. and Murdoch, J. (1993), 'Locality and Community: Coming to Terms with Place', *Sociological Review*, Vol. 41, pp. 82-111.

Department of the Environment and Ministry of Agriculture, Fisheries and Food (1995), *A Nation Committed to a Living Countryside*, HMSO, London.

Department of the Environment, Transport and the Regions and Ministry of Agriculture, Fisheries and Food (2000), *Our Countryside: The Future. A Fair Deal for Rural England*, HMSO, London.

Dilworth, R., Gowdie, T. and Rowley, T. (2000), 'Living Landscapes: the Future of the Western Australian Wheatbelt?', *Ecological Management and Restoration*, Vol. 1, pp. 165-175.

Duane, T.P. (1997), 'Community Participation in Ecosystem Management', *Ecology Law Quarterly*, Vol. 24, pp. 771-97.

Environment Agency (1998a), *Consensus Building for Sustainable Development*, Environment Agency, Bristol.

Environment Agency (1998b), 'Major New Partnership Push to Improve River Avon', Press Release 2 Dec 1998.

Environment Agency (1999), 'Increase in Agricultural Water Pollution Incidents "Disappointing" says Environment Agency', Press Release 18 Aug 1999.

Environment Agency (2000), 'Free Farmers Toolkit to Help Improve River Avon', Press Release 12 April 2000.

Environment Agency (2001a), *Rebuilding Agriculture*, Position Statement, Environemnt Agency, Bristol.

Environment Agency (2001b), 'New Group Sets Out Future for Wessex Pigs', Press Release 11 Sept 2001.

Environment Agency (2001c), 'Farmers Test Ways to Reduce Pollution and Flooding', Press Release 6 Sept 2001.

Environment Agency, South West Region (2002), *Landcare Baseline Monitoring Report*, Environment Agency South West Region, Blandford Forum, Dorset.

Environmental Protection Agency (1999), *EPA's Framework for Community-Based Environmental Protection*, EPA, Washington, D.C.

Environmental Protection Agency (2003a), www.epa.gov.

Environmental Protection Agency (2003b), *Evaluation of Community-Based Protection Projects: Accomplishments and Lessons Learned*, Office of Policy Economics and Innovation, EPA, Washington D.C.

Environmental Protection Agency (2003c), *Memorandum: Evaluation of Community-Based Protection Projects: Accomplishments and Lessons Learned*, Office of Policy Economics and Innovation, EPA, Washington D.C.

Ewing, S. (1996), 'Whose Landcare? Observations on the Role of "Community" in the Australian Landcare Programme', *Local Environment*, Vol. 1, pp. 259-276.

Freeman, J (1997), 'Collaborative Governance in the Administrative State', *UCLA Law Review*, Vol. 45, pp.1-98.

Haith, D.A. (2003), 'Systems Analysis, TMDLs and Watershed Approach', *Journal of Water Resources Planning and Management*, Vol. 129, pp. 257-260.

Halfacree, K. (1995), 'Talking About Rurality: Social Representations of the Rural as Expressed by Residents of Six Rural Parishes', *Journal of Rural Studies*, Vol. 11, pp. 1-20.

Harper, S. (1987), 'A Humanistic Approach to the Study of Rural Populations', *Journal of Rural Studies*, Vol. 3, pp. 309-320.

Harper, S. (1989), 'The British Rural Community: An Overview of Perspectives', *Journal of Rural Studies*, Vol. 5, pp. 161-184.

Higgins, V. and Lockie, S. (2002), 'Re-discovering the Social: Neo-liberalism and Hybrid Practices of Governing in Rural Natural Resource Management', *Journal of Rural Studies*, Vol. 18, pp. 419-428.

Jennings, S.F. and Moore, S.A. (2000), 'The Rhetoric Behind Regionalisation In Australian Natural Resource Management', *Journal of Environmental Policy and Planning*, Vol. 2, pp. 177-191.

Karamanos, P. (2001), 'Voluntary Environmental Agreements: Evolution and Definition of a New Environmental Policy Approach', *Journal of Environmental Planning and Management*, Vol. 44, pp. 67-84.

Lane, M. and McDonald, G. (2002), *Community-Based Environmental Planning: Operational Dilemmas and Practical Remedies*, Working Paper, Department of Geographical Sciences and Planning, University of Queensland.

Latour, B. (1993), *We Have Never Been Modern*, Harvester Wheatsheaf, Hemel Hempstead.

Leach, W.D. and Pelkey, N.W. (2001), 'Making Watershed Partnerships Work: A Review of the Empirical Literature', *Journal of Water Resources Planning and Management*, Vol. 127, pp. 378-385.

Leach, W.D., Pelkey, N.W. and Sabatier, P. (2002), 'Stakeholder Partnerships as Collaborative Policymaking: Evaluation Criteria Applied to Watershed Management in California and Washington', *Journal of Policy Analysis and Management*, Vol. 21, pp. 645-670.

Liepens. R (2000), 'New Energies For An Old Idea: Reworking Approaches to "Community" In Contemporary Rural Studies', *Journal of Rural Studies*, Vol.16, pp. 23-35.

Little, J. and Austin, P. (1996), 'Women and the Rural Idyll', *Journal of Rural Studies*, Vol. 12, pp. 101-111.

Lowe, P., Clark, J., Seymour, S. and Ward, N. (1997), *Moralizing the Environment: Countryside Change, Farming and Pollution*, UCL Press, London.

Lowe, P., Ray, C., Ward, N., Wood, D. and Woodward, R. (1998), *Participation in Rural Development: A Review of European Experience*, Centre for Rural Economy, University of Newcastle, Newcastle upon Tyne.

Marsden, T. and Murdoch, J, (1998), 'The Shifting Nature of Rural Governance and Community Participation', *Journal of Rural Studies*, Vol. 14, pp. 1-4.

Marsh, D. and Rhodes, R. A. W. (eds) (1992), *Policy Networks in the British Government*, Oxford University Press, Oxford.

Martin, P. and Ritchie, H. (1999), 'Logics of Participation: Rural Environmental Governance under Neo-liberalism in Australia', *Environmental Politics*, Vol. 8, pp. 117-135.

Martin, S. (1999), 'Democratising Rural Development: Lessons from Recent European Union and UK Programmes', in N. Walford, J. Everitt and D. Napton (eds), *Reshaping the Countryside: Perceptions and Processes of Rural Change*, CABI, Wallingford, pp.169-178.

Massey, D. (1991), 'A Global Sense of Place', reprinted in S. Daniels and R. Lee (eds), *Exploring Human Geography*, Arnold, London, pp.237-245.

Moseley, M.J. (2003), *Rural Development: Principles and Practice*, Sage, London.

Murdoch, J. (1997), 'The Shifting Territory of Government: Some Insights from the Rural White Paper', *Area*, Vol. 29, pp. 109-118.

Murdoch, J. and Marsden, T. (1994), *Reconstituting Rurality: Class, Community and Power in the Development Process*, UCL Press, London.

Newby, H. (1977), *The Deferential Worker*, Allen Lane, London.

O'Riordan, T. (2000), 'Environmental Science on the Move', in T. O'Riordan (ed), *Environmental Science for Environmental Management*, 2nd edn, Pearson, Harlow, Essex, pp. 1-27.

Phillips, M. (1998), Investigations of the British Rural Middle Classes. Part 2: Fragmentation, Identity, Morality and Contestation, *Journal of Rural Studies*, Vol.14, pp. 427-444.

Potter, C. (1998), 'Conserving Nature: Agri-environmental Policy Development and Change', in B. Ilbery (ed), *The Geography of Rural Change*, Longman, Harlow.

Rose, N. (1996), 'The Death of the Social? Refiguring the Territory of Government', *Economy and Society*, Vol. 25, pp. 327-356.

Sansom, A. (1999), *Farmer/Regulator Relationships and How to Improve Them*, Nuffield Farming Scholarships Trust, Uckfield, East Sussex.

Selman, P. (1996), *Local Sustainability*, Paul Chapman, London.

Selman, P. (1998), 'Local Agenda 21: Substance or Spin?', *Journal of Environmental Planning and* Management, Vol. 41, pp. 533-553.

Seymour, S., Turner, R., Gerber, J. and Kinsman, P. (1998), *Research into Cost Effective Methods of Influencing Attitudes within the Agricultural Community in the Upper Hampshire Avon Catchment: The Case of Non-Point Source Agricultural Pollution and 'Best Management Practices'*, Report commissioned by the Environment Agency.

Seymour, S., Lowe, P., Ward, N. and Clark, J. (1999), 'Moralizing Nature? The National Rivers Authority and New Moral Imperatives for the Rural Environment', in N. Walford, J.C. Everitt and D.E. Napton (eds), *Reshaping the Countryside: Perceptions and Processes of Rural Change*, CABI, Wallingford, pp. 39-56.

Short, B. (ed) (1992), *The English Rural Community: Image and Analysis*, Cambridge University Press, Cambridge.

Singleton, S. (2000), 'Co-operation or Capture? The Paradox of Co-management and Community Participation in Natural Resource Management and Environmental Policy-making', *Environmental Politics*, Vol. 9, pp. 1-21.

Singleton, S. (2002), 'Collaborative Environmental Planning in the American West: the Good, the Bad and the Ugly', *Environmental Politics*, Vol. 11, pp. 54-75.

Smith, M. J. (1992), 'The Agricultural Policy Community', in D. Marsh and R. A. W. Rhodes (eds), *Policy Networks in the British Government*, OUP, Oxford, pp. 27-50.

Thomas, C. (1994), 'Beyond UNCED: An Introduction', in C. Thomas (ed), *Rio: Unravelling the Consequences*, Frank Cass, Ilford, Essex, pp. 1-27.

Tsouvalis, J., Seymour, S., Watkins, C., Holloway, L. and Steven, M. (2000), *Farmers and Precision Farming: a Study of Attitudes and Practices in Lincolnshire and Suffolk*, Working Paper 34, School of Geography, University of Nottingham.

Walker, G. (1995), 'Social Mobilization in the City's Countryside: Rural Toronto Fights Waste Dump', *Journal of Rural Studies*, Vol. 11, pp. 234-254.

Warburton, D. (1998), 'A Passionate Dialogue: Community and Sustainable Development', in D. Warburton (ed), *Community and Sustainable Development: Participation in the Future*, Earthscan, London, pp. 1-39.

Ward, N., Lowe, P., Seymour, S. and Clark, J. (1995), 'Rural Restructuring and the Regulation of Farm Pollution', *Environment and Planning A*, Vol. 27, pp. 1193-1211.

WCE (1987), *Our Common Future*, Oxford University Press, Oxford.

Williams, R. (1983), *Keywords: A Vocabulary of Culture and Society*, Fontana, London.

Winter, M. (1996), *Rural Politics*, Routledge, London.

Winter, M. (2003), 'Responding to the Crisis: The Policy Impact of the Foot-and-Mouth Epidemic', *Political Quarterly*, Vol. 74, pp. 47-56.

Wright, S. (1992), 'Image and Analysis: New Directions in Community Studies', in B. Short (ed), *The English Rural Community*, Cambridge Univeristy Press, Cambridge, pp. 195-217.

Young, I. M. (1990), 'The Idea of Community and the Politics of Difference', in L. J. Nicholson (ed), *Feminism/Postmodernism*, Routledge, London, pp. 300-323.

PART IV
CULTURES OF FARMING
AND FOOD

Chapter 12

Lost Words, Lost Worlds? Cultural Geographies of Agriculture

Carol Morris

Whither the 'Culture' in Agricultural Geography?

In a recent collection of essays in cultural geography, *Cultural Turns/ Geographical Turns* (Cook et al, 2000), agriculture makes no appearance, at least, not in respect of an entire chapter devoted to farming issues. Meanwhile, agriculture's 'sister' land use of forestry does feature in one chapter of the volume (Tsouvalis, 2000). Similarly, in two reviews of the cultural turn within rural geography, neither makes explicit reference to this phenomenon in the context of geographical investigations of agriculture (Cloke, 1997; Little, 1999). Although Little (1999) refers, within her review of work concerned with nature-culture relations, to 'mad cows and hounded deer' (Woods, 1998), agriculture does not in fact constitute a starting point in this study. Meanwhile, Cloke (1997, p. 369) describes how 'culturally inspired studies are being overlain on very important existing topographies of behavioural and political-economic geographies of agriculture'. However, he does not go on to elaborate the type and scope of work that has been produced by this engagement, if indeed this is implied. It is unclear whether he is suggesting that geographies of agriculture are only behavioural and political economic, never 'cultural'. Furthermore, in an earlier book, *Contested Countryside Cultures*, edited jointly by the same authors, and evidence of rural geography's 'turn to culture', there are no chapters that deal with agriculture, in spite of a front cover depicting a 'traditional' agricultural scene (Cloke and Little, 1997).

This evidence appears to point towards an absence of research interest in the culture(s) of agriculture and their spatial manifestations. But is this actually the case? This chapter seeks to demonstrate that there is clear (if relatively limited) evidence of a concern for culture within geographical studies of UK agriculture. It approaches the task by firstly addressing some definitional issues, briefly reviewing the concepts of culture and the cultural turn within geography. The chapter then moves on to offer some suggestions as to why the cultural turn has had an uneven impact within British rural geography i.e. why culture has been of less interest to agricultural geographers than other researchers working in rural geography[1]. The various ways in which culture has been explored within geographical studies of agriculture are then discussed. The final section of the chapter considers the potential future content and contribution of geographical research into 'agri-culture(s)'[2].

Culture and the Cultural Turn in Geography

It is widely accepted that human geography, together with some of its sister disciplines within the social sciences, has recently experienced a 'cultural turn'. Within British rural geography specifically, researchers have enthusiastically embraced the cultural turn (Little, 1999) and it is this engagement, between the rural and the cultural, so Cloke (1997) suggests, which accounts for the recent resurgence of British rural geography. The 'moment' of the cultural turn within the discipline as a whole has been dated to the late 1980s / early 1990s, enabling it to be located within 'the wider set of debates that emerged in the late 1980s around postmodernism which in large part were the vehicles for geography's entry into new fields of cultural theory' (Barnett, 1998a, p. 381). Dissatisfaction with the theoretical hegemony of political economy is also identified as a cause of the turn to culture at this time. In simple terms, the cultural turn has entailed an increased sensitivity within geographical inquiry to the role of culture and cultural processes in the mediation of all aspects of social life. Culture has therefore been recognized as something that is central to the organization and operation of society, not a residual category that is marginal to the social, the political and economic (Crang, 1998).

While the precise meanings of culture and 'the cultural' are strongly contested, Crang (1997, p. 5) identifies two main types of understanding of culture in geography. The first conceptualizes culture as 'a "generic" facet of human life, bound up with human competencies to make the world meaningful and significant'. This view of culture takes seriously questions of beliefs and values that give meaning to ways of life and produce (and are reproduced through) material and symbolic forms. 'The cultural then concerns the meaningful mapping of the world and one's positionings within it' (Crang, 1997, p. 5). Culture as a 'differential' quality is the second understanding of culture that Crang identifies within geography. In this view culture marks out and helps to constitute distinctive social groups each with their own meaning and value systems. Key within these understandings is that culture is a process that we are all involved in, and not a thing that we possess, so that culture is 'best viewed as it occurs, rather than through the lens of some metaphorical end product' (Crang, 1997, op cit). However, as Crang moves on to argue, culture is often treated as if it were a thing. 'Hence it is precisely this construction of cultural things which needs explaining. Cultures as things are the starting point of analytical endeavour, not the end point. In and of themselves they explain nothing' (Crang, 1997, op cit). Another important dimension to definitions of culture within geography is contestation i.e. that cultural constructions are always 'up for grabs', both from those responsible for their production and from those outside the particular domain in which they have been produced. So,

> culture is [and our maps of meaning are] not something out there we seek to grasp, a text or hidden code. It is [they are] a relatively instable product of the practice of meaning, of multiple and socially situated acts of attribution (Friedman, 1994, p. 74, cited in Crang, 1997, p. 5).

By the mid 1990s, the cultural had become firmly embedded, but by no means always welcomed (e.g. Harvey, 2000; Philo, 2000), within British human geographical thinking and practice. This had profound implications for methodology, as evidenced in the increased interest in qualitative approaches such as ethnography and the use of techniques such as discourse analysis (Hughes et al, 2000). The cultural turn has also affected the choice of data sources, many of which would have previously been of little concern or interest to geographers e.g. an increasingly wide variety of texts and images have become the subject of interpretation. Furthermore, it is apparent that the cultural turn has not been uniform within and between geography's sub-disciplines. Within rural geography, the interest in culture appears to have been particularly marked in respect of the non-agricultural elements of rural spaces and places as the reviews by Cloke (1997) and Little (1999) testify. In the next section, some reasons are offered for this uneven impact of the cultural turn within rural geography.

An Agri-Cultural Turn in Geography: Exploring the Limiting Factors

The first reason that is advanced relates to 'research fashion'. During the 1990s, at the time when the cultural turn was beginning to impact on rural geography, research increasingly became concerned with those aspects of the rural beyond the agricultural. Larger numbers of new researchers have appeared to favour the rural rather than the agricultural[3]. This represents a reversing of, if not a deliberate reaction to, the situation in rural geography's 'fallow years', when the rural was conflated with the agricultural (Cloke, 1997). The limited impact of the cultural turn on studies of agriculture may in part therefore be a question of numbers of researchers within rural geography[4] working on agricultural and agri-food topics compared with those investigating the non-agricultural aspects of rural society. More fundamentally (and perhaps more convincingly) the greater emphasis upon 'rural' rather than 'agricultural' research has inevitably entailed a greater concern for the countryside as a place of consumption, as opposed to production (Marsden et al, 1993; Murdoch and Marsden, 1994). Rural geography might be broadly characterized therefore as increasingly consumption, rather than production, oriented. If, as some have claimed, the 'consumption turn' in social science is a parallel development (perhaps even synonymous) with the cultural turn[5] it is easy to see how those areas of research which have traditionally focused on matters of production, such as agriculture, have been less subject to the influence of the cultural turn than those areas of rural research (and also research on food consumption) that are broadly concerned with consumption.

Second, the hegemony of particular theoretical perspectives can be highlighted. Just at the time when the cultural was beginning to interest rural researchers, a number of key commentators working in, and arguably dominating, research on agricultural change were deploying, to considerable effect, political economy perspectives (see Marsden et al, 1996 for overview). As Morris and Evans (1999) have argued, political economic narratives have effectively overshadowed all other types of analysis in agricultural geography. The power of the political economy approach, at least in the early years of the 1990s, may have been such as to divert researchers away from the potential insights offered by culturally sensitive analyses. This is suggestive of the observations by Hughes et al (2000, pp. 10-11) that 'ethnographic methodologies have not been adopted ... widely in ... areas of rural research such as in agricultural geography'. They go on to argue that this 'may be partly explained by the interests of key "gate keepers" within these research areas', although they acknowledge that this is an issue which requires further unpacking.

The third explanation for a relatively limited agri-cultural turn lies in the policy orientation of a significant proportion of the agricultural research that is undertaken by geographers. As Morris and Evans (1999, p. 354) have argued, 'it is understandable that [agricultural] work has not been more culturally sensitive, because much of it has been delivered within a 'policy evaluation' mould. The monitoring brief demanded by government agencies leads inevitably to questionnaire-type approaches', rather than the more in-depth approaches required by cultural perspectives. These sentiments are echoed in Cloke (1997) and Milbourne (2000), who suggest that there is an incompatibility between qualitative rural research (characteristic of the cultural turn) and contemporary policy discourses which valorize numerical data.

A final explanation lies in the history of the positioning of sub-disciplines within human geography. Traditionally, agricultural geography was a branch of economic geography. This had an inevitable influence on the orientation of agricultural research. One example of this is the treatment of farm animals, which have tended to be conceptualized as 'homogeneous items of mass production, broad types within policy mechanisms or at best 'quality products' within the food production system' (Morris and Evans, 1999 p. 355). A consequence of this is that 'little has been written on the association of animals with local folklore and culture, so that a discourse is lacking on the importance of particular animals to particular locales, and to the construction of these locales' (Morris and Evans, op cit). The enculturing of economic geography has only very recently given added legitimacy to a culturally sensitive approach within agricultural research (Thrift and Olds, 1996; Crang, 1997).

Cultural Geographies of Agriculture: The Evidence

Having posited some reasons why agricultural geographers might have been more reluctant than other rural geographers to engage with the cultural turn, in this section evidence is presented of recent research into cultural geographies of agriculture together with some discussion and commentary on this work. However, it is important to acknowledge that this work cannot be solely attributed to the particular set of intellectual debates that have been circulating within human geography over the past decade. The 'valuable manoeuvres' of humanistic, and more particularly, behavioural studies of agriculture in the 1970s and 80s must also be recognized as 'paving the way for the cultural turn of more recent vintage' (Philo, 2000, p. 32). These precursors to the cultural geographies of agriculture identified herein entailed 'detailed investigations into the shadowy recesses of human perception, cognition, interpretation, emotion, meanings and values, creating a rich vein of inquiry' (Philo, op cit). The considerable body of behavioural investigations of agriculture, i.e. studies of farmers' goals, values and attitudes should be highlighted in this context (e.g. Gasson, 1973; Gillmor, 1986; Ilbery, 1985; Newby et al, 1977). Commentary on geographical agri-cultural research of the 1990s must also acknowledge the significant contribution of social anthropological studies of rural communities undertaken in the 1950s and 60s (e.g. Arensberg and Kimball, 1940; Davies and Rees, 1960; Emmett, 1964; Frankenburg, 1957; Littlejohn 1963; Rees, 1950) and also more recently in the 1980s and 90s (Bell, 1994; Bouquet, 1985; Middleton, 1986; Rapport 1993; Strathern 1982a and b) in which the agricultural members of these communities were a feature, and oftentimes the foci. Moreover, for these authors 'the cultural' was always a principal concern, not something that was 'discovered' as in the agri-cultural turn research of recent years.

In examining the 'cultural stuff' of those geographers who study agriculture and have a 'cultural interest' (Philo, 2000, pp. 27-28) it is apparent that relatively few of these studies have made deliberate efforts to define culture in this context (rather, they have simply donned a 'cultural cloak' and got on with the particular research task at hand without feeling the need to justify the terms or perspective adopted). As Young et al (1995, p. 73) observe in the context of environmental decision-making in agriculture, 'researchers have adopted "socio-cultural" or "socio-cognitive systems" approaches...Others have engaged in "actor-oriented research" which incorporates a cultural as well as a structural account of actions ... in order to investigate the complex web of values and aspirations extending far beyond purely economic rationales which make up agrarian ideology and culture'. However, while 'this research has introduced the consideration of culture into analyses... it has failed to give a clear definition of what culture actually is' (Young et al, 1995, p. 73). In spite of this definitional laxity, a range of studies can still be identified as illustrative of a body of work labelled here as 'cultural geographies of agriculture' because they concern themselves with exactly those processes identified by Crang (1997) and summarized above i.e. the meanings and value systems within agriculture and examinations of distinctive agricultural cultures.

The following subsections discuss these agri-cultural geographies. Three areas of work are identified: 'representations of agriculture'; 'nature-society relations'; and 'enculturing the agri-food economy'. The first two appear to have attracted more attention than the latter, which can be seen to represent an 'emergent' area of activity. These groupings are proposed as a means of organizing and commenting on the work undertaken, but it is acknowledged that any such categorization is inevitably subjective, the product of a particular 'reading' of research interpreted as 'agri-cultural'. Furthermore, the work referred to under each heading could be (and in some cases is) positioned under more than one heading. The review does not claim to be exhaustive, nor does it set out to give the impression of acting as a 'kind of satellite circling...[rural studies' agri-cultural]...turn, claiming the "scopic power" to see clearly all that is taking place' (Philo, 2000, p. 27).

Representations of Agriculture

The first area of work into the cultural geography of agriculture is concerned with the representation of aspects of agricultural life. As Fish and Phillips (1997, p. 1) have observed within the context of rural studies as a whole:

> increasing attention has been paid to images and representations of the countryside and how they are created, circulated and consumed, and also how they may influence material actions...There has also been....a concern with examining how these representations are embedded in relations of power.

Their observation can also be applied to the agricultural sphere, in which both written and pictorial texts have been analysed for the meanings they encapsulate, with media texts being a notably popular source. Gender issues in agriculture have been a particular focus of attention, supporting Barnett's (1998a) assertion that the study of the construction of social relations of gender is one exemplar of human geography's cultural turn. Within an agricultural context, authors have begun to explore the role of the agricultural and countryside media in the construction and reproduction of farming masculinities and femininities[6]. For example, the neglect of women in 'farming success stories' in the North American agricultural press (Walter and Wilson 1996) inspired Morris and Evans (2002) to examine the changing representations of farm women in a major UK farming publication. Portrayals of farm life were exposed as increasingly subtle in the communication of long-established patriarchal views about the roles of farm men and women.

The majority of this work is based upon academic readings of media texts, which is both interesting in itself and revealing of the ways in which powerful institutions such as the media draw upon and replicate ingrained constructions of masculinity and femininity in agriculture. However, there is a danger that such analyses lead researchers away from the 'patient excavation of the grain of component social lives, social worlds and social spaces' (Philo, 2000, p. 37)[7]. As such, analyses of these media representations need to be accompanied by study of the various ways in which these media products are consumed, as illustrated in a non-agricultural context by Jackson and Thrift (1995) and Jackson et al (1999), and, in a non-British but agricultural context, by Liepins (1996) who has detailed how media texts are actively resisted by Australian farm women and the various campaign methods that these women have adopted so that their views and contributions to agriculture are more accurately represented by the press. Moreover, examination of the social processes involved in the *production* of representations of agriculture has begun to add depth and critical insight to these analyses (e.g. Phillips et al, 2001).

Within this work on the representation of agriculture it is possible to locate a growing body of research that has focused upon 'discourse'. 'Discourse has become one of the most widely and often confusingly used terms in recent theories in the arts and social sciences, without a clearly definable single unifying concept' (Meinhof, 1993, p. 161). It is certainly evident within research into cultural geographies of agriculture that there is some variation in the interpretation and use of the term. Nonetheless, there is a measure of concurrence among discourse-centred accounts of agriculture, with discourse typically taken to mean a framework, embracing 'particular combinations of narratives, concepts, ideologies and signifying practices' (Liepins, 1996, p. 9), in which something e.g. agriculture, the environment, food quality etc., is made meaningful. Authors have been concerned to uncover the many competing discourses that exist to give meaning to specific agricultural phenomena and to explore the ways in which discourse structures experience and action. In this last respect, the relationship between discourse and power is highlighted. As Liepins (1996, p. 3) argues, 'discourse as a way of structuring knowledge and social practice is an important medium through which agricultural power can be studied'.

The discursive practices surrounding the agriculture-nature or agri-environment interface (Lowe and Ward, 1997; Clark et al, 1997; Buttel, 1998) and food (Morris and Young, 2000) have all been the subject of research attention, with interview transcripts and a variety of printed, e.g. media, texts used as sources. For example, Morris and Young's (2000) analysis of discourses of food quality and quality assurance schemes in the British agricultural press has revealed how quality is contested and a focus of resistance and disagreement amongst different agro-food system actors. Other documentary sources have also been the subject of 'discourse analysis'. European agricultural policy documents (together with the transcripts of interviews with policy makers) provided the focus of Clark et al's (1997) 'discourse approach' to the analysis of the 'greening' of European agricultural policy. They make a strong case for the examination of discourse in this context, observing 'agricultural policy researchers have been overly preoccupied with theorizing the role of interests, *at the expense of ideas*, in their analyses of national and supranational policy development' (Clark et al, 1997, p. 1870, emphasis added). By adopting a discourse approach they are able to demonstrate how the culture of European policy making has sustained a particular vision of agricultural and agri-environment policy in which agricultural land occupancy and the small scale, family farm are central concepts.

The relative merits of these discourse centred accounts of rural and agricultural change are summed up by Buttel (1998, p. 1152) who suggests that they have 'led to some notable advances, particularly by documenting the fact that discursive practices are important guides to social action, and pervasive and potentially efficacious resources in political struggles'. However, he goes on to identify two reasons why caution should be exercised in the analysis of 'the processes of rural and agricultural change that privilege the social origins and patterns of articulation of discourses' (Buttel, op cit). First, although the analysis of discourse potentially side-steps the problem of structural determinism, this is only achieved by putting in its place 'an equally one-sided voluntarism'. Second, Buttel (1998, op cit) warns against the study of discursive practices becoming 'a de facto form of structural determinism if it is presumed that the course of social action or fate of discursive exercises or struggles is essentially a reflection of the power of the groups that advance these narratives'. The combining of discourse based explanations of agricultural change with political economy, as advocated by Clark et al (1997) may be one way of avoiding the first of these difficulties. Meanwhile, examination of the complex ways in which discourses, that are seen as structuring the lives of particular agricultural groups, are negotiated by these groups may be a means of countering the second (Liepins, 1998).

Nature-Society Relations

'It is important that work [on the relationship between nature and society] is noted in the context of the "cultural turn" in rural studies' (Little, 1999, p. 440). This argument can be readily extended to agricultural geography where research concerned with nature-society relations can be identified. It is in this context that there has been a notable, if not the greatest, degree of engagement between agricultural geography and the cultural turn (Holloway, 2000). In some ways this is no surprise, given that agricultural activity is bound up with 'nature' and that considerable agricultural policy debate in recent years has centred upon the relationship between agriculture and the environment.

Social science research on nature-society relations has taken a number of distinct forms and this is also evident within agri-cultural research which has adopted various analytical entry points and methodological approaches. Many of these studies might be broadly characterized as 'social constructivist'[8], in that they reject realist perspectives on the environment and instead emphasize the inseparability of nature and society and the need to explore cultures of nature(s), i.e. the spatially and temporally contingent ways in which people come to understand and apply meaning to nature and the environment (Macnaghten and Urry, 1998; van Koppen, 2000). Thus, agri-cultural research has begun to uncover the different constructions of the environment amongst farmers, and particularly the contrasting meanings and understandings of nature, and the appropriate management of nature, between farmers and so-called environmental 'experts' e.g. conservation organizations and policy makers (Carr and Tait, 1991; McEachern, 1992; Walsh, 1997; McHenry, 1998; Holloway, 1999). As Burgess et al (2000, p.120) summarize, 'social and cultural research has shown that farmers and conservationists may view the same landscapes or species, but see them quite differently'. The context of these studies has been specific agri-environment schemes, localities of high natural value, and environmental issues such as habitat management and climate change[9]. McEachern's (1992, p. 168) ethnographic study of the Yorkshire Dales National Park, for example, detailed how the Park officers 'often perceived as 'dreadful' and 'ugly' the parts of the landscape which the farmers most admired: the pastures and fields extending far up the fellsides, destroying diversity and wildlife habitats'. In the exploration of these contrasting understandings of nature and the environment such studies have also been concerned with the (re)production of farmer and farming community identities[10].

A closely related entry point to those studies concerned with 'constructions of nature', is knowledge(s) about nature and the environment. Here, the locally specific knowledge of farmers, which is created in large part through the experience of working closely with the land, is contrasted with the scientific knowledge of agricultural and environmental experts. As Wilson (1997, p.307) asserts, in the context of assessment of the environmental impact of the Environmentally Sensitive Area Scheme:

> ... farmers usually know their land better than other actors, and are, therefore, in a good position to evaluate subtle changes over large areas on their farms that are intractable even with the most sophisticated...permanent monitoring plots...Yet, positivist quantitative approaches are often seen as providing more 'solid' information than the more 'intangible' knowledge that local actors may have about the ecology of their area.

The conflicts and negotiations that take place at the interface between different, i.e. local / lay and scientific / expert, knowledge forms and 'knowledge-cultures' (Tsouvalis et al, 2000) have become a significant feature of cultural geographies of agriculture (e.g. Wynne, 1996; Clark and Murdoch, 1997; Gerber et al, 1998; Harrison et al, 1998; Holloway, 1999; Burgess et al, 2000; Morgan and Murdoch, 2000). The scope of these analyses, in terms of the issues they consider, is similar to the 'cultures of nature' studies outlined above, but they also include examination of environmental risks, organic agriculture and precision farming technology (with the GM crop issue a notable omission, although see Rushbrook, 2003). All of this research has provided fresh, and increasingly sophisticated, insights into the relationships between the natural and social worlds in the context of agriculture. Moreover, work of this kind has clear, practical implications, as indicated by Burgess et al (2000, p. 131):

> Nature in general, and wetlands in particular, might be better aided if scientific conservation were to concede more ground to local knowledge and local specificity.

And if farmers were to give more recognition to the invisible wildlife that shares their space, but is not part and parcel of their everyday lives.

Further investigations of the relationship between local agricultural knowledges and understandings of nature / environment, local agricultural practices and the development of nature conservation and agri-environment schemes, plans and policies would therefore appear to be of value. Although agri-environment schemes such as the Countryside Stewardship Scheme and Environmentally Sensitive Areas, purport to be sensitive to some of the idiosyncrasies and traditions of local farming practice, there is limited evidence that policy evaluation work, and other research in this area has taken sufficient account of this dimension (nor perhaps has it been adequately addressed within the design and implementation of agri-environment schemes, at least in a British context). Similarly, the relatively new concept of biodiversity action plans (BAPs) includes biodiversity targets that are both national in scope and also reflect local people's values, local conditions, and local distinctiveness. The challenges involved in incorporating 'local natures' (including farming's natures) within BAPs have only just begun to be addressed (e.g. Harrison et al, 1998; Morris and Winter, 2002; Morris and Wragg, 2003).

The observation that particular (notably, scientific) agri-environmental knowledge forms are privileged over others (local knowledges) within the context, for example, of the conventional agricultural industry (Holloway, 1999; Morgan and Murdoch, 2000) and policy debates about agri-environment schemes and farmer enrolment (Burgess et al, 2000), has lead some commentators to adopt actor-network theory (ANT). This enables the different understandings and knowledges of the environment that are held by different actors, e.g. farmers and environmentalists, to be treated symmetrically, or on an equal basis, with no a priori distinctions between different categories (e.g. Clark and Murdoch, 1997; Lowe et al, 1997; Lowe and Ward, 1997; Burgess et al, 2000). ANT has also been advanced as a means of overcoming the human-centredness within studies of nature-society relations in social science generally and in the specific context of agriculture. Outside the agri-environmental domain, the impact of ANT also helps to explain recent interest in the geographies of farm animals, studies of which have begun to appreciate the distinctiveness of different breeds of farm livestock in the cultural landscape (Evans and Yarwood, 1995; Yarwood and Evans, 1998), and also to challenge the 'inevitable and universal farmer-conservationist orthodoxy' in an examination of the Rare Breeds Survival Trust (Evans and Yarwood, 2000). Some considerable effort has recently been accorded to discussing the relative merits of ANT (e.g. Murdoch, 1997; Goodman 1999; Marsden, 2000) which represents a distinct, but still largely embryonic, strand within research into the cultural geographies of agriculture.

Enculturing the Agri-food Economy

The final body of work discussed here as representing cultural geographies of agriculture is one that is possibly the least well defined and the least concerned with 'the farm'. It encompasses the import (both consciously and not) of the cultural into the analysis of the agri-food economy, and connects into a wider set of debates within human geography about the relationship between the economic and the cultural (e.g. Thrift and Olds, 1996; Lee and Wills, 1998; Ray and Sayer, 1999) in which culture is increasingly key to economic geography's research agendas (Crang, 1997). One exemplar of the 'enculturation' of research on the agri-food economy, is the investigation of the social construction of 'quality' food, i.e. the different meanings attached to quality by different actors within agro-food networks as they seek to gain economic advantage and protect their interests (e.g. Morris and Young, 2000; Murdoch et al, 2000; and Parrott et al, 2002). The complexities of these processes are highlighted by Ilbery and Kneafsey (2000, p. 218) who state,

> regulatory institutions may be concerned with the so-called objective indicators of quality, such as the application of hygiene requirements in the case of food products. The very objectivity of these indicators is in itself socially constructed and will vary according to political and economic pressures, scientific understandings and cultural contexts. Consumers may be interested in what have been traditionally described as subjective indicators of quality such as experiential phenomena which lie in the eye of the beholder, while producers may emphasize raw materials and methods of production.

The study of the meanings and ideologies attached to particular food production-consumption spaces, such as Farmers' Markets, is also indicative of an increased sensitivity to the role of cultural processes in the current reshaping of British foodscapes (Holloway and Kneafsey, 2000).

The cultural embeddedness of economic processes, which Barnett (1998a) identifies as one manifestation of human geography's cultural turn, has also begun to receive attention within agri-food research. As Ilbery and Kneafsey (2000, p. 218) explain:

> The current combination of demands and regulations [on the agro-food sector] could offer potential for a 'cultural relocalization' of food production in which locally produced SFPs [speciality food products] with designations of authenticity of geographical origin are transferred to regional and national markets. This does not entail a physical relocation of production but rather a conscious 'fixing' of local SFPs or regional cuisines to territory.

It is likely (and desirable) that the issue of embeddedness within the agri-food sector will attract further research attention particularly given current public and political concerns about various aspects of food production, together with the potential of re-embedded food chains to provide rural development and environmental protection opportunities (Murdoch et al, 2000). For example, in 2000, the Countryside Agency[11] launched its 'Eat the View' campaign, which is based on the belief that 'some products, because of the way they are produced, their area of origin, or other qualities, can help maintain the environmental quality and diversity of the countryside while at the same time bringing benefits to the rural economy and local communities' (Countryside Agency, 2000, unpaginated).

Although extending beyond the farm, the studies referred to above are, for the most part, concerned with questions of, and for, agro-food production. It is interesting to note that where the starting point has been food *consumption* cultural analysis has been both more explicit and more fully developed. The work by Cook, Crang and Thorpe is exemplary in this regard (e.g. Cook, 1994; Cook and Crang, 1996; Cook et al, 1998; Cook et al, 2000). A key focus of their research is the various meanings ascribed to food as it moves through the agro-food system and the 'work' (or material consequences) of these meanings. For example, in his analysis of the production and consumption of exotic fruit, Cook (1994, p. 232) suggests that,

> the meanings that companies attempt to ascribe to these fruits play a crucial role in the articulation of commodity systems. Just because they are produced and packed in one place and shipped, ripened and delivered fresh to a store in another, it does not necessarily follow that anyone will buy them. In short, there is a symbiotic relationship between the 'material' production of a fruit or vegetable and the 'symbolic' production of its meanings.

The meanings and understandings of food that are held by consumers are further explored by Cook et al (1998) who assert that in both academic and agro-food industry accounts, the consumer tends either to be constructed as knowledgeable (and therefore powerful) or ignorant (and, by implication, manipulated) of the origins of their food. Moving away from this 'blunt dichotomy' Cook et al (1998) begin to work with more subtle and sophisticated understandings of consumers' relations to systems of provision, identifying a 'structural ambivalence' within consumers' relationship with the rest of the food system. This involves both an impulse to forget and a need to know about the origins of the food they consume, suggesting a two-way relationship between producers and consumers in which consumer knowledges of the origins of food are both structured by and help to create relations with food provision.

What is the Future Contribution of Cultural Geographies of Agriculture?

The evidence presented above demonstrates a clear interest in the culture in agriculture within geographical studies of the sector. However, it is debatable whether this can be interpreted as a coherent *agri*-cultural turn that has been a self-conscious development. Furthermore, and unlike human geography as a whole, the agri-cultural turn can hardly be described as being too successful and too hegemonic, and suggests a somewhat different trajectory for agricultural geography over the last decade when compared with the sociology of agriculture, in which, so Buttel (2001, p. 167) suggests, a cultural turn represents a 'contender for intellectual dominance' with agrarian political economy.

In this final section the overall value of culturally informed work to date is assessed and consideration given to whether or not there is a need to further enculture geographical studies of agriculture. This seems all the more pressing in the light of Barnett's (1998b) observation that cultural analyses in geography have been haunted by doubts about political relevance, and in the context of rural studies, Little's (1999, p. 437) assertion that 'the application of the cultural turn has, at times, been simplistic and uncritical'. A key concern is:

> the theoretical and political purpose (or end point) of such research... It has been suggested that too strong an emphasis on the cultural construction of rural society and marginalization will encourage description and detract from an examination of the underlying causes and processes of disadvantage (Little, 1999, p. 439).

Suggestions for a culturally informed research agenda within agricultural geography have also been the subject of critique. With the economic pressures facing farmers both in the UK and abroad, cultural investigations of agriculture may be regarded as 'something of a luxury that we cannot yet afford, given the need to salvage rural life and provide food' (Battersby, 2000: pers. com.). However, such concerns do not inevitably lead to a complete rejection of cultural analyses of agriculture in geography and can be countered in a number of ways. In elaborating these responses, I assert that cultural geographical research will continue to have a value within analyses of the agricultural sector.

First, culturally informed analyses of agriculture balance and offer alternative insights to those provided by the rather deterministic and structural analyses of political economy. There is an acknowledged tension between cultural approaches and political economy (e.g. Cloke, 1994; Little, 1999), but the adoption of one does not necessarily entail the rejection of the other. In this sense, Cloke's (1997, p. 372) conclusion about the cultural turn in rural studies has much to recommend it. He asserts that there is a need 'to retain....important insights from political economy approaches and to place these *alongside* some of the exciting ways of seeing rurality and ways of doing rural studies which draw on aspects of the cultural turn' (emphasis added). This resonates with Lowe and Ward's (1997, p. 270) suggestion that 'the search for new, all encompassing paradigms is an insufficiently modest objective. Better to revel in theoretical and methodological diversity and be promiscuous in our interests and approaches'. In a similar vein, the conclusion drawn by Philo (2000, p. 44) about human geography's cultural turn can also be applied to research in agricultural geography: 'enquiry should embrace the material and the social (thereby resisting any dogmatic dematerializing or desocializing...) but also continue to draw inspiration from the whole sweep of the cultural turn'. A complementarity between cultural and other, e.g. political economic, perspectives mirrors the second response to the critics of an agricultural turn, namely that some of the agri-cultural work referred to above has deliberately set out to utilize methodologies which are sensitive to the interplay between textual representation and structural contexts (e.g. Clark et al's (1997) work on EU agricultural policy).

A third reason why agri-cultural research has a place within rural geography relates to the relationship between agri-cultural perspectives and policy. Little (1999) has also suggested that making links between cultural approaches to rural marginalization with political practice and policy is one means of negotiating some of the difficulties with (uncritical) cultural rural studies. Although 'policy orientation' was identified as one explanation for the relative paucity of culturally informed agricultural research, more often than not cultural geographies of agriculture have been situated quite deliberately within a distinct policy context (e.g. most of the agri-environmental work outlined above). Moreover, given the current re-directions in agricultural and rural policy, there is considerable value in promoting greater levels of cultural sensitivity among researchers and policy makers alike (see also Burgess, 2000). For example, as rural policy becomes increasingly regionalized[12], it would seem appropriate for research to map out the regional and sub-regional differences in agri-cultures[13], as an important means of informing decisions about the future shape and form of regionally specific policy. Other suggestions about the nature and scope of this type of 'policy informing' work have been outlined in the preceding 'review' sections of this chapter.

Cultural geographies of agriculture, therefore, have much to contribute to current agricultural and rural policy debates. An agri-cultural turn in this sense then implies a turn towards and embracing of contemporary policy and public concerns about the agricultural sector. However, this is not to advocate an agri-cultural research agenda designed exclusively for questions of policy and policy audiences. Agri-cultural research possibilities can, and should, be equally associated with academic questions about the future of agriculture and the food system. These possibilities are wide-ranging, from consideration of the cultural construction of different groups within the farming community (drawing on work on the rural 'other' - largely neglected by agricultural geographers to date) to 'political ethnographies' of those organizations responsible for the formulation of food and agricultural policy (Winter and Potter, 1999). What is important to bear in mind in pursuing any of these possibilities is the 'careful and critical deployment of culture concepts' (Crang, 1997, p. 5). If researchers choose to investigate the culture in agriculture then they must be clear about what they mean by this. In this respect, it is useful to make reference to Don Mitchell's concerns about cultural geography which omits materialist questions. Future cultural geographical investigations of agriculture may want to consider and apply Mitchell's (1995 and 2000) notion of the 'idea of culture', an outline of which has been provided by Young et al (1995) in the context of agriculture-environment relations in the UK. While a case can be made for the need to continue to make space for the cultural within research into contemporary geographies of agriculture, this research needs to be more explicit than it has been previously about what notion(s) of culture it is working with.

Notes

1. The scope of this paper is restricted to a particular geographical context (the UK) as this provides coherence through the particularities of the structure and organization of the farming and food industry, and those of the British academic community (institutional and career structures, research funding etc.). A geographical focus is also a means of making a review article manageable. This is not to deny the existence (and contribution) of non British 'agricultural' research in the developed market economies, conducted both by geographers and other social scientists, notably rural sociologists e.g. in the Wageningen School, and various researchers based in Northern Europe, Australasia and North America. Indeed, this work will be referred to in the review where it has provided inspiration to / influenced British research.

2. None of the assertions made in this chapter are meant to imply that the cultural turn is all encompassing and has eclipsed other, more traditional approaches to and concerns of research in agriculture (or the rest of the rural for that matter). So, cultural geographies of agriculture are understood as having developed alongside traditional behavioural and political economic investigations of the sector and are by no means dominant.

3. Evidenced by the relative balance of rural and agricultural papers presented at events hosted by the Rural Geography Study Group of the RGS-IBG and the Rural Economy and Society Study Group.

4. Many researchers in other disciplines are, of course, interested in agriculture, but my concern here is with geography. Agricultural economics has, of course, always been concerned with the agricultural sector, but this area of rural studies has arguably been the least susceptible to cultural perspectives due to the strength of neo-classical economic theory and the dominance of a quantitative methodology (although see Midmore, 1999 for an alternative view).

5. The strong co-existence, or correlation between consumption and the cultural has been observed by Jackson and Thrift (1995), in their review of geographies of consumption: 'There is still a tendency for studies of consumption to be exclusively "cultural", paying insufficient attention to other parts of the "circuit of culture" such as the social relations of production' (Jackson and Thrift, 1995, p.228).

6. This has been particularly noticeable in a non-British context, for example Brandth, 1995; Liepins, 1996; Brandth and Haugen, 1997.

7. This is not to deny those analyses e.g. by feminist scholars, that have continued to undertake 'on the ground' research with farming men and women, and which complement the text based accounts discussed here.

8. It is acknowledged that the constructivist perspective encompasses a diversity of approaches to examining environmental issues (e.g. Whatmore, 1999; van Koppen, 2000) and it is not without its critics (Peterson, 1999; Soule and Lease, 1995; Demeritt, 1998).

9. Although not an agricultural study, the contribution of Jackie Burgess and colleagues on the study of environmental meanings must be acknowledged (Burgess, 1990; Harrison and Burgess, 1994).

10. The study of the identities of various rural groups represents a significant element of the cultural turn within non-agricultural rural studies (Cloke and Little, 1997; Little, 1999).

11. The Countryside Agency defines itself as the statutory body in England working to: conserve and enhance the countryside; promote social equity and economic opportunity for the people who live there; and help everyone, wherever they live, to enjoy this national asset.

12. Notably, the England Rural Development Plan (the means of implementing the Rural Development Regulation) comprises a national framework document with nine regional chapters based on Government Office Regions. Each region has a Regional Planning Group

that is responsible for sensitizing the measures available under the RDR to local needs and conditions.

13. Examples exist of this type of work, which take as their starting point the cultures of particular groups within, or associated with, farming in specific localities e.g. McEachern (1992) in the Yorkshire Dales; Hermann and Shucksmith (1994) in Scotland and West Germany; and Gray (1996; 1998) in the Scottish borders.

References

Arensberg, C. and Kimball, S. (1940), *Family and Community in Ireland*, Harvard University Press, Harvard.

Barnett, C. (1998), 'The Cultural Worm Turns: Fashion or Progress in Human Geography?', *Antipode*, Vol. 30, pp. 379-94.

Barnett, C. (1998), 'Cultural Twists and Turns', *Environment and Planning D: Society and Space*, Vol. 16, pp. 631-4.

Bell, M. (1994), *Childerley: Nature and Morality in a Country Village*, University of Chicago Press, Chicago.

Bouquet, M. (1985), *Family Servants and Visitors: the Farm Household in Nineteenth and Twentieth Century Devon*, Geobooks, Norwich.

Brandth, B. (1995), 'Rural Masculinity in Transition: Gender Images in Tractor Advertisements', *Journal of Rural Studies*, Vol. 11, pp. 123-133.

Brandth, B. and Haugen, M. (1997), 'Rural Women, Feminism and the Politics of Identity', *Sociologia Ruralis*, Vol. 37, pp. 325-344.

Burgess, J. (1990), 'The Production and Consumption of Environmental Meanings in the Mass Media: a Research Agenda for the 1990s', *Transactions, Institute of British Geographers*, Vol. 15, pp. 139-161.

Burgess, J., Clark, J. and Harrison, C. (2000), 'Knowledges in Action: an Actor Network Analysis of a Wetland Agri-Environment Scheme', *Ecological Economics*, Vol. 35, pp. 119-132.

Buttel, F. (1998), 'Nature's Place in the Technological Transformation of Agriculture: Some Reflections on the Recombinant BST Controversy in the USA', *Environment and Planning A*, Vol. 30, pp. 1151-1163.

Buttel, F. (2001), 'Some Reflections on Late Twentieth Century Agrarian Political Economy', *Sociologia Ruralis*, vol. 41, pp.165-179.

Carr, S. and Tait, J. (1991), 'Farmers' and Conservationists' Attitudes', *Journal of Environmental Management*, Vol. 32, pp. 281-294.

Clark, J., Jones, A., Potter, C., and Lobley, M. (1997), 'Conceptualising the Evolution of the European Union's Agri-Environment Policy: a Discourse Approach', *Environment and Planning A*, vol. 29, pp. 1869-1885.

Clark, J. and Murdoch, J. (1997), 'Local Knowledge and the Precarious Extension of Scientific Networks: a Reflection on Three Case Studies', *Sociologia Ruralis*, Vol. 37, pp. 38-60.

Cloke, P. (1994), '(En)culturing Political Economy: a Life in the Day of a 'Rural Geographer', in P. Cloke, M. Doel, D. Matless, M. Phillips and N. Thrift, *Writing the Rural: Five Cultural Geographies*, Paul Chapman, London, pp. 149-190.

Cloke, P. (1997), 'Country Backwater to Virtual Village? Rural Studies and "The cultural turn"', *Journal of Rural Studies*, Vol. 13, pp. 367-375.

Cloke, P. and Little, J. (1997), *Contested Countryside Cultures: Otherness, Marginalisation and Rurality*, Routledge, London.

Cook, I. (1994), 'New Fruits and Vanity: Symbolic Production in the Global Food Economy', in A. Bonanno, L. Busch, W. Friedland, L. Gouveia and E. Mingione (eds), *From Columbus to ConAgra: the Globalisation of Agriculture and Food*, University Press of Kansas, Kansas, pp. 232-248.

Cook, I. and Crang, P. (1996), 'The World on a Plate: Culinary Culture, Displacement and Geographical Knowledges', *Journal of Material Culture*, Vol. 1, pp. 131-153.

Cook, I., Crang, P. and Thorpe, M. (1998), 'Biographies and Geographies: Consumer Understandings of the Origins of Foods', *British Food Journal*, Vol. 100, pp. 162-167.

Cook, I., Crang, P. and Thorpe, M. (2000), 'Regions to be Cheerful: Culinary Authenticity and its Geographies', in I. Cook, D. Crouch, S. Naylor and J. Ryan (eds), *Cultural Turns/ Geographical Turns*, Pearson Education, Harlow, pp.109-139.

Cook, I., Crouch, D., Naylor, S., and Ryan, J. (eds) (2000), *Cultural Turns/Geographical Turns*, Pearson Education, Harlow.

Countryside Agency (2000), *Eat the View*, Countryside Agency, Cheltenham.

Crang, P. (1997), 'Cultural Turns and the (Re)constitution of Economic Geography', in R. Lee and J. Wills (eds), *Geographies of Economies*, Arnold, London, pp. 3-15.

Crang, M. (1998), *Cultural Geography*, Routledge, London.

Davies, E. and Rees, A. (eds) (1960), *Welsh Rural Communities*, University of Wales Press, Cardiff.

Demeritt, D. (1998), 'Science, Social Constructivism and Nature', in B. Braun and N. Castree (eds), *Remaking Reality: Nature at the Millennium*, Routledge, London, pp.173-193.

Emmett, I. (1964), *A North Wales Village: a Social Anthropological Study*, Routledge and Paul Kegan, London.

Evans, N. and Yarwood, R. (1995), 'Livestock and Landscape', *Landscape Research*, Vol. 20, pp. 141-146.

Evans, N. and Yarwood, R. (2000), 'The Politicisation of Livestock: Rare Breeds and Countryside Conservation', *Sociologia Ruralis*, Vol. 40, pp. 228-248.

Fish, R. and Phillips, M. (1997), *Explorations in Media Country: Political Economic Moments in British Rural Television Drama*, Paper presented at the third British-French Rural Geography Symposium, Nantes, September 1997.

Frankenberg, R. (1957), *Village on the Border: a Social Study of Religion, Politics and Football in a North Wales Community*, Cohen and West, London.

Gasson, R. (1973), 'Goals and Values of Farmers', *Journal of Agricultural Economics*, Vol. 24, pp. 521-537.

Gerber, J., Holloway, L., Seymour, S., Steven, M. and Watkins, C. (1998), 'New Technologies and Old Knowledges: The Impact of Precision Farming on the Management of the English Countryside', in N. Croix (ed), *Environnement et Nature dans les Campagnes: Nouvelles Politiques, Nouvelles Pratiques*, Presses Universitaires de Rennes, Rennes, pp. 187-204.

Gillmor, D. (1986), 'Behavioural Studies in Agriculture: Goals, Values and Enterprise Choice', *Irish Journal of Agricultural Economics and Rural Sociology*, Vol. 11, pp. 19-33.

Goodman, D. (1999), 'Agro-Food Studies in the "Age of Ecology": Nature, Corporeality, Biopolitics', *Sociologia Ruralis*, Vol. 39, pp. 17-38.

Gray, J. (1996), 'Cultivating Farm Life on the Borders: Scottish Hill Sheep Farms and the European Community', *Sociologia Ruralis*, Vol. 36, pp. 27-50.

Gray, J. (1998), 'Family Farms in the Scottish Borders: a Practical Definition by Hill Sheep Farmers', *Journal of Rural Studies*, Vol. 14, pp. 341-56.

Harrison, C. and Burgess, J. (1994), 'Social Constructions of Nature: a Case Study of Conflicts Over the Development of Rainham Marshes SSSI', *Transactions, Institute of British Geographers*, Vol. 19, pp. 291-310.

Harrison, C., Burgess, J., and Clark, J. (1998), 'Discounted Knowledges: Farmers' and Residents' Understandings of Nature Conservation Goals', *Journal of Environmental Management*, Vol. 54, pp. 305-320.

Harvey, D. (2000), *Spaces of Hope*, Edinburgh University Press, Edinburgh.

Hermann, V. and Shucksmith, M. (1994), 'Habitus and Practise of Farmers in Scotland and West Germany', in A. Copus and J. Marr (eds), *Rural Realities, Trends and Choices*, Proceedings of the 35th EAAE seminar, Aberdeen, June, pp. 239-244.

Holloway, L. (1999), 'Understanding Climate Change and Farming: Scientific and Farmers' Constructions of "Global Warming" in Relation to Agriculture', *Environment and Planning A*, Vol. 31, pp. 2017-2032.

Holloway, L. (2000), '"Hell on Earth and Paradise all at the Same Time": The Production of Smallholding Space in the British Countryside', *Area*, Vol. 32, pp. 307-316.

Holloway, L. and Kneafsey, M. (2000), 'Reading the Space of the Farmers' Market: A Preliminary Investigation from the UK', *Sociologia Ruralis*, Vol. 40, pp. 285-299.

Hughes, A., Morris, C. and Seymour, S. (2000), *Ethnography and Rural Research*, Countryside and Community Press, Cheltenham.

Ilbery, B. (1985), *Agricultural Geography: A Social and Economic Analysis*, Oxford University Press, Oxford.

Ilbery, B. and Kneafsey, M. (2000), 'Producer Constructions of Quality in Regional Specialty Food Production: a Case Study From South West England', *Journal of Rural Studies*, Vol. 16, pp. 217-230.

Jackson, P. and Thrift, N. (1995), 'Geographies of Consumption', in D. Miller (ed), *Acknowledging Consumption: A Review of New Studies*, Routledge, London, pp. 204-237.

Jackson, P., Brooks, K. and Stevenson, N. (1999), Making Sense of Men's Lifestyle Magazines', *Environment and Planning D: Society and Space*, Vol. 17, pp. 353-368.

Lee, R. and Wills, J. (eds) (1997), *Geographies of Economies*, Arnold, London.

Liepins, R. (1996), 'Reading Agricultural Power: Media as Sites and Processes in the Construction of Meaning', *New Zealand Geographer*, Vol. 52, pp. 3-10.

Liepins, R. (1998), '"Women of Broad Vision": Nature and Gender in the Environmental Activism of Australia's "Women in Agriculture" Movement', *Environment and Planning A*, Vol. 30, pp. 1179-1196.

Little, J. (1999), 'Otherness, Representation and the Cultural Construction of Rurality', *Progress in Human Geography*, Vol. 23, pp. 437-442.

Littlejohn, J. (1963), *Westrigg: the Sociology of a Cheviot Parish*, Routledge and Paul Kegan, London.

Lowe, P., Clark, J., Seymour, S. and Ward, N. (1997), *Moralizing the Environment: Countryside Change, Farming and Pollution*, UCL Press, London.

Lowe, P. and Ward, N. (1997), 'Field-Level Bureaucrats and the Making of New Moral Discourses in Agri-Environmental Controversies', in D. Goodman and M. Watts (eds), *Globalising Food: Agrarian Questions and Global Restructuring*, Routledge, London, pp.256-272.

Macnaghten, P. and Urry, J. (1998), *Contested Natures*, Sage, London.

Marsden, T. (2000), 'Food Matters and the Matter of Food: Towards a New Food Governance?', *Sociologia Ruralis*, Vol. 40, pp. 20-29.

Marsden, T., Munton, R., Ward, N. and Whatmore, S. (1996), 'Agricultural Geography and the Political Economy Approach: a Review', *Economic Geography*, Vol. 72, pp. 361-75.

Marsden, T., Murdoch, J., Lowe, P., Munton, R. and Flynn, A. (1993), *Constructing the Countryside*, UCL Press, London.

McEachern, C. (1992), 'Farmers and Conservation: Conflict and Accommodation in Farming Politics', *Journal of Rural Studies*, Vol. 8, pp. 159-171.

McHenry, H. (1998), 'Wild Flowers in the Wrong Fields are Weeds! Examining Farmers' Constructions of Conservation', *Environment and Planning A*, Vol. 30, pp. 1039-1053.

Meinhof, U. (1993), 'Discourse', in W. Outwaite and T. Bottomore (eds), *The Blackwell Dictionary of 20th Century Social Thought*, Blackwell, Oxford, pp.161-162.

Middleton, A. (1986), 'Making Boundaries: Men's Space and Women's Space in a Yorkshire Village', in T. Bradley, P. Lowe and S. Wright (eds), *Deprivation and Welfare in Rural Areas*, Geobooks, Norwich, pp. 121-134.

Midmore, P. (1999), 'Towards a Postmodern Agricultural Economics', *Journal of Agricultural Economics*, Vol. 47, pp. 1-17.

Milbourne, P. (2000), 'Exporting "Other" Rurals: New Audiences For Qualitative Research', in A. Hughes, C. Morris and S. Seymour (eds), *Ethnography and Rural Research*, Countryside and Community Press, Cheltenham, pp.136-157.

Mitchell, D. (1995), 'There's No Such Thing as Culture: Towards a Reconceptualization of the Idea of Culture in Geography', *Transactions. Institute of British Geography*, Vol. 20, pp. 102-16.

Mitchell, D. (2000), *Cultural Geography: a Critical Introduction*, Blackwell, Oxford.

Morgan, K. and Murdoch, J. (2000), 'Organic vs. Conventional Agriculture: Knowledge, Power and Innovation in the Food Chain', *Geoforum*, Vol. 31, pp. 159-173.

Morris, C. and Evans, N. (1999), 'Research on the Geography of Agricultural Change: Redundant or Revitalized', *Area*, Vol. 31, pp. 349-358.

Morris, C. and Evans, N. (2002), 'Cheesemakers are Always Women: Gendered Representations of Farm Life in the Agricultural Press', *Gender, Place and Culture*, Vol. 8, pp. 375-390.

Morris, C. and Winter, M. (2002), 'Barn owls, Bumble-bees and Beetles: UK Agriculture, Biodiversity and Biodiversity Action Planning', *Journal of Environmental Planning and Management*, Vol. 45, pp. 653-671.

Morris, C. and Wragg, A. (2003), 'Talking about the Birds and the Bees: Biodiversity Claims-Making at the Local Level', *Environmental Values*, Vol. 12, pp. 71-90.

Morris, C. and Young, C. (2000), '"Seed to Shelf", "Teat to Table", "Barley to Beer" and "Womb to Tomb": Discourses of Food Quality and Quality Assurance Schemes in the UK', *Journal of Rural Studies*, Vol. 16, pp. 103-115.

Murdoch, J. (1997), 'Inhuman-Nonhuman-Human: Actor Network Theory and the Prospects for a Nondualistic and Symmetrical Perspective on Nature and Society', *Environment and Planning D: Society and Space*, Vol. 15, pp. 731-756.

Murdoch, J. and Marsden, T. (1994), *Reconstituting Rurality*, UCL Press, London.

Murdoch, J., Marsden, T. and Banks, J. (2000), 'Quality, Nature and Embeddedness, Some Theoretical Considerations in the Context of the Food Sector', *Economic Geography*, Vol. 76, pp. 107-125.

Newby, H., Bell, C., Saunders, P. and Rose, D. (1977), 'Farmers' Attitudes to Conservation', *Countryside Recreation Review*, Vol. 2, pp. 23-30.

Parrott, N., Wilson, N., and Murdoch, J. (2002), 'Spatialising Quality: Regional Protection and the Alternative Geography of Food', *European Journal of Urban and Regional Studies*, Vol. 9, pp. 241-261.

Peterson, A. (1999), 'Environmental Ethics and the Social Construction of Nature', *Environmental Ethics*, Vol. 21, pp. 339-353.

Phillips, M., Fish, R. and Agg, J. (2001), 'Putting Together Ruralities: Towards a Symbolic Analysis of Rurality in the British Mass Media', *Journal of Rural Studies*, Vol. 17, pp. 1-27.

Philo, C. (2000), 'More Words, More Worlds: Reflections on the "Cultural turn" and Social Geography', in I. Cook, D. Crouch, S. Naylor and J. Ryan (eds), *Cultural Turns/Geographical Turns*, Longman, London, pp.26-53.

Ray, L. and Sayer, A. (1999), *Culture and Economy After the Cultural Turn*, Sage, London.

Rapport, N. (1993), *Diverse World-Views in an English Village*, Edinburgh University Press, Edinburgh.

Rees, A. (1950), *Life in the Welsh Countryside*, University of Wales Press, Cardiff.

Rushbrook, L. (2003), *Models of Controversy: Reflections on Cultural Theory and the GM Crop Debate*, Unpublished PhD Thesis, University of Gloucestershire.

Soule, M. and Lease, G. (eds) (1995), *Reinventing Nature? Responses to Postmodern Deconstruction*, Island Press, Washington DC.

Strathern, M. (1982a), 'The Place of Kinship: Kin, Class and Village Status in Elmdon, Essex', in A. P. Cohen (ed), *Belonging: Identity and Social Organisations in British Rural Culture*, Manchester University Press, Manchester, pp. 72-100.

Strathern, M. (1982b), 'The Village as an Idea: Constructs of Villageness in Elmdon, Essex', in A. P. Cohen (ed), *Belonging: Identity and Social Organisations in British Rural Cultures*, Manchester University Press, Manchester, pp. 247-277.

Thrift, N. and Olds, K. (1996), 'Refiguring the Economic in Economic Geography', *Progress in Human Geography*, Vol. 20, pp. 311-337.

Tsouvalis, J. (2000), 'Socialised Nature: England's Royal and Plantation Forests', in I. Cook, D. Crouch, S. Naylor and J. Ryan (eds), *Cultural Turns/Geographical Turns*, Pearson Education, Harlow, pp.288-312.

Tsouvalis, J., Seymour, S. and Watkins, C. (2000), 'Exploring Knowledge-Cultures: Precision Farming, Yield Mapping, and the Expert-Farmer Interface', *Environment and Planning A*, Vol. 32, pp. 908-924.

van Koppen, K. (2000), 'Resource, Arcadia, Lifeworld. Nature Concepts in Environmental Sociology', *Sociologia Ruralis*, Vol. 40, pp. 300-318.

Walsh, M. (1997), 'The View from the Farm: Farmers and Agri-environmental Schemes in the Yorkshire Dales', *The North West Geographer*, Vol. 1, pp. 24-35.

Walter, G. and Wilson, S. (1996), 'Silent Partners: Women in Farm Magazine Success Stories, 1934-1991', *Rural Sociology*, Vol. 61, pp. 227-248.

Whatmore, S. (1999), 'Culture-Nature', in P. Cloke, P. Crang and M. Goodwin (eds), *Introducing Human Geographies*, Arnold, London, pp. 4-11.

Wilson, G. (1997), 'Assessing the Environmental Impact of the ESA Scheme: a Case For Using Farmers' Environmental Knowledge?', *Landscape Research*, Vol. 22, pp. 303-326.

Winter, M. and Potter, C. (1999), *CAP Reform, Liberalisation and Farming Futures*, Paper presented to Rural Economy & Society Study Group Conference, Royal Agricultural College, Cirencester.

Woods, M. (1998), 'Mad Cows and Hounded Deer: Political Representations of Animals in the British Countryside', *Environment and Planning A*, Vol. 30, pp. 1219-1234.

Wynne, B. (1996), 'May the Sheep Safely Graze? A Reflexive View of the Expert-Lay Knowledge Divide', in S. Lash, B. Szerszynski and B. Wynne (eds), *Risk, Environment and Modernity: Towards a New Ecology*, Sage, London, pp.44-83.

Yarwood, R. and Evans, N. (1998), 'The Changing Geographies of Domestic Livestock Animals', *Society and Animals*, Vol. 6, pp. 137-66.

Young, C., Morris, C. and Andrews, C. (1995), 'Agriculture and the Environment in the UK: Towards an Understanding of the Role of "Farming Culture"', *Greener Management International*, Vol. 12, pp. 63-80.

Chapter 13

Producing-Consuming Food: Closeness, Connectedness and Rurality in Four 'Alternative' Food Networks

Lewis Holloway and Moya Kneafsey

Introduction

This chapter aims to explore some key ideas and relationships associated with what we categorize as 'alternative' food production-consumption networks. We use the problematic term 'alternative' here to suggest two inter-related things. First, it implies that food production-consumption is undertaken within an ethical framework contrasting with those networks regarded as 'conventional'. An ethic of care (e.g. for people, livestock animals, 'nature', a locality), which might take very different forms, is frequently of central importance. As Sage (2003, p. 29) suggests, 'Recovering a sense of morality within the food and agriculture sector is arguably one of the most important emerging characteristics of alternative food networks'. Second, 'alternative' networks and ethical relations are allied with spatialities which distinguish them from the conventional, although clearly both encompass a diversity of geographies. These spatialities are often associated with the desire to foster relations of 'closeness' or 'connectedness'. Additionally, in many cases, specific conceptions of rurality are significant. Thus, in examining such 'alternatives', we are particularly concerned with two broad issues; first, the production of notions and practices of 'close' relationships and 'connectedness' between production and consumption; and second, implications of this for understandings and uses of rural space. The chapter begins by outlining some key currents of debate surrounding contemporary agro-food networks. It then deploys four case studies to take different perspectives on the key issues of closeness and connectedness in particular forms of alternative food production-consumption networks. We thus illustrate the ways in which closeness, connectedness and rurality are negotiated in different contexts. As a result, as suggested in the concluding section, an agenda for future research into the relationships between food production-consumption networks, 'alternative' or otherwise, needs to maintain a critical sensitivity towards the specificities of particular production-consumption relations, while simultaneously maintaining a capacity to conceptualize, and perhaps argue for changes in, larger-scale production-consumption relations transcending the local, the ephemeral and the specific. Similarly, differences in the discursive and material spaces of the 'rural' need to be recognized in research into 'alternative' production-consumption networks.

'Alternative' Agro-Food Networks

Recent discussion of 'alternative' food production-consumption has raised a series of issues surrounding the relationships between production and consumption, and producers and consumers. Debates have been influenced by a diversity of theoretical perspectives. For example, many (e.g. Murdoch et al, 2000) have engaged with the economic sociology of Granovetter (1985), focusing on the ways in which economic relations are embedded in social relations of trust. Critiques of this position (e.g. Winter, 2003), however, suggest that such an approach simply reproduces a problematic separation of economy and society. In combination with an increasing interest in considering the nonhuman entities which mediate human relations (e.g. Stassart and Whatmore, 2003), a desire to incorporate ideas of nature into studies of food production-consumption (Goodman, 2001), and a heightened awareness of the ecological implications of different ways of producing-consuming food, such critique has led researchers to work with various forms of network theory (e.g. Lockie and Kitto, 2000; Murdoch, 2000). These network approaches are capable of allowing a reassessment of notions of human and nonhuman agency within food production-consumption relations, implying the inclusion of nonhuman entities as material-semiotic actants (Whatmore, 2002). In addition, some have begun to engage with theories of aesthetics in relation to food, arguing, for example, that the aesthetic needs to be understood as a key part of the notion of food quality (Miele and Murdoch, 2002).

Academic debate surrounding food production-consumption has intensified in response to the particular conditions of contemporary society-food relations. Much has been made of the increasing separation of producers and consumers in modern Western food supply systems. Agriculture has become an increasingly specialized activity undertaken by relatively few people, and remote from the experience of most urban, and many rural, dwellers. Industrial-scale food processing transforms food commodities and re-presents them in forms which may render the primary inputs unrecognizable. Finally, food has been increasingly retailed through large supermarkets which, again, seem to increase the perceptual distance between consumers and both producers and the food itself. Nevertheless, and perhaps as a result, there has been increasing consumer interest in 'alternative' modes of food production-consumption which apparently bring consumers and producers of food into different ethical and spatial relationships.

This movement has been reinforced by consumers' growing awareness of the health risks, such as vCJD or salmonella, associated with particular foods, and as such, the ethical status of current conventional food production-consumption systems has been called into question. Here, as Whatmore (2002, p. 119) suggests, 'Public anxieties around industrial foodstuffs and growing consumer participation in alternative food networks, from Fairtrade to organics, suggest that food is a ready messenger of connectedness and considerability that is fleshing out the spaces and practices of a relational ethics, even as academic and policy analysts struggle to register or make sense of them'. Food has thus become an issue and a substance of heightened academic and public concern.

Current geographical interest in food *per se* is associated with increasing interest in consumption. Buying and eating food is acknowledged as occurring within, and constitutive of, social relations, in ways which ascribe cultural value to particular foods. Thus, Bell and Valentine (1997, p. 3) write that 'For most inhabitants of (post)modern Western societies, food has long ceased to be merely about sustenance and nutrition. It is packed with social, cultural and symbolic meaning'. They argue, drawing on Bourdieu's (1984) concept of distinction, that the socio-cultural significance of food is related to social differentiation, as people seek to distinguish between themselves and others by what, how and where they eat. Simultaneously, food consumption can be regarded as an ethical act, since it draws the consumer into relationships with human and nonhuman others that are always ethically charged (Jones, 2000). One's food consumption practices might thus impact on, *inter alia*, animal welfare, agro-chemical use, farm labour relations and the balance of power between supermarkets and small independent retailers. Falk (1994, p. 81) thus suggests that it is important for people that their food has positive associations; food should be '"good to taste", it must act as a *positive* representation. In other words it must *stand* for something valued ...'. An ethical dimension to individual acts of consumption, as well as to collective social and political debates over what forms of food supply system are desirable, is thus important in culturally-oriented understandings of food.

Yet, despite this upwelling of interest in consumption in wider fields of geographical study, commentators such as Goodman (e.g. 2002, 2003) and Goodman and DuPuis (2002) have argued that studies of agro-food networks have tended to focus predominantly on production. For these authors, much of the work which purports to bring consumption into the frame of reference ends by reconfirming production as the locus of power and agency in food supply systems. Thus, for example, the recent focus on speciality or regional foods amongst European rural geographers has tended to examine them as part of the adaptation of the farming sector to changing economic circumstances, and of efforts to encourage 'rural development'. While appeals to the cultural are made, for example in the emphasis on the symbolic significance of 'the local' or the contested semiotics of 'quality', thus recognising that food is more than a material commodity, consumers have seldom been ascribed agency or power in such accounts. Goodman (2002) acknowledges that network approaches begin to address this issue (e.g. Lockie, 2002; Whatmore and Thorne, 1997), but suggests that a more concerted effort to overcome the binary opposition of 'producer' against 'consumer' is required. The challenge, then, is 'to move beyond the theoretical asymmetries and linearities of this framework, with its implicit alignment of power relations and assignment of agency, and acknowledge consumers as relational actors in recursive, mutually constituted food circuits' (Goodman, 2002, p. 272). Thus, ascriptions of power and agency in food networks should be redistributed so that consumers, alongside heterogeneous other actors, are conceptualized as active in the relational constitution of material-semiotic food networks (Goodman, 2001; Whatmore, 2002). Further, dualistic categories of 'consumer' and 'producer' may prove inadequate in such a conceptualization (Hassanein, 2003). Such categorization can be seen as an attempt to maintain a division between the worlds of production and consumption, which can instead be blurred to allow individual human actors to negotiate shifting relational positions within hybrid networks of production-consumption.

Further debate in relation to 'alternative' food networks concerns whether such networks are part of a radical movement with a politicized agenda for restructuring the global food system, or whether they are better regarded as marginal activities taking up specialist niche markets in the interstices of a powerful, globalized agro-food industry (see, for example, Allen et al, 2003; Goodman, 2002; Hassanein, 2003). A related question is how far individual consumer engagement with specific, often small-scale and local networks, can be understood as political action - that is, as related to a wider movement for change in modes of food production-consumption (Goodman and DuPuis, 2002). Returning to approaches influenced by actor-network theory, even those activities which are not overtly political in a conventional sense might be regarded as politicized, since the expression of agency in a consumption context locates the consumer in a relation of power within the production-consumption system. 'Consumer actions can be seen as political when they exercise "the capacity to act" in any way that affects the future form of society' (Goodman and DuPuis, 2002, p. 18). Consumers are thus able, even in unpremeditated, unconscious ways, to influence relations of food production-consumption.

In this context, consumers' engagements with alternative food systems might be understood as part of a broader set of reflexive responses to the late-modern 'risk society' identified by Ulrich Beck (1992; 1999). For Beck, hazards produced by modern industrial systems (e.g. pollution) assume a primary importance in human consciousness. Industrial farming and food processing systems have been recently identified with similar hazards, and thus food consumption has increasingly been seen as a 'risky' activity. He argues that one response has been the emergence of new forms of 'cosmopolitan' geographical imagination, in which human subjects exhibit greater awareness of their mutually being affected by global issues. Thus, 'World-wide public perception and debate of global ecological danger or global risks of a technological and economic nature ("Frankenstein food") have laid open the cosmopolitan significance of fear' (Beck, 2000, p. 79). Returning to what constitutes political action, Beck defines a stratum of 'subpolitical' activity. Subpolitics take place in 'sites which were previously considered unpolitical' (Beck, 1999, p.93), and implicate individuals and a range of non-governmental institutions in new forms of political agency and practice:

> The concept of 'subpolitics' refers to politics outside and beyond the representative institutions of the political systems of nation-states ... Subpolitics means '*direct*' politics - that is, *ad hoc* individual participation in political decisions, bypassing the institutions of representative opinion-formation (political parties, parliaments) and often even lacking the protection of the law. In other words, subpolitics means the shaping of society from below. Economy, science, career, everyday existence, private life, all become caught up in the storms of political debate ... Crucially, however, subpolitics sets politics free by changing the rules and boundaries of the political so that it becomes more open and susceptible to new linkages - as well as capable of being negotiated and reshaped (Beck, 1999, pp. 39-40).

In these terms, spaces for subpolitical agency are clearly being carved out within the conventional food supply system by 'alternative' practices of food production-consumption. These practices are associated with a re-emerging public sphere; a space for more democratic dialogue about reshaping society (Beck, 1999), including the important arena of food production-consumption.

Yet, despite this potential for a radical subpolitics, Beck suggests that subpolitical action and cosmopolitan imagination may exist alongside reactionary turns to the perceived reassurances of a conservative social order. The idea of 'the local', figuring alongside ideas of rurality, community and so on, can play an important part here and is reflected in the localist or nationalist sentiments seen in the contexts of some 'alternative' food production-consumption networks (Holloway and Kneafsey, 2000). For some, the politics of the local is a 'militant particularism' deploying specific forms of small-scale, production-consumption as a radical challenge to the globalized food system (Allen et al, 2003). Yet, it might be that appeals to the local are instead part of what Allen (1999) refers to as a 'defensive localism' (see also Winter, 2003). Thus, 'buying local' might be seen as a reactionary (re)turn to a parochialism in which the local is imagined as a fixture against the anxieties of change, rather than as a cosmopolitan engagement with global ecological and social issues. Even so, a 'subpolitics of food' might be identified in contemporary, and perhaps also in historical (see Holloway, 2003), production-consumption networks which seek, in different ways, to challenge, subvert or bypass conventional food supply systems by attempting to re-establish relations of closeness and connectedness, and by engaging with specific notions of rurality.

We now consider four case studies which in different ways exemplify how food production-consumption networks involve the negotiation of connectedness, closeness and rurality. The case studies can be read as two pairs. The first pairing focuses on networks which maintain a discursive and practical distinction between food producers and food consumers while seeking a connection between them which might be strongest at the point of food purchase. Simultaneously ethical and economic relations are, in this pair, expressed through what Sage refers to as relations of 'regard' between buyers and sellers of food, as 'a sense of "entanglement" arises from the hybridity of moral and money economies that impose certain obligations and responsibilities on both transacting parties' (Sage,

2003, p. 49). First, foods marketed on the grounds that they are the specialities of particular places or regions rely on processes of certification and the circulation of symbols of authenticity and 'quality', apparently establishing relations of closeness and trust while allowing the possibility of a physical distance and relations of anonymity between producer and consumer - such products might be purchased at a supermarket physically distant from the site of production. In contrast, Farmers Markets are seen to bring producers and consumers into direct, face-to-face contact, associating the local with socially-constructed ideas of quality in consumers' searches for closer relationships with the food they consume and with those people and places involved in its production. The second pairing focuses on modes of production-consumption in which the categories 'producer' and 'consumer' are significantly blurred. Those engaged in small-scale 'alternative' farming enter into direct material relations with the foods they eat and the means of growing those foods, simultaneously being immersed in semiotic contexts linked to particular ethical positionings and notions of rurality. Finally, Internet businesses which have given clients a say in the growing of food which is then delivered to their door create more ambiguous relations between 'production' and 'consumption', by giving clients the opportunity to engage in food production over the Internet, in a virtual space which nevertheless gives access to a material food product, is linked to a food production assemblage of soil, plants, animals, farmers and so on, and engages with ethical and cultural associations of connectedness and rurality.

The 'Quality Turn'

The 'turn' to quality (Goodman, 2003) is associated with the proliferation of 'alternative' food networks operating at the margins of mainstream industrial food production. This is having direct impacts on the socio-cultural and economic restructuring of rural areas. Firstly, research in rural regions in England (Kneafsey and Ilbery, 2000a), Wales (Kneafsey et al, 2001) and Ireland (Sage, 2003; Tovey, 1997) has shown that the producers of quality foods are often counter-urbanites and non-nationals seeking a means of sustaining a rural lifestyle and often driven by environmental and ethical motivations. Quality food netwroks might thus be linked to changing population profiles in rural areas. Secondly, throughout Europe, the production of high quality regional 'speciality' foods has been heralded by some as the basis for a new economic dynamic in areas by-passed by the productivist logic of the 'conventional' food system (Ilbery and Kneafsey, 1998; Marsden et al, 2002). This dynamic can be seen as a form of resistance to the disembedding forces of globalization, allowing regions to carve out niches for food products which appeal to consumers not on the basis of price competitiveness, but in terms of their ecological, moral and aesthetic qualities. These qualities are in turn embedded within producer-consumer relationships in which notions of trust, regard, authenticity and 'connectedness' are given prominence, even though the discursive and practical distinction between producers and consumers is maintained.

In this case, we want to consider how 'connectedness' may be communicated through spatially extended short food supply chains, which often incorporate conventional retail structures including supermarkets. Such chains aim to reduce the perceived gap between producers and consumers, yet may extend over large physical distances (Marsden et al, 2000; Renting et al, 2003). As defined by Marsden et al (2000), they depend on information about the place and nature of production being transmitted to consumers outside the region of production. A crucial characteristic of these supply chains is that products reach consumers embedded with information, enabling connections to be made between product characteristics and place of production. The embedding of information may be more or less formalized and regulated. At the less formalized end, producers of speciality foods may attempt to use regional imagery to appeal to pre-existing aesthetics concerning regional identities and lifestyles. This potentially creates a sense of connection between the consumer and specific places of production and works because, as Bell and Valentine (1997, p. 205) note, '[T]he region ... is still articulated as a place of tradition - local, place specific tradition'. Moreover, this articulation often has a rural dimension and taps into broader societal notions of idyllic rural lifestyles and values. For instance, research into food marketing strategies amongst speciality food producers in the South-West of England has shown that many producers attempt to make links between the rural context of production and the products' 'farm fresh', 'traditional', 'wholesome' and 'high quality' characteristics (Kneafsey and Ilbery, 2001). This is achieved on the product packaging, through the use of place names and imagery combined with text descriptions. However, the extent to which producers have developed these connections varies and certainly is not always aligned to radical personal or collective political action related to challenging the dominant agro-food system. For instance, many producers use a combination of 'conventional' and more 'alternative' market outlets for their produce, and few rely totally on direct sales to consumers. For those distributing their foods through conventional wholesale and retail networks, the need to create strong connections between food quality and place is not always evident, especially if the products are destined for national and international markets. Moreover, for small producers there is always a risk of such connections being appropriated by larger organizations such as supermarket retailers who can tap into strong but vaguely defined popular associations between rurality, region of origin and quality and who have far more extensive marketing resources at their disposal.

One way of resisting this potential for appropriation is for smaller producers to participate in more formalized, regulated methods of creating connectivity between product and place of origin. This can be achieved through the use of labels of origin which, in the European context at least, can be carried by 'traditional' or 'typical' products. Several authors have identified labels of origin as important expressions of 'local', 'quality' or 'endogenous' food systems (Marsden et al, 2000, Ilbery and Kneafsey, 2000b) facilitating a reconnection of people, products and place in the context of rural development (Barham, 2003). For example, modelled on the French *appellation d'ôrigine controlée* (AOC), the European Union's system of Protected Designations of Origin (PDO) and Protected Geographical Indications (PGIs) enables regions to claim collective ownership of locally-embedded modes of food production. Unlike trademarks which are owned by corporations, labels of origin 'belong' to regions and are administered by state governments to prevent fraud and oversee certification systems. They 'hold the potential of re-linking production to the social, cultural or environmental aspects of particular places, further distinguishing them from anonymous mass produced goods' (Barham, 2003, p. 129). AOCs, PDOs and PGIs are built on the French concept of *'terroir'* which historically, refers to an area or terrain (usually rural) whose soil and micro-climate impart distinctive qualities to food products. Importantly, these natural factors are combined with human factors such as *savoir faire* and techniques confined to that area and historically established public knowledge of the product and its association with the area (Barham, 2003). These certification schemes are regulated by extra-local institutions and are designed to be accessible to groups of producers rather than sole individuals. As such, they provide what may be described as a radical potential to resist the appropriating impulses of larger scale agro-food production systems by enabling smaller producers to combine and achieve protection for their unique products. Such labelling schemes also allow for the establishment of 'closer' relationships between producers and consumers through the formalization and certification of trust in production methods and resulting food quality. However, they are not necessarily easily transplanted from one cultural context to another. For instance, the uptake of PDO/PGI certification in the UK is lower than in France, Spain and Italy, reflecting both the relatively weak tradition of producer co-operation in the UK and the loss of consumer appreciation of food origins and production which has been a feature of the post-war British foodscape.

To summarize, quality regional speciality foods can be exchanged through a variety of supply chain relationships which may or may not challenge conventional structures and the distinctions between producers and consumers. In this section we have focused on relationships which are often conventional in nature but which are mediated through efforts to promote a greater sense of connectedness between food products and place of origin. These efforts range from unregulated appeals to popular perceptions and understandings of rurality through to highly regulated certification schemes which authenticate product origins and qualities. In the following section, we turn to an alternative space of exchange in which the conventional distinction between producer and consumer is maintained but moments of reconnection are experienced through face-to-face contact between buyer and seller.

Farmers' Markets

Farmers' Markets (FM) have recently become something of a *cause célèbre* in the UK and USA, seen as a potential solution to the problems of economic marginality faced by small-scale food producers, mechanisms for local economic development, and means of generating local food economies and cultures. As such, they have been perceived as valuable by diverse interest groups, including food activists, suggesting their potential for challenging the supremacy of the global food supply system (see, for example, Hinrichs, 2000; Norberg-Hodge, 1999; Norberg-Hodge et al, 2002; Pretty, 2002), and local authorities with responsibilities for economic development. Motivations for supporting FM are variable, ranging from demands for local, sustainable and ethical food production-consumption, to conventional regional economic competitiveness and some food producers' search for niche markets and direct sales. The growth of FM in the UK and US has as a result been significant. For example, in the UK, numbers expanded from none in the mid-1990s, to over 270 by the end of that decade (Norberg-Hodge et al, 2002). Similarly, there were around 2900 FM in the US by 2000, rising from 1700 in 1994 (Pretty, 2002). In an earlier paper (Holloway and Kneafsey, 2000), we explored a new FM in Stratford-upon-Avon, Warwickshire, detailing a series of issues raised by the nascent FM movement in the UK, including the formation of a National Association of Farmers' Markets (NAFM). Here, we briefly discuss negotiations of closeness and connectedness between production and consumption, and the relationships between constructions of rurality and this particular 'alternative' food network.

One of the founding premises of FM is their role in re-establishing connections between 'local' production and consumption. The construction of closeness is mediated in several ways, illustrated by the moments of connection which occur at FM, and by the institutions forming a context within which FM have emerged, and which mediate moments of encounter. An emblematic image of the FM is the face-to-face contact between producer and consumer at the moment of purchase. Such moments are social as much as economic, seemingly establishing a closeness between the consumer and the person producing the food. The significance of the 'local' is also apparent, as closeness to a place (an identifiable location where food has been grown, with its specificities of climate, seasonality, soils etc. influencing what can be produced as 'local' foods) is established within the FM. The sense of direct connection through the food one has purchased to an individual grower, a specific site and conditions of production (e.g. a named farm), and the broader sense of the local, might be associated with an ethical relation of trust between consumer and producer. Food thus purchased carries social and geographical meaning. Its material existence (perhaps later, as cooked and eaten) retains meaning beyond the moment of encounter and mediates a sense of socio-spatial closeness and connection, which additionally becomes embodied within the consumer through eating (see Bell and Valentine, 1997; Lockie, 2002).

Surrounding these encounters, relations of closeness and connectedness are also institutionally mediated. For example, FM are closely regulated by local authorities: at Stratford, what counted as 'local' was defined as a 30 mile (c.48km) radius around the centre of the town (Stratford District Council, undated). Rules also stipulated that food was to be sold only by the producer themselves, by members of their family or an employee. Food could, therefore, only be sold by people with their own 'close' relations to production and the site of production. Similarly, indicators of 'quality', such as organic status, are authenticated by institutions such as the Soil Association (SA), while the more recent NAFM regulates and legitimizes what counts as a 'proper' FM. The relations of closeness established at FM are thus institutionalized as well as inter-personal; the institutional mediation of the relationship acts to guarantee the inter-personal trust.

Understandings of 'rurality' here are multi-layered. There is an association of 'the rural' with a series of other positive ideas, including the local, family farming, community and 'quality'. Growing and processing food are presented as rural 'crafts' or 'traditions', positioning FM against the industrial manufacturing of conventional food systems. In this sense FM might be seen as part of a reactionary 'defensive' (re)turn to the (rural) local. Simultaneously, however, FM can be understood as part of a more radical agenda for the reshaping of rural space, involving the transformation of food supply systems on the basis of small-scale farming, ecological sustainability, local production-consumption, etc. For others, an understanding of the countryside as threatened informs perceptions of FM. Consumers at Stratford spoke of their desire to support local, or British, farmers, recognising the sense of crisis which currently pervades British farming. Here, FM might be understood more simply as a response of some producers to difficult economic conditions, attempting to find specialist niches in the interstices of conventional food production-consumption. These different senses in which FM might be understood can be complementary. Norberg-Hodge, for instance, discusses how 'local food economies' are essential to the sustainability of rural economies and communities. She argues that a strong local food economy is 'the root and fibre of the entire rural economy and efforts to strengthen it thus have systemic benefits that reach far beyond the local food chain itself' (1999, p. 214). Moreover, 'social life often flourishes when like-minded suppliers and consumers meet as friends' (1999, p. 212). For writers like Norberg-Hodge, then, events like FM have collective economic and social benefits.

This discussion suggests the potential for FM to be understood simultaneously in different ways. First, as potentially part of a radical movement for change in the way food is produced and consumed. The SA, for example, has been involved with the establishment of the NAFM, seeing it partly as having a lobbying function for smallholders and allotment societies, as well as a national 'movement' for change in food production-consumption (Soil Association, 1999a; 1999b). In this understanding, food producers are seen as pioneers of new (or revisited) modes of food production-consumption, and the idea of the 'movement' suggests something extending away from the 'local' focus of a FM towards more far reaching change. Second, for many consumers, FM are simply a supplement to conventional shopping practices, and are encouraged by local authorities as part of rural or local development strategies. They are used for the purchase of speciality foods associated with the rural local, perhaps as part of a process of 'positional consumption' (Urry, 1990), and are unlikely to be seen as challenges to the dominant food supply system. Specialist producers fill economic niches opened by both consumers' interest in traditional or speciality foods, and heightened public sensitivity towards some of the health and other issues which have become associated with industrialized food supply systems. Consumer agency is thus expressed in its effect on what producers can successfully sell at FM, and in the trading practices (such as friendly face-to-face contact) associated with that success. Finally, FM also engage with a reactionary or nostalgic turn to parochial or defensive rural localism, or nationalism.

In summary, FM are associated with ephemeral moments of production-consumption reconnection, in some cases leading to longer-term relationships of friendship (Norberg-Hodge et al, 2002) or intermittent connection as face-to-face encounters recur on a monthly or biweekly basis. They act as a nexus of the mundane and the alternative, with ordinary acts of food purchase and consumption being overlain with ideas of the local, the rural, reconnection and closeness. However, FM, like 'quality' food networks, in practice reproduce categorical distinctions between producers and consumers, despite their (differently mediated) constructions of closeness. While consumer agency might be understood as having some influence on farm practice, this is likely to be indirect and non-intentional. The second pairing of 'alternative' food production-consumption networks in different ways disrupt or blur more thoroughly the boundaries of production and consumption and the agencies and identities of producer and consumer.

Small-Scale 'Alternative' Farming

While this section covers an immense diversity of form, motivation and practice, the production-consumption networks included share an enacted hybridity of production-consumption. UK examples of such production-consumption include (but are not restricted to) the following. First, communal and individual attempts to go 'back to the land' in the nineteenth and early twentieth centuries were often associated with specific political and/or moral ideologies (such as anarchism, utopian socialism or 'organicism'), or Christian philosophies, seeing small-farming as part of a solution to the moral, social and economic problems of increasingly urbanized and industrialized society (Hall and Ward, 1998; Hardy, 1979, 2000; Marsh, 1982; Matless, 1998). Second, the later twentieth century saw the (re)emergence of interest in ideas of 'self-sufficiency', associated, for example, with the writing of John Seymour (e.g. 1961, 1975; Seymour and Seymour, 1973). Third, the more recent emergence of groups such as The Land is Ours (see Halfacree, 1999) is associated with a radical, campaigning demand for access to land for alternative lifestyles which sometimes include food production-consumption, and which are often marginalized or excluded by current planning practices. Finally, a different form of alternative small-scale agriculture is practised by 'hobby-farmers', whose participation in food production-consumption is part of a set of 'lifestyle' choices.

Clearly, the most significant differences between this diverse set of alternative food systems and those already discussed lie in the convergence of food production and consumption within an individual household or community. Production and consumption are likely to be located in the same place, giving significance to the home as a site of food growing, processing and eating. There are a number of dimensions to the closeness and connectedness which can be associated with these modes of production-consumption. First, there is a sense of reconnection with the land. The persistent motif of going 'back to the land' implies a regaining of a lost closeness. Second, food production-consumption here involves physical work. Such work is ethically-weighted, consistently seen as morally, spiritually and physically improving. John Seymour, for instance, describes the 'health of body and peace of mind which come from hard varied work in the open air' (1975, p. 7). Work is important in the making of place, food and identity. Third, re-establishing connections with 'nature' and wildlife is considered important by many. For example, for some respondents to a survey of hobby-farmers (Holloway, 2000), their practices allowed them to 'commune with nature' and take 'care of animals, plants and wildlife'. Undertaking small-scale production-consumption is also regarded as a more 'natural' lifestyle, and is often positioned against an 'unnatural' urban existence. Seymour, again, states this position, arguing that 'what I am interested in is *post*-industrial self-sufficiency: that of the person who has gone through the big-city-industrial way of life and ... wants to go on to something better' (Seymour and Seymour, 1973, p. 9). Fourth, a closeness to livestock animals is important. In many instances, working with animals in small-scale production-consumption produces close, inter-subjective relations between humans and livestock which contrasts with a presumed objectification of animals in conventional agriculture. An implicitly or explicitly 'ethical' approach to livestock care is often taken, while animals might be imagined as embedded in natural systems (the rhythms of season, birth, growth, death, nutrient recycling and so on) which themselves consist of relations of closeness and connectedness (Holloway, 2001, 2003). Paradoxically, the 'reality' of the closeness of these relationships, in comparison with the distancing which is reproduced in the previous case studies, means that they are often more problematic for those engaged in them. In particular, subjective relations with animals in production-consumption contexts can raise ethical and emotional questions if they are slaughtered and eaten. A fifth relation of closeness and connectedness, that to food, is linked. Food is particularly significant as something one has produced oneself. It is food which is known intimately, connecting the consumer to their land, animals, plants and work at the moment of eating. The closeness produced by this process is again contrasted with the disconnection (and, perhaps, an increasing sense of risk) experienced in relation to purchased foods and the conventional food supply system.

As suggested above, motivations for engaging in these forms of production-consumption vary widely (Holloway, 2002a). They range from an inward-looking 'lifestyle' choice in which producing and consuming one's own food is an important part of a 'country life', to engaging in food production-consumption as part of an outward-looking, sometimes radical, agenda for social change. Explicitly adopted ethical frameworks may or may not be present. These varying motivations for involvement are associated with differing conceptions of rurality. One is an association of the rural with farming. It is suggested, for example, that participation in food production-consumption is more authentically rural than simply living in the countryside. One guide for would-be smallholders states;

> Many [rural] houses are now lived in by people who work in town and do not know how agriculture really operates. These people do not always contribute properly to the community they dwell in ... However, as an agriculturist [sic] yourself, however new to the game, you will be expected to participate a little more than that. Do so willingly, and you will find a society of a deeper and different kind from what you knew in the city (Humphreys and Gabriel, 1988, p. 15).

Growing food is thus rendered part of a 'real' rurality and community. A duality of potentially simultaneous understandings of rurality related to engaging in small-scale food production-consumption is identifiable. First, the smallholding or self-sufficient home can be seen as part of a rurality which is a retreat from a threatening, ugly and alienating urban-industrial world. Second, the spaces and practices of small-scale food production-consumption can be regarded as a microcosm of potential alternatives for wider society. Here, particular rural sites might be understood as space for revolutionary change in, for example, food production, landownership and modes of social life. These positions may be connected in specific ways in specific cases, varying with the objectives of those involved, and may combine elements of the reactionary and radical in similar ways to, for example, FM. A common theme, however, as suggested above, is the idea of a 'return' to the rural. While this apparently reflects a reactionary, even nostalgic, perspective, for some it takes on a radical or utopian edge as part of a particular vision for a future rurality. Seymour argues that this 'going back' is instead a 'going forward', deploying different notions of 'progress' and 'improvement' to those commonly held. This position is articulated in the following;

> The time is ripe for change. Thousands of people who have been through the big-conurbation stage of development ... want to come out of it. They want to get back to their birthright - the land. They want to plant orchards and woodlands and plantation, and build good-looking homes, and till fields which ring again with the voices of children and not just the roar of tractor engines (Seymour, 1977, pp. 114-15).

There is thus a debate about the use of rural space, about the ethics of human relationships with food, land and each other, which is played out in specific instances of small-scale 'alternative' farming. In contrast to the two earlier case studies, the reconnection between people and food in this instance is a sustained engagement, eliding production with consumption and necessitating a situated, material, earthy closeness to food production, contrasting with moew ephemeral, momentary connection that reproduces a dualism of production and consumption.

The final case study extends this notion of a production-consumption elision, but in relation to situations where the materiality of the engagement is more complex, and technologically-mediated over geographical space.

Virtual-Material Food Production-Consumption

The possibilities for marketing food using the Internet are increasingly recognized by companies ranging from large scale manufacturers to specialist niche-market suppliers (see, for example, Pritchard, 1999; Ray and Talbot, 1999). Most of these supply routes, however, are simply extensions of existing modes of marketing and delivering food, and retain strong distinctions between producer and consumer. In contrast, this final case study examines Internet-based schemes which give clients input into the production of food which they then consume. Two such schemes are outlined below (see Holloway, 2002b, for a fuller appraisal), and questions raised about how closeness, connectedness and rurality are problematized within the context of schemes which blur distinctions between the 'virtual' and the 'real', and 'production' and 'consumption'.

'My Veggie Patch' was an on-line service offering urban customers in West London the opportunity to grow/have grown for them vegetables on a farm in rural Suffolk. Although the scheme was short-lived, opening in 2000 and closing in 2002, it usefully illustrates the ideas suggested here. The complex, hybrid nature of production-consumption associated with the scheme was demonstrated in the opening text on the website.

> What is My Veggie Patch? - it is about *you* having *your* vegetables delivered to *your* door - *you* decide what *you* want to grow and how they are to be grown - we do all the hard work and deliver the produce *direct to you. My Veggie Patch - controlled by you, grown for you!*
>
> At myveggiepatch we will deliver fresh to your door on a weekly basis vegetables grown per your instructions in your own vegetable patch. We will provide on a weekly basis a photograph and narrative ... showing the development of your veggie patch. You will decide what action is required to any problems which may arise during the growing of your vegetables and we will action it (My Veggie Patch, 2001).

Here, food consumers could engage with the materiality of food production through the 'virtual' medium of the Internet, allowing participation in production without physically working or being at the production site. Ultimately, the customer (re)gained material contact with the food, when it was delivered to their door, then cooked and eaten.

A very different, and current, scheme, with the material site of production based in Italy, extends its network internationally, while still suggesting that a closeness and connectedness can be constructed between customers and the food, production process and location. Clients 'adopt' a sheep, have some input into its management, and receive, along with adoption documents, produce including sheep's cheese, woollen socks and salami from the yearly lamb. The website text and photographs attempt to instil a sense of connectedness within customers, evoking the smells, sounds, sights and textures of the region, and suggesting that adoption of sheep contributes to the conservation of 'nature', local sheep breeds, and 'traditional' ways of life. Reconnection was emphasized in an interview with one of the scheme's owners in a UK national newspaper (Carroll, 2000);

> The idea is to give people in the cities a new sense of faith in what we are doing ... Clients will once again have direct contact with the origins of what they eat ... this project recreates a direct contact between the producer and the buyer, and restores the client's confidence in the quality of produce – something that was lost with mass production and distribution.

Similarly, the website deploys language of connection and assurance in advertising the scheme:

> The objective is to offer certified biological [organic] products directly to the public without expensive intermediaries, and also to guarantee the consumer the possibility of a direct control of production and rearing ... The rearing stages can be controlled from a distance by email or by conventional correspondence, and a periodical bulletin with articles on the important phases of production ... will be drawn up and sent (Porta dei Parchi, 2001).

The website also focuses on the crafts and traditions of food manufacture, which customers will be able to tap into through participation. Speciality cheeses and meats are mentioned along with an evocation of the craft, sensuality and place of their production. Thus, for example, readers are invited to use their:

> Sense of smell that brings you close to the fruits of the earth: to moss, mushrooms, truffle, and aromatic herbs that are so abundant in the pastures around us and whose fragrance is transferred directly into the products we transform (Porta dei Parchi, 2001).

Growing virtual vegetables and rearing 'adopted' sheep seem to offer customers a potential closeness-at-a-distance, a technologically-mediated, clean, connectedness to the rural, the earthy, and food. Such production-consumption seemingly occurs 'between' the (urban) location of the customer and the (rural) site of physical food production. Growing and eating food here moves between different registers, not fitting easily into distinctions between real and virtual, close and distant, production and consumption. Such production-consumption raises complex questions about what closeness and connectedness might mean in virtual-material networks and about the idea of agency and participation in food networks. For example, customers of the schemes mentioned above were or are warned of the 'natural' phenomena which might interrupt them - the seasonality of vegetable production, severe weather, even wolf attacks on sheep. For customers, these disruptions offer a chance to re-engage with 'natural' cycles and risks which they feel separated from, again in a relationship which is at once distanced and close. A return to such risks might also be understood as displacing anxieties over the late-modern risks identified by Beck (1992), perhaps as part of a safely nostalgic imaginitive retreat to lives where natural hazards were significant threats. Further research into schemes which offer forms of participation is needed, particularly focusing on the experiences of the different actors involved. Yet this final case study is valuable in drawing attention to the complexity of closeness and connectedness in 'alternative' food production-consumption networks, raising questions which might be equally applicable to some of what appear to be more clear-cut distinctions employed in relation to the other examples.

Conclusions

The four case studies of 'alternative' food production-consumption networks discussed above suggest that the notions of reconnection and closeness which are often used in relation to such networks can be played out very differently in specific examples - differences are evident both between and within the case studies. It is possible, for example, for closeness to be simulated over a distance rather than to be taken literally, and for connectedness to be mediated in different material, semiotic and virtual ways. Similarly, the role and understanding of rurality is variable. Rurality might be deployed, in one instance, as a marketing signifier to associate a food product with a local authenticity, or in another, as something to move 'back' to. This range of differences emphasizes that research into food production-consumption networks needs to be able to account for the specificities of particular cases. Two implications follow. First, research needs to remain critical of terms such as closeness, connectedness and rurality which might be used unreflexively in relation to alternative production-consumption networks; 'closeness', for example, does not always mean the same thing. Second, if case-specific difference and complexity is recognized, it is difficult to make general statements about alternative production-consumption networks, and problematic for researchers and activists to critically analyse, or even to urge, changes in food networks at wider scales. The tension between particularism and global vision mentioned previously is reconfirmed in specific cases. A case could therefore be made for focusing only on the particular. Yet larger-scale theorization and empirical work, concentrating on the unequal power relations inherent in the conventional food supply system, and on the tactical responses of those involved in 'alternative' networks, should also be of value in critical examination of proposed and actually existing 'alternative' food production-consumption networks. Given this, research into food production-consumption networks also needs to focus on ways of negotiating between scales of analysis. We conclude by raising several other issues needing further consideration in relation to specific 'alternative' networks and the broader context of changing food production-consumption.

First, food production-consumption takes place simultaneously across a number of registers - the material, cultural, social, economic and political. These are increasingly considered as porous, discursively constituted categories, and the use of hybrid concepts such as the material-semiotic points to alternative ways of understanding the complexity of production-consumption networks. Food is something which has materiality, economic value and cultural value which are simultaneously important within production-consumption networks. Wheat, for example, grows in material relations with its farmed environment and may be part of an aesthetically valued landscape, while as bread purchased at FM it enters corporeal relations with human consumers and may be valued for its semiotic qualities - perhaps through an association with organic production or a specific rural location. Similarly, at different stages it is part of farm and household economies. Materiality, economic value and cultural value differ at different nodes within the network and are transformed through the network, but are always conjoined and mutually-constitutive.

Second, in considering 'alternative' food networks, consumer agency needs to be further understood. Studies need to focus on the fluid and relational construction of consumption identities, and on the different motivations and desires which structure consumers' social, economic and cultural engagement with food production-consumption. In addition, the multiplicity of possible expressions of consumer agency needs to be recognized. Consumers' positions within networks are negotiated, contested and able to shift. Again, specificity is important; consumption practices and identities will vary within and between different types of production-consumption network, and across time and space. Consumers too are able to effect change in food production-consumption, by participating in particular ways in alternative networks. Such participation might be mundane, such as the desire to engage in face-to-face contact with a vegetable grower, or more radical, such as going 'back to the land'. These forms of agency are complex, relying on relations with other constituent parts of networks, and associated with complex motivations (such as a desire to protest against the global food industry coupled with a nostalgic yearning for rural connectedness). Research into food networks needs to account for the multiple forms of agency, and multiple actors, which constitute them. The ability to do things should thus be seen as distributed through the network and as the product of relationships between the actors involved. As yet, there is little research on consumers' understandings of 'alternative' food networks (but see Weatherell et al, 2003), and more work needs to be done to account for consumer agency in such contexts.

Finally, the political nature of engagement in 'alternative' food production-consumption needs to be accounted for, even where engagement is not overtly or intentionally political. Following Beck (1999), notions of food subpolitics can be further explored, suggesting that growing social unease with the risks produced by modern, industrial modes of food production-consumption is leading people to make alternatives. These alternatives politicize the sites, spaces and practices of alternative food production-consumption in the ways that they implicitly or explicitly criticize, challenge, substitute for or subvert conventional food networks. Participation can thus be seen as a form of political action in various real and metaphorical spaces of alternative encounters with food. Engagement in at least some alternative production-consumption networks might thus be understood as 'the shaping of society from below ... changing the rules and boundaries of the political so that it becomes more open and susceptible to new linkages - as well as capable of being negotiated and reshaped' (Beck, 1999, p. 40). The rural is an important part of these subpolitics, the idea of rurality becoming involved in the negotiation and reshaping of food production-consumption and the creation of new connections between, or hybrid versions of, producer and consumer. Although some alternative modes of production-consumption use urban space to grow food, rural space continues to be significant both as an actual site of food production, but also as part of a social imaginary which gives it powerful symbolic and moral associations. The reshaping of food production-consumption might thereby involve the material reshaping and cultural re-imagining of rural spaces in a variety of different ways.

References

Allen, P. (1999), 'Reweaving the Food Security Net: Mediating Entitlement and Entrepreneurship', *Agriculture and Human Values*, Vol. 16, pp. 117-129.

Allen, P., FitzSimmons, M., Goodman, M. and Warner, K. (2003), 'Shifting Plates in the Agrifood Landscape: the Tectonics of Alternative Agrifood Initiatives in California', *Journal of Rural Studies*, Vol. 19, pp. 61-75.

Barham, E. (2003), 'Translating Terroir: the Global Challenge of French AOC labelling', *Journal of Rural Studies*, Vol. 19, pp. 127-138.

Beck, U. (1992), *Risk Society: Towards a New Modernity*, Sage, London.

Beck, U. (1999), *World Risk Society*, Polity Press, Cambridge.

Beck, U. (2000), 'The Cosmopolitan Perspective: Sociology of the Second Age of Modernity', *British Journal of Sociology*, Vol. 51, pp. 79-105.

Bell, D. and Valentine, G. (1997), *Consuming Geographies: We Are Where We Eat*, Routledge, London.

Bourdieu, P. (1984), *Distinction: a Social Critique of the Judgement of Taste*, Routledge, London.

Carroll, R. (2000), 'Adopt a sheep on the Internet', *The Guardian*, 7th November, p. 17.

Falk, P. (1994), *The Consuming Body*, Sage, London.

Goodman, D. (2001), 'Ontology Matters: the Relational Materiality of Nature and Agro-food Studies', *Sociologia Ruralis*, Vol. 41, pp. 182-200.

Goodman, D. (2002), 'Rethinking Food Production-consumption: Integrative Perspectives', *Sociologia Ruralis*, Vol. 42, pp. 271-277.

Goodman, D. (2003), 'The Quality 'Turn' and Alternative Food Practices: Reflections and Agenda', *Journal of Rural Studies*, Vol. 19, pp. 1-7.

Goodman, D. and DuPuis, E. (2002), 'Knowing Food and Growing Food: Beyond the Production-consumption Debate in the Sociology of Agriculture', *Sociologia Ruralis*, Vol. 42, pp. 5-20.

Granovetter, M. (1985), 'Economic Action and Social Structure: the Problem of Embeddedness', *American Journal of Sociology*, Vol. 91, pp. 481-510.

Halfacree, K. (1999), 'A New Space or Spatial Effacement? Alternative Futures for the Post-productivist Countryside', in N. Walford, J. C. Everitt and D. E. Napton (eds), *Reshaping the Countryside: Perceptions and Processes of Rural Change*, CABI Publishing, Wallingford, pp. 67-76.

Hall, P. and Ward, C. (1998), *Sociable Cities: The Legacy of Ebenezer Howard*, Wiley, Chichester.

Hardy, D. (1979), *Alternative Communities in Nineteenth Century England*, Longman, London.

Hardy, D. (2000), *Utopian England: Community Experiments 1900-1945*, E&FN Spon, London.

Hassanein, N. (2003), 'Practicing Food Democracy: a Pragmatic Politics of Transformation', *Journal of Rural Studies*, Vol. 19, pp. 77-86.

Hinrichs, C. (2000), 'Embeddedness and Local Food Systems: Notes on Two Types of Direct Agricultural Market', *Journal of Rural Studies*, Vol. 16, pp. 295-303.

Holloway, L. (2000), '"Hell On Earth and Paradise All At The Same Time": the Production of Smallholding Space in the British Countryside', *Area*, Vol. 32, pp. 307-315.

Holloway, L. (2001), 'Pets and Protein: Placing Domestic Livestock on Hobby-farms in England and Wales', *Journal of Rural Studies*, Vol. 17, pp. 293-307.

Holloway, L. (2002a), 'Smallholding, Hobby-farming and Commercial Farming: Ethical Identities and the Production of Farming Space', *Environment and Planning A*, Vol. 34, pp. 2055-2070.

Holloway, L. (2002b), 'Virtual Vegetables and Adopted Sheep: Ethical Relation, Authenticity and Internet-mediated Food Production Technologies', *Area*, Vol. 34, pp. 70-81.

Holloway, L. (2003), '"What a thing, then, is this cow ...": Positioning Domestic Livestock Animals in Texts and Practices of Small-scale "Self-sufficiency"', *Society and Animals*, Vol. 11, pp. 145-165.

Holloway, L. and Kneafsey, M. (2000), 'Reading the Space of the Farmers' Market: a Preliminary Investigation from the UK', *Sociologia Ruralis*, Vol. 40, pp. 285-299.

Humphreys, P. and Gabriel, T. (1988), *Practically in the Country*, Comma, Abergavenny.

Ilbery, B. and Kneafsey, M. (2000a), Producer Constructions of Quality in Regional Speciality Food Production: a Case Study from South West England', *Journal of Rural Studies*, Vol. 16, pp. 217-30.

Ilbery, B. and Kneafsey, M. (2000b), 'Registering Regional Speciality Food and Drink Products in the United Kingdom: the Case of PDOs and PGIs', *Area*, Vol. 32, pp. 317-25.

Jones, O. (2000), '(Un)ethical Geographies of Human-nonhuman Relations. Encounters, Collectives and Spaces', in C. Philo and C. Wilbert (eds), *Animal Spaces: Beastly Places: New Geographies of Human-Animal Relations*, Routledge, London, pp. 1-16.

Kneafsey, M., Ilbery, B., and Jenkins, T. (2001), 'Exploring the Dimensions of Culture Economies in Rural West Wales', *Sociologia Ruralis*, Vol. 41, pp. 296-310.

Kneafsey, M. and Ilbery, B. (2001), 'Regional Images and the Promotion of Speciality Food and Drink in the West Country', *Geography*, Vol. 86, pp. 131-40.

Lockie, S. (2002), '"The Invisible Mouth": Mobilising "the Consumer" in Food Production-consumption Networks', *Sociologia Ruralis*, Vol. 42, pp. 278-294.

Lockie, S. and Kitto, S. (2000), 'Beyond the Farm Gate: Production-consumption Networks and Agri-food Research', *Sociologia Ruralis*, Vol. 40, pp. 3-19.

Marsden, T., Banks, J. and Bristow, G. (2000), 'Food Supply Chain Approaches: Exploring their Role in Rural Development', *Sociologia Ruralis*, Vol. 40, pp. 424-438.

Marsden, T., Banks, J. and Bristow, G. (2002), 'The Social Management of Rural Nature: Understanding Agrarian-based Rural Development', *Environment and Planning A*, Vol. 34, pp. 809-825.

Marsh, J. (1982), *Back to the Land*, Quartet Books, London.

Matless, D. (1998), *Landscape and Englishness*, Reaktion Books, London.

Miele, M. and Murdoch, J. (2002), 'The Practical Aesthetics of Traditional Cuisines: Slow Food in Tuscany', *Sociologia Ruralis*, Vol. 42, pp. 312-328.

Murdoch, J. (2000), 'Networks: a New Paradigm of Rural Development?', *Journal of Rural Studies*, Vol. 16, pp. 407-419.

Murdoch, J., Marsden, T. and Banks, J. (2000), 'Quality, Nature and Embeddedness: Some Theoretical Considerations in the Context of the Food Sector', *Economic Geography*, Vol. 76, pp. 107-125.

My Veggie Patch (2001), (http:www.myveggiepatch.com), Accessed 2001.

Norberg-Hodge, H. (1999), 'Reclaiming Our Food: Reclaiming Our Future', *The Ecologist*, Vol. 29, pp. 209-214.

Norberg-Hodge, H., Merrifield, T. and Gorelick, S. (2002), *Bringing The Food Economy Home: Local Alternatives To Global Agribusiness*, Zed Books, London.

Porta dei Parchi (2001), (http:www.asca.dimmidove.com), Accessed 2001.

Pretty, J. (2002), *Agri-Culture: Reconnecting People, Land And Nature*, Earthscan, London.

Pritchard, W. (1999), 'Local and Global in Cyberspace: the Geographical Narratives of US Food Companies', *Area*, Vol. 31, pp. 9-17.

Ray, C. and Talbot, H. (1999) 'Rural Telematics: the Information Society and Rural Development', in M. Crang, P. Crang and J May (eds), *Virtual Geographies: Bodies, Space and Relations*, Routledge, London, pp. 149-163.

Renting, H., Marsden, T. and Banks, J. (2003), 'Understanding Alternative Food Networks: Exploring the Role of Short Food Supply Chains in Rural Development', *Environment and Planning A*, Vol. 35, pp. 393-411.

Sage, C. (2003), 'Social Embeddedness and Relations of Regard: Alternative 'Good Food' Networks in South-west Ireland', *Journal of Rural Studies*, Vol. 19, pp. 47-60.

Seymour, J. (1961) [1991], *The Fat of the Land*, Metanoia Press, New Ross, Ireland.

Seymour, J. (1975) [1996], *The Complete Book of Self Sufficiency*, Dorling Kindersley Ltd, London.

Seymour, J. and Seymour, S. (1973), *Self-sufficiency: the Science and Art of Producing and Preserving Your Own Food*, Faber and Faber, London.

Stassart, P. and Whatmore, S. (2003), 'Metabolising Risk: Food Scares and the Un/re-making of Belgian Beef', *Environment and Planning A*, Vol. 35, pp. 449-462.

Soil Association (1999a), 'National Association of Farmers Markets Now Established', *Press Release* 8th March 1999.

Soil Association (1999b) Summary: Soil Association Seminar, National Association of Farmers Markets, 10th February 1999.

Tovey, H. (1997), 'Food, Environmentalism and Rural Sociology: on the Organic Farming Movement in Ireland', *Sociologia Ruralis* Vol. 37, pp. 21-37.

Urry, J. (1990), *The Tourist Gaze*, Sage, London.

Weatherell, C., Tregear, A. and Allinson, J. (2003), 'In Search of the Concerned Consumer: UK Public Perceptions of Food, Farming and Buying Local', *Journal of Rural Studies*, Vol. 19, pp. 233-244.

Whatmore, S. (2002), *Hybrid Geographies. Natures Cultures Spaces*, Sage, London.

Whatmore, S. and Thorne, L. (1997), 'Nourishing Networks: Alternative Geographies of Food', in D. Goodman and M. Watts (eds) *Globalising Food: Agrarian Questions And Global Restructuring*, Routledge, London, pp. 287-304.

Winter, M. (2003), 'Embeddedness, the New Food Economy and Defensive Localism', *Journal of Rural Studies*, Vol. 19, pp. 23-32.

Chapter 14

Winners and Losers?
Rural Restructuring, Economic Status and Masculine Identities among Young Farmers in South-West Ireland

Caitríona Ní Laoire

Introduction

Agricultural society in Ireland is undergoing significant restructuring, associated with the continued rationalization of agriculture and a declining farming population. As is the case across the EU, the trend in agriculture in Ireland is towards fewer farms, less farm employment, larger farm units and the specialization and concentration of farm production (Frawley and Commins, 1996). Agricultural employment in the Republic of Ireland declined by 63 per cent between 1961 and 1995 (Department of Agriculture, Food and Rural Development, 1999). Between 1985 and 1994, the total number of farms in Ireland declined by 30 per cent, this decline occurring mainly in small farms (Frawley and Commins, 1996).

The economic implications of these changes for the sector as a whole include, arguably, a more efficient agricultural sector, together with a rise in pluri-activity and in non-agricultural rural economic activities. However, the social and cultural implications, in particular at the level of the farm and the locality, have not been explored in great depth. At the level of the farm, the competitiveness of the contemporary agricultural sector has given rise to a situation where farms need to expand in order to survive, with farmers competing with each other for available production quota allocations and land. As a result, farmers' livelihoods often depend on their neighbours' losses. The competitive process is driven by a dominant rhetoric that constructs farming as a modern business with expansion as a key objective and entrepreneurialism as a key attribute. Given that some farmers can succeed better than others in expanding and modernising, to what extent do farmers themselves relate to this expansionist discourse?

This research question is addressed here with reference to the construction of male farmer identities, given the male dominance of the farming sector. It is argued that the masculine nature of dominant farming discourses and values needs to be recognized in seeking to understand the social and cultural responses to contemporary rural restructuring.

Using evidence from interviews with young male farmers in a locality in the south-west of Ireland, this chapter explores the ways in which male farmer identities are constructed in relation to the nature of their farm work and economic status. It is argued that economic status is an important element of farmer identity, and that economic restructuring processes that involve producing 'winners' and 'losers' have important implications for the construction of farming identities. The chapter explores the implications of increased commercialism and the rationalization of the agricultural labour force for the ways in which farmers construct their work identities. A number of dominant discourses of farming masculinity among both economically successful and marginal farmers are identified. The complexities of the relationships between farming masculinities and economic success are unpacked, and it is suggested that material circumstances, together with perceptions of potential threat to one's livelihood and one's sense of self, contribute to shaping farmer values and identities.

Research Context and Methodology

This chapter explores constructions of masculinity in the 21st century among a particular group of farmers, that is, young farmers in a locality in north Cork in the south-west of Ireland. The locality is one of medium to large farms in terms of economic size, by national standards, according to analysis by Lafferty et al. (1999). Agriculture is the largest employer in the locality (62.5 per cent of those 'at work'), and farmers form the largest social group in the area (figures from Small Area Population Statistics for 1991). Dairying is the most important enterprise, although there is also some beef and sheep farming. However, the milk production sector (and the agricultural sector generally) in the area has been experiencing problems associated with the quota system and increased competitiveness (ó Donnchadha and Ní Shé, 1994).

The research aimed to explore the implications of contemporary rural restructuring processes for the experiences of young farmers in the area. It comprised a series of in-depth interviews with eleven young farmers, exploring their views on farming as a way of life and as an occupation. All were full-time farmers, identified through the snowballing technique, all were male and were aged between 25 and 42 years. (Contact was made with one female farmer, but unfortunately she was unable to participate). Most interviews were conducted in the farmers' homes, although two were conducted by telephone. The interviews were semi-structured, and focussed upon the interviewees' current farming practices, intergenerational change, views on agricultural policy, their own biographies and their attitudes towards farming and rural life. One of the aims was to explore underlying values and goals through the stories the farmers told and the languages and discourses they drew on. The most dominant themes that emerged from the interviews in relation to constructions of farm work and perceptions of economic status are explored here. In particular, the analysis focuses on the degree to which the farmers drew upon traditional or more entrepreneurial discourses of farming in constructing their own identities as farmers.

Producing 'Winners' and 'Losers'

The economic reality of the contemporary agricultural sector in western Europe is one in which growing competitiveness contributes to a situation where farms need to expand in order to survive. In Ireland, this manifests itself in a reduction in numbers of farmers and farms, and a simultaneous increase in average farm size (Frawley and Commins, 1996). Agriculture becomes more efficient through improved technologies as fewer farmers produce higher levels of output (Lafferty et al, 1999). The most recent Census of Agriculture figures show a fall of 17.0 and 17.5 per cent respectively in the total number of farms and number of farmers in the state between 1991 and 2000 (Central Statistics Office, 2003) (More detailed regional or local data from the 2000 Census of Agriculture is not yet available). Young farmers are at the centre of these restructuring processes, making decisions to become or to stay farmers in the context of competing economic and social pressures. The pressures to leave farming include financial pressures as well as perceptions of the social isolation and harder working conditions of farming relative to other occupations.

The rationalization of the agricultural sector in Ireland varies regionally in intensity, with the decline in the number of holdings greatest in the north and north-west, and less marked in the south-west (Lafferty et al, 1999). In the study area (in the south-west), there was a decline of 9.7 per cent in numbers directly employed in agriculture between 1981 and 1986, while numbers of milk suppliers fell by between 22 and 25 per cent in the late 1980s and early 1990s (ó Donnchadha and Ní Shé, 1994). Ó Donnchadha and Ní Shé (1994) argue that the sharp decline in milk suppliers in the area was a direct result of the introduction of the milk quota system, which resulted in a reduction in production levels. Figures from Lafferty et al (1999) suggest that the decline in the south-west of Ireland, as elsewhere, occurred mainly in the small farm (less than 20 ha.) sector, while the 20 to 60 ha. category increased in numbers, reflecting a gradual process of enlargement. Of the remaining farms in the state as a whole, 29 per cent can be considered to be economically viable, and 71 per cent are economically non-viable, according to analysis of 1994 figures by Frawley and Commins (1996). Of the non-viable farms, approximately one-third are categorized as 'residual', with a very doubtful future in farming, while two-thirds are categorized as 'marginal', considered to be moving towards either viability or residual status.

The dynamics of how these processes of enlargement and rationalization are worked out at the local scale merit attention. Dudley (2000) has explored the social implications of agricultural rationalization in rural America in the 1980s, and found that high levels of debt and ultimate dispossession were key factors in the movement out of farming. In contrast, the shift away from agriculture in Ireland is a more gradual process, involving transitional part-time farming and short-term land-leasing. According to data presented in Lafferty et al (1999), leasing of land tends to be associated particularly with areas of medium to large farms and is quite a significant phenomenon in north Cork. In the case of the dairying sector in particular, occasionally, a shift to beef farming may represent a transitional step towards going out of production altogether. In general however, holdings that are non-viable economically, or demographically (with no heir), tend to be passed on to a neighbouring farmer via a short-term lease, or may be bequeathed to a relative on the death of the landholder. Gradually, these holdings become incorporated into neighbouring or nearby holdings where there might be sufficient capital to buy or rent additional land. In the case of the dairying sector, milk production quotas have become a marketable commodity in their own right, which (until recently[1]) have tended to move around in a similar manner. The result is that due to a combination of economic and demographic factors, some farmers are in a position to expand production and become more efficient, some are unable to expand due to lack of capital to purchase or rent land or quota, and others gradually move out of production.

Survival for one farmer is often a result of land becoming available through a neighbouring farmer selling or leasing land. This means that some farmers emerge as 'winners' through their ability to expand and thereby secure the future of their family holding. The second group, those who are unable to expand, are in a progressively weaker position from which to do so, as a result of the expansionist dynamic inherent in the agricultural sector.

> The pressure to maintain economic viability in farming obliges farm operators to enlarge the scale of their farm business by acquiring extra land and/or intensifying the scale of their farm operations (Commins, 1999, p. 6).

The ability to expand, then, can mean the difference between survival and loss in farming. Clearly, economic success in these terms is not simply a function of the entrepreneurial or managerial skills of the individual farmer, but also relies very heavily on purchasing or borrowing power. The financial power with which to lease or purchase land or quota is a key factor in determining economic success. In other words, expansion tends to be easiest for the larger or more commercial farmers. The economic processes at work here are similar, clearly, to those operating elsewhere in the capitalist free market economy. However, the social and cultural framework in which these processes are set is quite distinctive. The loss of a farm is more than the loss of a business enterprise, but instead, can represent the loss of a way of life and of a family inheritance, and thus can be represented as a failure in upholding one's responsibilities. Indeed, the close relationship between farm work and masculine identity means that the loss of the farm can have a devastating effect on one's very sense of self. Identity, then, is bound up to a certain extent with economic success.

Farm Work and Masculine Identities

The relationship between economic success and farming identity is mediated through the construction and reproduction of masculinity. It is generally agreed that work is a central source of masculine identity and power (Collinson and Hearn, 1996). For example, traditional working class masculinities are based on notions of the male breadwinner (Willott and Griffin, 1996), while middle class masculinities are often based on men's domination of management (Collinson and Hearn, 1996). Male pride is invested not just in participation in these work spheres, but also in 'success' achieved through work, whether that is measured in physical or financial terms.

> Typically, it seems, men's gender identities are constructed, compared and evaluated by self and others according to a whole variety of criteria indicating personal 'success' in the workplace (Collinson and Hearn, 1996).

Masculine values surrounding success in the workplace have tended to be those that are taken-for-granted as the norm, thus perpetuating male dominance of many workplaces. However, it is increasingly being recognized that male dominance is a complex phenomenon. Recent research on masculine power has involved moving beyond a conventional homogeneous concept of masculinity that associates privilege and power with maleness in an unproblematic way, and moving towards the recognition of the existence of multiple masculinities in different social and spatial contexts. Connell (1995), among others, argues that some masculinities occupy more dominant positions than others, in other words, that there are hegemonic masculinities. These are defined as masculinities that occupy 'the hegemonic position in a given pattern of gender relations' (Connell, 1995, p.74). Hegemonic masculinities are those ideologies that privilege some men (and women) by associating them with particular forms of power (Carrigan et al, 1985, cited in Cornwall and Lindisfarne, 1994). They are those masculinities which are the most powerful, or most 'honoured' (Connell, 2000) within a social context, and substantial research exists on the ways in which dominant masculinities marginalize women and less powerful masculinities (Cornwall and Lindisfarne, 1994).

In recent years, there has been a growing recognition in rural gender research of the need to understand the experiences of both men and women, and to interrogate the highly gendered nature of rural life from both sides of the male-female binary. This has given rise to an expanding body of literature focusing on rural masculinities, much of which explores the power relations inherent in constructions of masculinity in the rural sphere (for example, Brandth, 1995; Woodward, 1998; Evans, 2000; Campbell, 2000; Little and Jones, 2000; the special issue of *Rural Sociology*, 2000; Saugeres, 2002a; Saugeres, 2002b). These critical perspectives on rural masculinities contribute to understandings of the gendered nature of rural society and space, by highlighting the visible and invisible ways in which masculine practices and preferences are normalized and legitimized. Campbell (2000), for example, analyses men's behaviour in rural pub space, while Little and Jones (2000) focus on the space of rural development activity.

There is a growing body of research exploring the relationship between masculine identities and work in particular rural contexts, such as forestry (Brandth and Haugen, 2000), the armed forces (Woodward, 1998), trucking (Evans, 2000) and the farm. The farm is a highly significant workspace in the construction and performance of rural masculinities. The highly gendered nature of work and social relations in the farming sector is well-documented (Brandth, 1994; Leckie, 1996; Liepins, 1998; Shortall, 1999; Saugeres, 2002a, 2002b). Research shows that farming is still an overwhelmingly masculine occupation, culturally as well as numerically. For example, Brandth (1995), Liepins (2000) and Saugeres (2002a, 2000b), working in different farming contexts, highlight the dominance of representations of masculinity in popular imagery of farmers. The masculinist nature of agriculture has material consequences for women involved in farming. Leckie (1996, p. 309) found in her research in southern Ontario that 'farm girls do not share in the occupational inheritance of agriculture - they are frequently excluded or marginalized from important agricultural resources, including information'. Shortall (1992) relates the subordinate position of farm women in Ireland to the highly gendered nature of rural power structures such as patrilineal inheritance patterns, information structures, the farm media, farm organizations and prevalent beliefs and ideologies.

Farming masculinities then are grounded in unequal material relations of land-ownership and control of property. This highlights the materiality of unequal gender relations, an aspect emphasized by Connell (2000, p. 25) in accounting for men's gender conduct. He calls for greater attention in research to what he terms the 'patriarchal dividend', that is 'the benefits accruing to men from unequal shares of the products of social labour'. In the emphasis on representation and performance in much recent masculinities research, there may be a tendency to under-emphasize the material dimensions of identity constructions and performances. In other words, there is a need to reiterate that gendered practices such as the performance of masculine identities are often grounded in material interests. In this chapter, it is argued that farmer identities and values cannot be separated from their economic and material situations, and the complexities of this relationship are explored in some detail.

Material inequality is reproduced through the normalization and hegemony of masculinist values and discourses that construct farming masculinities in particular ways. Discourse analysis can reveal the socially constructed nature of masculine identities. For example, Liepins' (1998, 2000) research on agricultural discourses in Australia and New Zealand emphasizes the existence of dominant constructions of masculinity in farming, based on key qualities of 'toughness' and 'physical activity'. This is based on an association of farming with outdoor work, control over the environment and tenacity in the face of difficult physical and economic conditions. Brandth's (1995) research on tractor adverts and Saugeres' (2002a, 2002b) on farming masculinities in France also highlight the hegemony of constructions of masculinity based on control of nature. Similarly, Peter et al (2000) assert the dominance of a conventional masculinity among farmers in Iowa, one that is based on an association of farming with dirt, a denial of bodily comfort and the control of nature. Evans (2000) critiques the over-emphasis in much rural masculinities literature on the importance of physical strength as a key attribute. He argues instead that the dominant ethos of masculinity associated with rural staples economies (such as, farming, forestry and fishing) is based on 'getting the job done', in other words, on achievement. There is a strong emphasis on the mental skills required to get work done, and the attributes that are highly honoured include a sense of resilience, independence and 'not complaining'. Therefore conventional hegemonic farming masculinities involve constructions of men's farm work as tough work, requiring both mental and physical strength and resilience.

Hegemonic masculinities are constantly being re-formed and re-shaped as they compete for dominance with alternative masculinities. Recent research points to the existence of competing discourses of masculinity in farming. Brandth's (1995) research on technology and farming masculinities highlights a tension between traditional discourses based on qualities of physical toughness, on the one hand, and more modern discourses based on efficiency and the use of high tech machinery, on the other. Brandth and Haugen's (2000) research on forestry workers identifies a similar tension between discourses of masculinity among manual forestry workers and forestry managers. Saugeres (2002b) found in her research in the south of France that farmers distinguished between 'true' farmers, who worked in harmony with nature, and modern farmers whose values were those of profit and exploitation of nature. There is therefore a recurring theme of an opposition between traditional and modern constructions of the male farmer.

Bryant's (1999) analysis of farmers' occupational identities is useful as a framework within which to identify discourses of farm work. She argues that, for the 'traditional' farmer, self-image is based on pride and a pleasure in physical work, and farming is seen as a way of life, usually involving rigid gender roles. On the other hand, the 'managerial' or 'entrepreneurial' farmer emphasizes planning, has a pride in organizational skill and efficiency, and sees farming as a business. This identity also involves a certain distancing from the physical labour of farm-work. In some cases, this also incorporates a pride in market responsiveness, progress and risk-taking, and a view of farming as a business enterprise (Bryant, 1999). The aim in this chapter is not, as Bryant (1999) has done, to classify farmers as either 'traditional' or 'managerial/entrepreneurial', but to identify the discourses used by farmers that idealize one or other set of values.

Constructions of Farming Masculinity in South-west Ireland

In Ireland, traditionally, farming masculinities have been rooted in an agrarian ideology in which the family farm was a key element. The concept of an 'agrarian ideology' is based here on Beus and Dunlap's (1994) conceptualization of agrarianism, as a set of beliefs and values that reflect the notion that agriculture is an important and valuable segment of society, that farmers should be independent and that family farms are important to democracy. The self-sufficiency and independence of the family farm was central to the version of agrarianism that prevailed in Ireland from the 19th century onward. The traditional rural family life that was idealized in this involved a highly masculinist relationship to place, based on patriarchal systems of land-ownership and inheritance. The agrarian ideology was, therefore, grounded in unequal material relations. The modernization of Irish agriculture from the 19th century onwards was intimately bound up with the establishment of patterns of impartible and patrilineal inheritance, and gendered divisions of labour in agriculture. This meant the construction of an Irish rural masculinity that was closely associated with land-ownership, control of property and the authority of a powerful father figure (Martin, 1997). Toughness, self-reliance and the ability to provide for the family were also important elements of this construction. Femininity became closely associated with domesticity and motherhood, while formal outdoor work became the domain of the masculine. This highly gendered social system was perpetuated by discourses of agrarianism and idealized notions of masculinity and femininity. Although the structures of Irish rural society have changed enormously throughout the twentieth century, gender inequalities remain persistent. Patrilineal inheritance patterns are still the norm, men predominate among landholders and women still occupy subordinate positions on the farm (Shortall, 1992). Women are still severely under-represented in the formal agricultural labour force, representing only 11 per cent of farm holders in the Republic of Ireland (Central Statistics Office, 2003), although their contribution to farm work is increasingly being recognized. Contemporary Irish farming masculinities have some of their roots in the agrarian ideology of family farming that prevailed in Ireland in the early twentieth century. Some aspects of this construction of masculinity are evident among farmers in north Cork in the 21st century, and are imbued with notions of farm work as physical men's work. These are enabled through patrilineal systems of land-ownership that have survived into the 21st century.

Conventional farming masculinities involve constructions of the farmer as a hard worker battling against environmental and economic obstacles and exerting his authority over the natural landscape (Liepins, 2000; Peter et al, 2000). This type of construction is one that is evident among farmers in the study area. There is a strong sense among them that farming as an occupation involves 'hard work' and a denial of comfort, in a way that other occupations do not. For some farmers, it is related to growing pressure to increase productivity. For many, the long and unsociable hours distinguish farming from other occupations.

> Having to work in different conditions; if 'tis raining we just have to go and work, even if the cows are all inside in one big shed, 'still have to go out and work... 'still have to work ten hours a day... hard sometimes when the weather is bad. [...]. You have to be very efficient and work hard. People always work hard like but you have to work that bit harder (Tom[2]).

This farmer recognizes the difficulties of his job, and he uses this to contribute to his own self-image as 'tough', distinguishing himself from others who work in less physically demanding jobs. In fact, there is an element of disdain among the farmers towards young people who turn down the rigours of farming in favour of the easier way of life and greater financial remuneration associated with life off the farm. The 'true' farmer is distinguished from those who are tempted away by the attractions of a perceived easier lifestyle.

This sense of masculine pride in the nature of farm work is further reinforced by reference to visible signs of their achievements in the landscape. Visible signs of their hard work, such as healthy animals or secure fences are identified by the farmers as sources of satisfaction and pride. The fruits of one's labour are pointed to as signs of the rewards of hard work.

> 'Tis nice to walk out in the summertime or late spring and to see all the cattle out in the fields and the cows grazing. That's rewarding because a good bit of work goes on indoors in the wintertime and it's rewarding to get through all that work and see them all out and healthy... (Tom)

> Job satisfaction? [To] do your day's work and say that's a job well done. Compared to a factory [where] you come out and you get your 50 or 60 pounds or whatever and you never see what you've done (Denis).

The ability to work the land and to provide for their families or themselves is a source of pride and status. As in many other sectors of society, 'breadwinning' status and hard work are symbols of masculinity.

> There were six of us here - it [the farm] reared them all, put them through college without help from anyone - a great achievement -you wouldn't do it now (Denis).

There is a sense among some, however, that making a living or supporting a family from farming is no longer possible. This is clearly a source of disappointment and disillusionment with farming for young farmers.

> Young farmers aren't taking it up. [They've] seen the reality - parents working day and night - seeing nothing out of it at the end of the day (Michael).

This is expressed by some in terms of the importance of the viable family farm, as opposed to the commercial business enterprise or the non-viable holding maintained by direct subsidies. The ability to support a family is seen to be the primary goal, and for that reason, direct subsidization is not a source of pride.

> And I think government policy should be directed more at those who want to maintain a viable holding and those who want to develop their holdings for their families particularly where there would be a successor... I think they should look after family farms better (Joe).

> The subsidy area is demoralising farmers... Farmers should be paid for produce. Subsidies are only a roundabout way (Pat).

This view reflects the historical importance of the agrarian ideology of family farming, self-sufficiency and independence. Pride is invested in the ability to support a family without external aid. Masculine pride then is clearly vested in particular aspects of farm work, such as physical strength, tenacity and the ability to earn a living. Dominant discourses of farming are used by farmers to reinforce a sense of self that is built on key male attributes and is intimately tied to the nature of farm work and also to success as a farmer.

Competing Discourses of Successful Farming

Success as a farmer can of course be defined and interpreted in different ways. Dominant rhetoric as well as economic realities define success in agriculture in economic terms, based on expansion, modernization and efficiency (Dudley, 2000). This represents a significant shift away from the traditional values of agrarianism, tradition and self-reliance. To what extent do farmers themselves relate to these competing values? What does the hegemony of the expansionist rhetoric mean for the ways in which farmers construct and perform their own identities? Of particular interest is the ways in which marginal farmers relate to the ideologies of expansionism and efficiency, given that these ideologies by their nature define marginal farmers as unsuccessful farmers. These questions are explored through an analysis of the discourses of farm work that are utilized by the farmers interviewed in the research.

There was no direct question on farm size in the interviews, but some volunteered information on acreage, numbers of livestock and/or quota size. All gave some indication of whether they considered their holdings to be small, medium-sized or large by local standards, and also of the financial state and future prospects of their enterprise. Based on these indicators, it is clear that five of the interviewed farmers are working on relatively marginal or struggling holdings. The remaining six consider their holdings to be viable; at least five of them are holdings of 100 acres or more, and at least four have expanded their enterprises. This is not an ideal method of assessing farm size and viability, due to the lack of complete information and also due to a possible tendency for farmers to underplay their economic assets for a number of reasons. However, it is useful as an indicator of the farmers' own perceptions of their economic status by local standards and of the viability of their holdings.

It could be expected that the five farmers who present a relatively successful image of their farm businesses would espouse views that reflect the values of economic rationality, modernization and efficiency. Certainly they tend to talk about farming more as a business than as a way of life, valuing expansion, progress and change. A prominent goal among them is expansion:

> A lot [of farmers] haven't changed with the times. Farming changes, you must go with it. If you don't keep expanding, you're going backwards. ... (Brian).

> I can see the herd increasing in the next ten years and probably taking more land, near home, as near home as we can get I presume. We had one farm that is bounding us on one side and there's a possibility we might get the one bounding us on the other side, so we might be farming three farms. Where there was three families before, there'd be one family living off it now (Tom).

A 'good' farmer is seen to be someone who is willing to make changes to become more efficient:

> Being able to think ahead. Doesn't have to be this way because your father did it this way. Maybe there's a better way... I would say... it was very easy for me to change a lot of systems because I was trained, I did a course, I am in a discussion group. I can see different ways of approaching things, cutting down expense, making the environment more operator-friendly. When I started making these changes, my father couldn't understand, [he] would say why are you doing it that way?, wasn't it alright?... (Brendan).

Some of these farmers would consider leaving farming if it become unprofitable or non-viable, or at least they talk about farm succession as a rational decision to be made, rather than a taken-for-granted step.

> I'm committed to a 15 per cent return on investment. If farming can't provide, I will pull out. I'm not interested in plodding along for the sake of being a farmer. There is room for the top ten per cent of farmers in the country. I'm certain that the returns will be good for those who can hang on for the next five years (Brian).

When asked if he would like to see any of his children become farmers, 'Tom' replied:

> I couldn't see anything wrong with it; of course I wouldn't pressurize them into it. If they make the decision to go into farming, I'd support them... as long as there's a good income in it. At the moment there is – I'm not sure about the future. We've always survived and I presume we will survive. I would recommend it but there's a lot of things to take into account.

However, these values are not so prominent among those farmers who present a relatively unsuccessful image of their enterprises. Emerging from the interviews with these farmers is a more traditional approach to farming, involving a strong emphasis on hard work, a view of farming as a way of life rather than a business, and a strong emotional or almost biological connection with farming and land.

> I suppose I was destined to be a farmer. 'Twas in my blood. I was always going to be farmer; that's the way it is.
> *Destined? (Interviewer)*
> Kind of... 'twas... destined in a way yes, because I'd a love for farming ever, always since I was knee high. I'd just a love for farming, I suppose that's why (Seamus).

This has strong resonances with Saugeres' (2002b) findings from her research in southern France, where there was a strong discourse of the 'good' farmer as having a natural, almost embodied, connection to the land. This is part of a general idealization of traditional peasant farming, a trend which is also evident among some farmers in north Cork, where a traditional approach to farming involves a certain romanticization of farming in the past. The 'community spirit' of the past is idealized, although the community that is referred to is mainly one of men and associated with the outdoors.

> Those people had more to do with each other in years gone past. There was more of a community spirit really. People worked with each other at harvest time. They used go to the creamery every day, meet other people; that doesn't happen at all now... (Seamus).

This type of discourse of masculinity values dedication to farming above all else and a commitment to 'sticking with it', although at times it may not be financially rewarding to do so. Again, there is a sense of tenacity and of struggling with external pressures.

> Even if I didn't make a profit, I'd still be at it. I'll be at it till the day I croak. I'm not there for the money (Kevin).

One young farmer, 'John', says that his income has dropped dramatically in recent years and, in his own words, he does not have a future. In order to get back to the income level he had ten years ago, he needs to quadruple in size. Despite this, he claims, he is going to stay in farming, as he likes it and it is the only thing he knows. Some farmers, like 'John', expressed the belief that, economically, things would improve for the agricultural sector in the medium-term, and that those who stayed in farming when times were difficult would then be successful. In other words, there is a sense among them that tenacity, resilience and 'staying power' are honourable qualities, which will ultimately be rewarded.

The research, then, identifies two dominant discourses of farming among the young male farmers in the area - a business-oriented managerialist discourse, prominent among (although not exclusive to) the more economically successful farmers, and a more traditional romantic discourse, prominent among (although not exclusive to) the more marginal farmers. All interviewees had a certain tendency (some more than others) to utilize a number of different discourses, but a degree of coherence could be identified between the ways in which they presented the viability of their own holdings and their constructions of what it meant to be a farmer. Both discourses provide different perspectives on the dominant notions of farming masculinity as tough men's work, one emphasising the managerial aspects of it, and the other emphasising the relationship with the land and nature. At the heart of both is a strong association between economic status, farm work and masculinity.

The findings here differ from those of Saugeres' (2002b) research, however. She found the traditional discourse of farming masculinity, emphasising the importance of harmony between farmer and nature, to be a hegemonic local discourse, while there was some resistance from a subordinate discourse of 'reluctance' among some farmers. The latter was associated with a critical attitude towards farming and the process of becoming a farmer, reflecting disappointments and frustrations with the reality of farming. In other words, the 'unsuccessful' farmers (as defined in commercial terms) denaturalized the process by which they became farmers and did not romanticize their relationship to farming. Similarly, in south-west Ireland, one farmer spoke bitterly of being put under pressure to enter farming and now finding it difficult to pay the bills. More common, however, in Ireland, among marginal farmers, was the tendency to naturalize one's decision to be a farmer and to romanticize farming, while the hegemonic construction of farming masculinity was one based on values of commercialism and economic success. The following section explores the tension between these two competing discourses in some more detail.

Economic Status and Farming Masculinities

The utilization of an entrepreneurial/managerial discourse among farmers must be set within the context of a wider tendency in contemporary society to associate the world of business, competitiveness and managerial positions with maleness (McDowell, 2001). Managerial cultures serve to reproduce gendered power structures in the workplace. These involve constructions of masculinity that tend to value competitiveness and entrepreneurialism as well as 'professional, competent and rational self-images' (McDowell, 2001, p. 4). Such images of masculinity have become quite potent in society generally, associated with the hegemony of values of competition and professionalism in work, and reproduced through media constructions of the business executive (Connell, 2000). In the context of agriculture, the relatively large and successful farmers can draw on these discourses to justify their farming methods and to shape their own identities. There is obvious pride in economic rationality and a professional approach to farming, built on a conventional sense of masculinity involving a pride in hard work. This represents a relatively unproblematic adaptation of conventional farming masculinity to the values of a changing society and economy. Therefore some farmers can continue to portray or to perform certain selected aspects of conventional Irish farming masculinity, while simultaneously pursuing a highly commercial and competitive approach to farming.

On the other hand, for more marginal or struggling farmers, many of whom are farming 'non-viable' farms, this is not an option. Perhaps for these farmers, valuing tradition and dedication to farming is a way of coping with a lack of economic status. Similarly, Willott and Griffin (1996) have found that unemployed men in urban England, faced with a loss of status due to the loss of their breadwinning status, use discursive strategies to take on alternative masculine identities or re-establish traditional forms of masculinity. Farmers who are not commercially successful can pride themselves on 'not being in it for the money' but for much deeper and almost spiritual reasons. This discourse of farming also often involves a biological metaphor among interviewed farmers, that is, the notion of farming being 'in the blood'. The 'true' farmer is portrayed as being selected for farming by the invisible hand of nature. Saugeres (2002b) also found that among farmers in her study area in the south of France, the notion of a 'true' farmer involved the concept of a natural predisposition for farmwork. The physical nature of farm work is idealized in this highly corporeal construction. It involves a construction of the farmer as being close to nature, or even as part of the natural world itself.

This contrasts with the more conventional construction of the farmer as controlling nature, and therefore is to a large extent an anti-modernist ideal. The 'true' farmer, according to this discourse, is not necessarily the ideal farmer portrayed in the popular rhetoric of commercial agriculture. This discourse, then, may be a form of resistance to some of the values of contemporary commercial agriculture. It may be a strategy for coping with the imposition of a set of values that define farming success in terms of the commercial viability of holdings. This may represent a subordinate discourse of successful farming, one that values affinity with nature, and love of farm work, of land and of animals, above commercial success. It may also represent an alternative construction of masculinity, one that incorporates roles that are often defined as 'feminine' that is, caring and nurturing, in relation to nature and animals. This points to the possibility of the emergence of alternative farming masculinities involving less rigid gender constructions than those of conventional hegemonic masculinities (Ní Laoire, 2002).

Some farmers, then, may resist the dominant values of commercialism and expansion, not necessarily for philosophical reasons, but because such institutions have not benefited them. Indeed, in many cases they have impoverished and disempowered them. The competitiveness of the contemporary agricultural economy produces 'winners' and 'losers'. Small or marginal farms tend to become non-viable or are phased out, while large farms are in a better position to expand and maintain viability. Marginal farms are clearly 'unsuccessful' in those terms due mainly to economic forces, and not necessarily due to lack of ability on the part of the farmer.

At a superficial analytical level, one might infer that an entrepreneurial/managerial farming ethos implies the ability to expand and modernize, while a traditional/romantic farming ethos implies a lack of innovation or poor management skills, resulting in an inability to expand. This is an assumption that underlies much agricultural literature and policy. According to Dudley (2000), the underlying logic of neo-liberal approaches to agriculture is that those farmers who are commercially successful are the best or most efficient farmers, while those who experience financial failure or loss of farm are the least able farmers. This logic is also manifest in common assumptions regarding farmers who do not expand or modernize, as being 'conservative', resistant to change or even 'idiosyncratic' (Battershill and Gilg, 1996). This stereotyping of farmers based on their behaviour reflects, in part, the prominence of behavioural analyses in agricultural geography. These tend to focus on the motives, values and attitudes determining farmer decision-making, and to explain farmer uptake of innovation as a function of their attitudes (Morris and Potter, 1995). Innovation-adoption research, for example, classifies farmers along a spectrum from active adoption to passive resistance, labelling farmers as 'adopters' or 'resistors' (for example, Potter and Gasson, 1988; Morris and Potter, 1995). Such analyses invariably tend to conclude that what is needed for progress is a change in farmer attitudes, but the underlying structural factors shaping farm modernization or diversification are often not given equal weight.

It is argued here that the underlying material forces that shape values and attitudes need to be recognized, in particular economic factors such as farm size. For example, Wilson (1996), in his study of farmers' participation in the ESA scheme, found that farm size and other material factors were as important as attitudes in determining participation in a scheme. He argues that a unique focus on farmer attitudes is futile, and instead the ways in which attitudinal and structural factors interact needs to be understood. Similarly, others such as Pile (1992) and Daskalopoulou and Petrou (2002) recognize that farmers develop survival strategies based on their readings of external conditions and how they inter-relate with internal considerations. Daskalapoulou and Petrou (2002) draw on Whatmore et al's (1987) conceptualization of the family farm, which emphasizes the importance of household reproduction and survival strategies in farm practices. This highlights the complexity of factors involved in farm decision-making. As Pile (1992, p. 79) states,

> How the restructuring of the family farm takes place is the result of the intersection of the farm's structural location and wider capitalist forces, and the family's feelings and wider social aspirations.

Farm practices and production patterns, then, are the outcome of decisions made in this way, and not simply a result of inherent farmer attitudes.

Decisions reflect sets of values that are formed in the context of changing material realities, in that some ideologies become redundant or contradictory, and some emerge or are adapted in the context of changing material or social circumstances. In the case of the growing commercialization and rationalization of agriculture in north Cork, some farmers can adapt conventional notions of 'farming as hard work' to modern entrepreneurial constructions of the 'farmer as business-man'. Others become aware of the inherent contradiction in a discourse that defines them as unsuccessful and threatens their sense of masculinity. One strategy for dealing with this is to re-turn to a more traditional or alternative discourse of farming masculinity. This discourse fulfils a very important function for the more marginal farmer, in that it reinforces 'his'[3] sense of pride and 'his' status as a farmer in the household and the locality. He may be losing out to more commercial or larger farmers in the struggle to expand or intensify, and therefore to survive as a farmer, but he can still be considered a good farmer, according to an alternative set of values.

Furthermore, this discourse justifies or reinforces the decision to stay in farming even when it is not economically rational to do so.

> 'Tis a farmer I am, 'tis a farmer I'll be. A conscious decision [to become a farmer]?, no. 'Tis in the blood. I can't explain it.... I suppose I love what I do... I stick with it... the land and cattle – that's my way of life (Denis).

If farming is constructed almost in terms of a biological imperative, then for some people, there is no choice but to be a farmer. If farming is a way of life and there is a strong emotional attachment to that way of life, then the 'true' farmer will display his love of farming by staying in it despite the obstacles. Therefore at a time when there are strong economic and social reasons for leaving farming, particularly for the marginal farmer, it is still possible to justify the decision not to do so. This 'staying power' is constructed as a sign of one's commitment and dedication to farming, as well as the invisible biological imperative, which is seen to select some individuals simply to be farmers. It may be necessary to present the decision to stay in farming in this way because the possibility of giving up the farm or phasing out production is incomprehensible. To someone who is running a family farm that has been in the family for generations, who may not have many transferable skills, and for whom farm work is central to one's very identity, then leaving farming is often a very last resort.

Conclusions and Future Research

Contemporary processes of rural restructuring in north Cork involve increased competitiveness in the agricultural sector and have resulted in the production of 'winners' and 'losers' as defined in terms of the commercial viability of holdings. Because of the close association of notions of 'success' at work with masculine identities, this has contributed to a certain shaping and re-shaping of male farmer identities, as some farmers become economically dominant and some are marginalized. Conventional masculine identities based on values of physical strength, hard work and land ownership are re-worked in a context where commercial success becomes the dominant measure of success as a farmer. The more economically powerful farmers can adapt conventional hegemonic rural masculinities in a relatively unproblematic way to the demands of the marketplace and contemporary society. However, more marginal farmers need to adapt in different ways, one of which is to draw on traditional non-materialist values to resist the de-valuation of their way of life and their status and to compensate for their lack of economic power. This challenges the idea that entrepreneurialism at the farm level simply reflects farmer attitudes to change and modernization. Instead it is argued that decisions to modernize, expand and diversify, or not, are made in the context of particular material and financial circumstances which may or may not allow such changes. Furthermore, farmer attitudes reflect values and identities that are constantly being shaped and re-shaped in the context of their material circumstances. Therefore, farm practices, attitudes and identities are related in complex ways, involving the interaction of external, household and internal factors in farm decision-making, and the adaptation of values to suit particular material circumstances and perceptions of threat to one's livelihood and identity.

The challenge for future research on farmer behaviour, then, is to contextualize behaviour by recognising the role of material circumstances in decision-making, and to recognize that farmer attitudes and identities are not formed in a vacuum, but are constantly being shaped in response to economic, social and cultural demands. This means that farmer identities are shaped by their economic, social and cultural contexts and will reflect particular local, regional or national processes. There are many parallels between farming masculinities in south-west Ireland and in southern France, for example, but the historical, cultural and economic contexts in which they are constructed may differ significantly. There is a role, then, for cross-cultural research into the construction of farmer identities, to explore interregional and international comparisons in constructions of identities.

The role of masculinism is central to this, as farmer identities are constructed in selective ways that emphasize the maleness of farm work. This is maintained despite the changing nature of constructions of farming masculinity. The power of masculinity to adapt to a changing context is associated with the privileged material position of men in farming society generally, intimately bound up in complex ways with the very powerful role of land-ownership, farm work and economic status in farmers' identities. A hegemonic construction of farming masculinity based on core values of hard work and control of nature is maintained, despite changing economic and cultural contexts. This contributes to the reproduction of patriarchal systems of land-ownership and the male dominance of the farming sector. Future research on rural masculinities, then, if it is to maintain a critical perspective on rural gender relations, needs to recognize the complex interrelationships between material circumstances and identity construction. This becomes particularly interesting in the current context of falling farm incomes, and future research could explore the implications of this for hegemonic masculinities and rural gender relations. The challenge is to recognize the complexity of relations of dominance and marginalization among farmers as well as between men and women in order to continue to provide a critical perspective on changing rural gender relations.

Notes

1. Recent changes in regulations now require available milk quota to be distributed centrally rather than sold on the open market.
2. All names have been changed to protect interviewees' anonymity.
3. The masculine pronoun is used deliberately to highlight the close association between masculinity and farming identities.

References

Battershill, M. and Gilg, A. (1996), 'Traditional Farming and Agro-Environment Policy in Southwest England: Back to the Future?' *Geoforum*, Vol. 27, pp. 133-147.

Beus, C. and Dunlap, R. (1994), 'Endorsement of Agrarian Ideology and Adherence to Agricultural Paradigms', *Rural Sociology*, Vol. 59, pp. 462-484.

Brandth, B. (1994), 'Changing Femininity: the Social Construction of Women Farmers in Norway', *Sociologia Ruralis*, Vol. 34, pp. 127-149.

Brandth, B. (1995), 'Rural Masculinity in Transition: Gender Images in Tractor Advertisements', *Journal of Rural Studies*, vol. 11, pp. 123-133.

Brandth, B. and Haugen, M. (2000), 'From Lumberjack to Business Manager: Masculinity in the Norwegian Forestry Press', *Journal of Rural Studies*, Vol. 16, pp. 343-355.

Bryant, L. (1999), 'The Detraditionalization of Occupational Identities in Farming in South Australia', *Sociologia Ruralis*, Vol. 39, pp. 236-261.

Campbell, H. (2000), 'The Glass Phallus: Pub(lic) Masculinity and Drinking in Rural New Zealand', *Rural Sociology*, Vol. 65, pp. 562-581.

Collinson, D. and Hearn, J. (1996), '"Men" at "Work": Multiple Masculinities/Multiple Workplaces', in M. Mac an Ghaill (ed), *Understanding Masculinities*, Open University Press, Buckingham, pp. 61-76.

Connell, R. W. (2000), *The Men and the Boys*, Polity Press, Cambridge.

Commins, P. (1999), 'Structural Change in the Agricultural and Rural Economy', Final Report, Project No. 4004, Teagasc, Dublin.

Central Statistics Office (1999), *Agricultural Labour Input*, (www.cso.ie/publications/agriculture/aglabinput.pdf); Accessed October 2002.

Central Statistics Office (2003), *Agricultural Labour Input 2000*, (www.cso.ie/principalstats/princstats.html#agriculture); Accessed March 2003.

Daskalopoulou, I. and Petrou, A. (2002), 'Utilising a Farm Typology to Identify Potential Adopters of Alternative Farming Activities in Greek Agriculture', *Journal of Rural Studies*, Vol. 18, pp. 95-103.

Department of Agriculture, Food and Rural Development (1999), *Statistical Compendium*, (www.irlgov.ie/daff/etable15.xls); Accessed 2000.

Dudley, K.M. (2000), *Debt and Dispossession: Farm Loss in America's Heartland*, University of Chicago Press, Chicago.

Evans, R. (2000), 'You Questioning my Manhood, Boy? Masculine Identity, Work Performance and Performativity in a Rural Staples Economy', Arkleton Research Paper No. 4, University of Aberdeen.

Frawley, J. and Commins, P. (1996), 'The Changing Structure of Irish farming: Trends and Prospects', Rural Economy Research Series 1, Teagasc, Dublin.

Lafferty, S., Commins, P. and Walsh, J.A. (1999), *Irish Agriculture in Transition: a Census Atlas of Agriculture in the Republic of Ireland*, Teagasc, Dublin.

Leckie, G. (1996), '"They Never Trusted Me to Drive": Farm Girls and the Gender Relations of Agricultural Information Transfer', *Gender, Place and Culture*, Vol. 3, pp. 309-325.

Liepins, R. (1998), 'The Gendering of Farming and Agricultural Politics: a Matter of Discourse and Power', *Australian Geographer*, Vol. 29, pp. 371-388.

Liepins, R. (2000), 'Making Men: the Construction and Representation of Agriculture-based Masculinities in Australia and New Zealand', *Rural Sociology*, Vol. 65, pp. 605-620.

Little, J. (2002), *Gender and Rural Geography: Identity, Sexuality and Power in the Countryside*, Pearson Education Ltd, Essex.

Little, J. and Jones, O. (2000), 'Masculinity, Gender and Rural Policy', *Rural Sociology*, Vol. 65, pp. 605-620.

Martin, A. (1997), 'The Practice of Identity and an Irish Sense of Place', *Gender, Place and Culture*, Vol. 4, pp. 89-119.

McDowell, L. (2001), 'Men, Management and Multiple Masculinities in Organizations', *Geoforum*, Vol. 32, pp. 181-198.

ó Donnchadha, G. and Ní Shé, N. (1994), 'Duhallow Farm Families 1994: Challenge and Prospect', Report for IRD Duhallow, Newmarket, County Cork.

Morris, C. and Potter, C. 'Recruiting the New Conservationists: Farmers' Adoption of Agri-Environmental Schemes in the UK', *Journal of Rural Studies*, Vol. 11, pp. 51-63.

Ní Laoire, C. (2002), 'Young Farmers, Masculinities and Change in Rural Ireland', *Irish Geography*, Vol. 35, pp. 16-27.

Peter , G., Mayerfeld Bell, M., Jarnagin, S., Bauer, D. (2000), 'Coming Back Across the Fence: Masculinity and the Transition to Sustainable Agriculture', *Rural Sociology*, Vol. 65, pp. 215-34.

Pile, S. (1992) '"All Winds and Weathers": Uncertainty, Debt and the Subsumption of the Family Farm', in A. Gilg, (ed), *Restructuring the Countryside: Environmental Policy in Practice*, Avebury, Aldershot, pp. 69-82.

Potter, C. and Gasson, R. (1988), 'Farmer Participation in Voluntary Land Diversion Schemes: Some Predictions from a Survey', *Journal of Rural Studies*, Vol. 4, pp. 365-375.

Potter, C. and Lobley, M. (1992), 'The Conservation Status and Potential of Elderly Farmers: Results from a Survey in England and Wales', *Journal of Rural Studies*, Vol. 8, pp. 133-143.

Saugeres, L. (2002a), 'Of Tractors and Men: Masculinity, Technology and Power in a French Farming Community', *Sociologia Ruralis*, Vol. 42, pp.143-159.

Saugeres, L. (2002b), 'The Cultural Representation of the Farming Landscape: Masculinity, Power and Nature', *Journal of Rural Studies*, Vol. 18, pp. 373-384.

Shortall, S. (1992), 'Power Analysis and Farm Wives: An Empirical Study of the Power Relationships Affecting Women on Irish Farms', *Sociologia Ruralis*, Vol. 32, pp. 431-451.

Shortall, S. (1999) *Women and Farming: Property and Power*, Macmillan Press Ltd, Basingstoke.

Whatmore, S., Munton, R., Marsden, T., and Little, J. (1987), 'Interpreting a Relational Typology of Farm Business in Southern England', *Sociologia Ruralis*, Vol. 27, pp. 103-122.

Willott, S. and Griffin, C. (1996), 'Men, Masculinity and the Challenge of Long-term Unemployment', in M. Mac an Ghaill (ed), *Understanding Masculinities*, Open University Press, Buckingham, pp. 77-92.

Wilson, G. (1996), 'Farmer Environmental Attitudes and ESA Participation', *Geoforum*, Vol. 27, pp. 115-131.

Woodward, R. (1998), 'It's a Man's Life!: Soldiers, Masculinity and the Countryside', *Gender, Place and Culture*, Vol. 5, pp. 277-300.

Index